Models of Adaptive Behaviour sets out a framework for explanations of behaviour in terms of the maximisation of fitness. Aspects of an organism's state, such as size, energy reserves or temperature may be important for survival and reproduction. The framework adopted takes this into account, and provides a common currency for comparing diverse actions. This approach is used to analyse a wide range of biological issues such as energetic gain, energy-predation trade-offs, dynamic games, state-dependent life histories, annual routines and fluctuating environments. In each case, the book summarises previous work, presents new material and suggests directions for future research. This book provides a unification and overview of much of the authors' previous work on state-dependent models.

ALASDAIR HOUSTON is Professor of Theoretical Biology at the University of Bristol, and JOHN McNAMARA is Professor of Mathematics and Biology, also at the University of Bristol. They first met as undergraduates at the University of Oxford and have been working together since 1978 developing theoretical approaches to behaviour and applying them to various systems.

Models of Adaptive Behaviour

Alasdair Houston & John McNamara
University of Bristol

PUBLISHED BY THE PRESS SYNDICATE OF THE UNIVERSITY OF CAMBRIDGE
The Pitt Building, Trumpington Street, Cambridge CB2 1RP, United Kingdom

CAMBRIDGE UNIVERSITY PRESS
The Edinburgh Building, Cambridge CB2 2RU, UK   www.cup.cam.ac.uk
40 West 20th Street, New York, NY 10011-4211, USA   www.cup.org
10 Stamford Road, Oakleigh, Melbourne 3166, Australia
Ruiz de Alarcón 13, 28014 Madrid, Spain

First published 1999

Printed in the United Kingdom at the University Press, Cambridge

Typeface Computer Modern 11/14pt   *System* LaTeX [EPC]

A catalogue record for this book is available from the British Library

Library of Congress Cataloguing in Publication data

ISBN 0 521 38480 X hardback
ISBN 0 521 65539 0 paperback

Contents

Contents

Acknowledgements

In writing this book, we have benefited from the comments of the following colleagues at Bristol who have read one or more chapters: Sean Collins, Innes Cuthill, Cathy Gasson, Jon Heron, Alison Lang, Sean Rands, Tamas Székely, James Webb, Bob Welham and Nicky Welton. John Hutchinson deserves special mention for reading the whole book and providing very detailed suggestions. We are grateful to the following for giving us feedback on particular chapters: Peter Abrams (Chapters 6 and 8), Troy Day (Chapter 8), Jan Ekman (Chapter 6), Alan Grafen (Chapter 7), Patsy Haccou (Chapter 10), Alex Kacelnik (Chapter 5), Olle Leimar (Chapter 7), Steve Lima, Carin Magnahagen and Andy Sih (Chapter 6), Marcel Visser and Thomas Weber (Chapter 10) and Franjo Weissing (Chapter 7). Figure 4.3 is based on computations carried out by Jon Heron and Figures 9.3–9.8 are based on computations carried out by Bob Welham. Our thanks to all of you. Thanks also to staff at CUP, particularly Tracey Sanderson for advice and assistance and Susan Parkinson for the thorough and perceptive suggestions in copy editing the book.

We are especially grateful to Rhoda Rees not only for typing the book, but for typing our papers for the past eighteen years with remarkable speed and efficiency.

A. Houston and J. McNamara
Bristol
December 1998

1

Introduction

1.1 The focus of the book

In this book we are concerned with the use of models based on optimisation to understand the behaviour of organisms. Natural selection produces organisms that are good at surviving and reproducing, so we can hope to make predictions about the products of natural selection by constructing optimisation models. The pros and cons of this approach have been widely discussed, e.g. Maynard Smith (1978), Gould & Lewontin (1979), Stephens & Krebs (1986), Mitchell & Valone (1990), Reeve & Sherman (1993), Orzack & Sober (1994). (Work on long-term evolution, e.g. Eshel 1996, Hammerstein 1996, Weissing 1996, Eshel *et al.* 1998 is also relevant.) Of course what evolves may be constrained by phylogeny and genetics. In this book we adopt what Grafen (1984) calls the phenotypic gambit, that is to say we take certain plausible constraints as given and find the optimal phenotype within these constraints. We do not expect organisms to be perfectly adapted to their environment, and hence do not expect them to be optimal in terms of any model. Nevertheless, optimality models should give reasonable predictions, especially when deviations from the optimum are costly in terms of fitness.

In this book we do not review the whole of behavioural ecology. We do not even review all the theoretical issues in behavioural ecology. Rather, we concentrate on state-dependent and dynamic models of behaviour. A central theme of the book is that at any given time an organism can be characterised by its current state, and that behaviour may depend on state. We are also often concerned with organisms that are making a sequence of decisions over time, i.e. with dynamic models. There is an obvious link between dynamic models and state-dependent models, in that an organism's state will typically change over time. State-dependent dynamic models incorporate both components; they model sequences of decisions over time where each action chosen can depend on current state and may influence future state.

In this book we have concentrated on models from areas in which we have worked. Within these areas, we have sought to expose the logic of various explanations of behaviour; given the limitations on space, we have not attempted to provide an exhaustive review of the data. Many of the models in the book are analysed using a technique known as dynamic programming. Although we describe this technique in Chapter 3, the book is primarily not about dynamic programming *per se*. For a complementary account of dynamic programming models of behaviour and how to construct them, see the excellent book of Mangel & Clark (1988).

State and its importance

We characterise an organism in terms of a set of state variables. These variables may represent aspects of the organism, such as its size or body temperature, or aspects of the organism's environment, such as the quality of the organism's territory. An organism's state may constrain the organism's possible actions, e.g. the nutrients available to a bird may limit its clutch size (e.g. Bolton *et al.* 1992). Within the set of possible actions, state may influence an organism's performance. For example, the body temperature of a lizard may affect its ability to catch and digest prey. A discussion of which state variable should be included in a model is given in Chapter 2.

The costs and benefits of an action may depend on state. If we consider an animal that is about to reproduce, then the condition of its immune system and the level of its nutrient reserves could be relevant components of state. The consequences of reproducing may depend on these state variables. In particular, the animal's state is likely to influence the probability that its young survive, and to influence the cost to the animal in terms of its own survival and future reproductive success. In studying a foraging animal, energy reserves might be a suitable state variable. If the animal must expose itself to predation in order to obtain a food item, then the advantage of obtaining the item is likely to decrease as energy reserves increase and the cost of being killed is likely to increase as energy reserves increase. This is discussed in detail in Chapter 6.

Given the importance of state, we can expect animals to make state-dependent decisions. An organism that ignored its state when making decisions would typically have a much lower fitness than an animal that based its decisions on its state (see Sections 3.6 and 5.8).

Many biological phenomena can only be understood in terms of differences between individuals in a population, i.e. in terms of state. Quality differences are central to understanding mate choice and signalling. State variables such as gut contents and energy reserves can account for satiation, and changes in state that represent information about the the world form the basis of learning.

The common currency for behavioural decisions

An organism's behavioural strategy is a rule that specifies how the organism deals with every possible circumstance. In this context, the organism's circumstances are characterised by its state. Thus a strategy is a rule specifying the dependence of behaviour on state (and time).

Fitness is a measure of the performance of a strategy based on the number of descendants left by individuals following the strategy. In the simplest case, the fitness of a strategy can be taken as the number of offspring produced over the lifetime of an individual following the strategy. This measure is not adequate in general. When offspring may be in different states (e.g. have different qualities) or may be produced at different times of year, we cannot evaluate the success of a strategy simply by counting offspring. Instead, fitness is measured as the long-term growth rate in descendant numbers given that the focal individual and all its descendants follow the strategy. This topic is discussed in detail in Chapter 8. Natural selection favours strategies that maximise fitness. We will refer to a strategy that maximises fitness as an optimal strategy.

Fitness is a measure that can be used to compare the performance of strategies, but organisms following a given strategy may differ in state and hence in their potential to produce offspring in the future. For every possible state, the reproductive value of the state is a measure of this potential. Thus reproductive value is a within-strategy measure that compares individuals. In the simplest case, in which fitness is measured in terms of lifetime production of offspring, the reproductive value of an organism in a particular state is the expected future number of offspring produced by the individual over the remainder of its lifetime. In the general case in which offspring may be in different states, reproductive value is a relative measure. It compares the number of descendants left at some time far into the future by an individual in one state and the number of descendants left at the same time by an individual in another state, given that both individuals follow the same strategy (see Section 8.4).

Reproductive value is central to the optimality analysis presented in this book. We can take the payoff from performing a particular action to be the sum of the reproductive values of any offspring produced as a direct result of the action and the expected reproductive value of the focal individual after it has taken the action. Under an optimal strategy, an organism always chooses the action that maximizes its payoff. In this way, reproductive value provides a common currency for comparing possible actions. Actions with very different consequences (e.g. feeding and searching for a mate) can be compared via their effect on the organism's reproductive value. These intuitive notions are made precise in Chapter 2, in which we also give some simple examples. This approach based on reproductive value is used in all chapters except Chapter 10.

The importance of the future

In finding an optimal strategy we often cannot consider actions in isolation; it is necessary (either explicitly or implicitly) to consider sequences of actions. This is so because the value to an organism of its current action may depend on the actions that it takes in the future. For example, the value of obtaining an item of food may depend on whether the animal then chooses to reproduce. It is important to realise that we may need to consider not just the future actions of the focal organism but also the future actions of any offspring that it produces. For example, when there is a trade-off between the number of offspring and their size, the value of producing large offspring depends on the subsequent growth and reproductive strategies of these offspring.

The fact that an action cannot be considered in isolation is an important reason for analysing sequences of actions and hence for using dynamic models. Because the value of current actions depends on future actions, it is natural to work backwards when finding optimal strategies. Once optimal behaviour from some time onwards is known, it is possible to find the optimal behaviour at the immediately previous decision time. The technique of working backwards that we explain in this book is known as dynamic programming. The systematic use of this general approach was introduced by Mangel & Clark (1986) and McNamara & Houston (1986). Rather than working explicitly with sequences, this technique uses reproductive value to summarise the consequences of future actions. Once reproductive value is known at some time, it can be used as a common currency to find the optimal behaviour at the immediately previous decision time. From this we obtain reproductive value at the previous decision time, and so on working backwards. This still leaves us with the problem of where to start the process of working backwards. This is discussed in detail in Chapters 3, 8 and

9. In some cases it may be reasonable to assume that reproductive value is known at a certain time in an organism's life, and hence analyse optimal behaviour before this time in isolation from behaviour after this time. If this is not possible, it may be necessary to analyse behaviour over the whole of the organism's lifetime. Even this may not be sufficient when the offspring produced may be in different states and hence have different reproductive values. It may then be necessary to calculate reproductive value by working backwards from some time far in the future, analysing the behaviour not only of the focal animal but also of its descendants. Details are given in Chapter 8.

Of course, if we knew reproductive value at a given time then a simple static optimisation could be used to find the best action at the preceding decision point. Even if the exact value is not known, it can be instructive to assume some plausible form of the dependence of reproductive value on state and carry out the static analysis. The problem with such an approach lies in justifying the chosen functional form. The true functional form may not be obvious, and using an incorrect function may produce erroneous conclusions. In performing a dynamic optimisation, we work backwards from some final time. It is necessary to specify reproductive value at this time. The form assumed will certainly influence reproductive value and optimal behaviour shortly before the final time, so again there could be problems if an incorrect functional form were chosen. We may, however, be concerned with behaviour a long way before the final time. Then reproductive value and hence optimal behaviour at this time of interest may be essentially independent of assumptions about the form of reproductive value at final time (see Section 3.5). In this context, reproductive value at the time of interest is not assumed but emerges from the analysis. We regard this as an advantage of the dynamic modelling approach. The technique of choosing a final time that is a long way after the time of interest is taken to the limit in Chapters 8 and 9, in which state-dependent life histories are analysed. In these chapters, the final time is essentially taken to be infinitely far into the future, and reproductive value at the time of interest is independent of any assumptions about its value at final time.

Some people build models by making assumptions about the costs and benefits of an action. Our approach is to specify the consequences of an action, in particular, the number of offspring produced and their states, the probability that the focal organism survives and its state if it survives. The disadvantage of an action may be that it reduces the survival of the focal organism or has a deleterious effect on the organism's state. The optimal strategy is found directly from consequences alone. Once this strategy has been found, it is possible to derive costs and benefits that explain optimal behaviour, but it is not necessary to do so.

Types of model and their uses

There are various reasons for constructing models. For example, it may be desired to make predictions about the outcome of experiments. Alternatively, models may be used to clarify assumptions and expose the logic behind a situation. The latter is an important use of models; indeed we feel that it is dangerous to make predictions without an understanding of the underlying logic. If we do not understand the basis of the predictions, then not only have we failed to explain the biology but also we cannot be sure of how robust our predictions are to changes in the assumptions.

As we have outlined above, the logic of adaptive decision-making leads naturally to state-dependent dynamic models. Typically, we have to find solutions to such models by a numerical technique such as dynamic programming. This means that quite a lot of work must be done to establish general conclusions (e.g. Gladstein *et al.* 1991, Houston *et al.* 1992). Because of this, we believe that it is better to start an analysis by constructing a relatively simple model. Such a model need not be dynamic, and may be solvable analytically. Analytical models are very valuable because they provide an excellent understanding of the underlying logic and the generality of their conclusions can be established. Examples of the value of a proper analysis of analytic models are provided by Abrams (1990, 1991, 1993a, b). If it is felt that a simple model is unrealistic, a more realistic model can be constructed. This might be a complex state-dependent dynamic model, and as such would be hard to analyse. Such an analysis would, however, be facilitated by insights obtained from the simple model.

1.2 The structure of the book

The basic approach adopted in this book is spelled out in Chapters 2 and 3. In Chapter 2 we introduce the basic decision-making framework, explaining how trade-offs are mediated by state variables. The analysis is static in that only a single decision is considered. In Chapter 3 we then describe how the framework expounded in Chapter 2 can be extended to look at a sequence of decisions. Chapters 2 and 3 form the basis of our analysis in subsequent chapters. These later chapters have only a loose dependence on each other.

Chapters 4, 5 and 6 are all concerned with foraging behaviour. In Chapter 4 we start off with the simplest case, in which mean net energy gain is all that matters, and go on to consider constraints on energy expenditure and the possible influence of state variables other than energy reserves. Chapter 5 is concerned with stochastic effects in foraging and the non-linear dependence of reproductive value on energy reserves. This chapter also touches on an issue of wide importance – the need for rules that cope with a changing environment. In Chapter 6 we analyse the trade-off between gaining energy and avoiding predation. The chapter is organised around a unifying framework based on a simple criterion for optimal choice. This criterion allows an analysis of many situations without the need for a complex dynamic model.

Chapter 7 is specifically about dynamic games and can be read independently of the other chapters, although this chapter is relevant to other chapters that contain game-theoretic models (e.g. Chapters 6, 8 and 10).

Chapter 8 is concerned with the whole life history of an organism, and the importance of state variables within these life histories. Up to this chapter, fitness and reproductive value are treated in a simple and intuitive manner. In Chapter 8, though, we define these quantities rigorously. This is necessary because when there are intergenerational effects, simple measures based on lifetime production of offspring are no longer adequate. Chapter 8 plays a key rôle in modelling sequences of behaviour. In Chapter 3 we are concerned with a sequence of decisions by an organism; in Chapter 8 we are also concerned with a sequence of decisions by an organism, but because we quantify the reproductive value of offspring, we are then implicitly concerned with the decisions made by both the organism and its descendants. Chapter 8 concentrates on annual decisions; in Chapter 9 we extend the analysis of Chapter 8 to consider the sequence of decisions during a year (i.e. we analyse annual routines).

Like Chapter 8, Chapter 10 is concerned with life-history decisions, but now the fluctuations in the environment influence all population members. This situation requires a fundamentally different definition of fitness. In contrast to the analysis in the rest of the book, there is no longer individual optimisation, nor is the concept of individual reproductive value of central importance.

1.3 Directions for future research

With the hindsight obtained from writing this book, we are now aware of various issues where more work is required. Three of the major issues are discussed below.

Physiology.   We believe that many trade-offs are mediated by the effect

of behaviour on physiological state variables. In this book we have made a
number of schematic models in which the underlying physiological variable is
referred to as 'condition'. To give more biological realism to these models, we
need much more empirical evidence about what constitutes condition, how
behaviour influences condition and the effect of condition on the organism's
ability to survive and reproduce.

There is evidence that animals avoid hard work if possible. In Chapter
4, we suggest that metabolic expenditure decreases condition, but we do
not know what physiological variables constitute condition. Both the state
of the immune system and the state of the muscles could be components
of condition, but other physiological variables could be important too. We
need to know not only which physiological variables are relevant but also
how they are influenced by energy expenditure and how they influence the
animal's ability to survive and reproduce. If we knew such details, we would
be in a better position to construct models that could make predictions
about foraging behaviour. Similarly, we need information about the effect
of reproduction on condition in order to understand the trade-off between
current and future reproduction. If aging is taken to be a change in condition
rather than simply an increase in age, then information about the relevant
aspects of condition is required if we are to model the aging process.

What problem is the animal built to solve?   Throughout this book we
have tended to assume that we can correctly characterise the problem that
selection has enabled the animal to solve. This assumption may be too
optimistic. There are many reasons why our view of the world may not
correspond to the aspects of the world that have shaped behaviour.

In particular it may be difficult to construct the correct model for an
animal in a simple laboratory environment. If we do no more than represent
what goes on in the laboratory, it is unlikely that we will produce a realistic
model (Houston & McNamara 1989). For example, the laboratory may not
contain predators, and the animal will not be allowed to starve. We cannot
assume, however, that its behaviour will reflect these facts. The animal may
have evolved in an environment in which predation always poses a threat
and in which starvation is a real danger. We expect that evolution will have
given it a rule for choosing between behavioural options that take these
possibilities into account. If the animal follows this rule in the laboratory
then we can expect that

  (i) the animal's behaviour will show consistent patterns, and
  (ii) the animal's behaviour may not appear optimal in terms of what we
       know about the laboratory environment.

If we wish to make predictions about behaviour in laboratory experiments, it may be necessary to put a considerable effort into understanding the environment in which the animal's rules have evolved. Even if we have this understanding, there will still be problems to overcome. First, we have to construct a model that captures the essence of what is likely to be a complex environment. For example, we may have to specify the range of circumstances that the animal encounters and the frequency with which they are encountered. Then we have to determine the behavioural strategy that is likely to evolve. In some cases it might be reasonable to assume that the optimal strategy evolves. In other cases, the optimal strategy may be so complicated that it is unreasonable to assume that the animal will evolve this strategy. Instead, it is plausible that relatively simple and robust rules will evolve. To model this evolution, it is necessary to choose a class of simple rules on which evolution acts. The choice of an appropriate class may not be straightforward, but one approach is to base the class of rules on known mechanisms. In the context of risk-sensitive foraging, some work has been done on optimisation in a wider context than the laboratory environment (see Chapter 5), and Kacelnik and his co-workers have explored the underlying mechanisms (e.g. Kacelnik & Bateson 1996, 1997), but even within this area further work needs to be done.

Interactions between animals. As we point out in Chapter 7, many game-theoretic models of interactions between animals are so schematic as to ignore most of the details of the interaction process. In fact, such details may make a considerable difference to the outcome of the interaction, and much more attention needs to be given to them. We believe that, in order to understand a variety of social interactions, it will be necessary not only to include a representation of the process but also to allow for individual differences and errors in decision-making. We also believe that certain phenomena can only be understood by recognising that animals show consistent patterns of behaviour analogous to personalities. This is briefly discussed in the context of pair bonds in Section 7.10.

2

States, actions and trade-offs

2.1 The decision framework

In this chapter we introduce the framework for optimal decision making based on state. We show how a single decision can be analysed; in the next chapter we look at sequences of decisions. We begin by describing the components of a state-based model.

State variables

Individual members of a species can differ in a variety of ways. An individual's type or condition may limit its behavioural repertoire and may also influence the consequences of behaviour for reproductive success. Some differences between individuals may be relatively short-lived, whereas others may be a fixed property of the individual. We represent these differences by using a set of state variables. Each individual is characterised by the values of its state variables. Such variables might include an organism's age and size, together with various aspects of its physical condition, such as its level of energy reserves, its parasite load and its body temperature. There may also be variables that represent what the organism 'knows' about its environment. Another category of variables describes what Dawkins (1982) refers to as an individual's extended phenotype. The extended phenotype is not part of the individual's body but is part of the environment with which the individual interacts. It may be inanimate (e.g. the condition of a bird's nest or territory) or animate (e.g. the quality of an individual's mate or the state of its young).

A complete description of an individual might involve a large number of state variables. However, in modelling an organism's behaviour it is desirable to concentrate on just a few components of the state, in the hope of capturing the essence of the problem. Any factor that has a significant

10

effect on an organism's ability to survive or reproduce is a potential state variable. In some cases such a factor may not change over the time period of interest and hence can be regarded as a parameter rather than a state variable. To illustrate this, suppose that we are modelling the foraging behaviour of a fish. The size of the fish is likely to influence both its rate of energetic gain and its probability of being killed by a predator. If we are looking at behaviour over a short time period, size will be approximately constant and appropriate state variables might be energy reserves and gut contents. In contrast, if we are looking at growth over a period of years, then we might ignore the fine detail of foraging behaviour and assume some relationships between foraging effort and predation, and between foraging effort and growth. In this model, the appropriate state variable is size. For a further discussion of how to choose state variables, see Mangel & Clark (1988).

In general we denote a single state variable by x and a vector of variables by $\boldsymbol{x}$. In the wild, food, the weather, attacks by predators and many other phenomena may be unpredictable. Thus an individual's state in the future cannot be known with certainty and we regard it as a random variable. We then denote the state by a capital letter: X will denote a single state variable and $\boldsymbol{X}$ a vector of state variables.

An organism's behaviour is often dependent on time; e.g. the time of day or time of year. Although it is possible to include time as a component of the state, we prefer to treat time separately, denoting time by t.

Reproductive value

The current state of an organism influences the reproductive success of the organism in its future. This could be, for example, because current state directly affects the amount of resources that can be put into current reproduction, or because current state affects future state and hence affects resources in future reproductive bouts, or because current state affects future survival and hence affects the expected number of future reproductive bouts. Similarly, future reproductive success may depend on the current time of day or year. To quantify the dependence of future reproductive success on current state and time we need to define the *reproductive value*, $V(\boldsymbol{x}, t)$, of an organism. In simple cases $V(\boldsymbol{x}, t)$ is the expected number of offspring produced from time t onwards (until death) by an organism in state $\boldsymbol{x}$ at time t. Various modifications to this definition are needed in some circumstances. If reproduction is sexual then the number of offspring produced needs to be discounted by the relatedness of parent to offspring. If relatedness r is fixed and V is taken to be the number of offspring produced, the factor r just rescales V and does not usually affect optimality analyses. The simple definition in terms of offspring number also fails to take into account the fact that it may be better to produce offspring sooner than later and to take into account quality differences between offspring. The precise definition of $V(x, t)$ when these two complications occur is delayed until Chapters 8 and 9.

Two simple examples illustrate the dependence of $V(x, t)$ on x and t.

Small bird in winter. Consider a small bird attempting to survive over the winter. We might take the state of this bird to be its level of energy reserves x, although other aspects of its physiology may well also be important. Clearly, at a given time t the reproductive value $V(x, t)$ of a bird with reserves x increases with increasing x. Indeed, it can be shown that for times well before the end of winter $V(x, t)$ is simply proportional to the probability that the bird survives the winter (see Section 3.5). Thus $V(x, t)$ will increase with x for given t, and may increase with t for given x since survival prospects increase as the time to go until the end of winter decreases.

Indeterminate growth. For organisms such as a fish or tree the most important state variables might be size, y, and age, n. If we are concerned with the organism at a particular time of year only (say when reproductive decisions are made by an organism that breeds annually), we might drop the dependence on t from our notation and let $V(y, n)$ be the reproductive value of an organism of size y and age n at our focal time. Large size is likely to increase the current reproductive success of an organism and may well increase its survival probability. Thus $V(y, n)$ is likely to increase with y. For given y, $V(y, n)$ will increase with n before the age of sexual maturity, and may decrease with n for large n because of senescence.

Actions

The idea of state as used in system theory to provide a framework for studying behaviour was pioneered by McFarland (1971). McFarland argued that an organism's behaviour could be viewed as being chosen to control the organism's state. We adopt this perspective and model an organism's behavioural choice by assuming that it has a set of possible actions. The set of actions will depend on the context under consideration. For example, if we are concerned with annual reproductive decisions in a large mammal, we might regard the animal as choosing between the actions $u_1 = $ 'reproduce' and u_2 = 'do not reproduce'. If we were concerned with the vigilance behaviour of a foraging animal, an action u might correspond to the proportion of time that an animal spends feeding as opposed to being vigilant.

Behaviour is modelled as a sequence of distinct actions taken at times $t = 0, 1, 2, \ldots$. Let the organism's state at time t be $\boldsymbol{x}$. The organism has a set of possible actions available to it at this time. This set may depend on $\boldsymbol{x}$. For example, an animal can only defend its territory if it has a territory. Each action has three sorts of consequence.

(i) *Direct contribution.* The action taken determines the direct contribution to lifetime reproductive success that is made between times t and $t + 1$. This contribution is the total reproductive value (at time $t + 1$) of all offspring produced between time t and time $t + 1$ that survive until $t + 1$ and are independent at that time. If an organism in state $\boldsymbol{x}$ at time t performs an action u then the resulting direct contribution to reproductive success is denoted by $B_{\text{off}}(\boldsymbol{x}, t; u)$.

In the simplest case the states of independent offspring alive at time $t + 1$ do not depend on parental state and action at time t. All offspring then have equal value and B_{off} is just the total number of these offspring. Examples in which it is not adequate simply to count offspring number are given in Chapters 8 and 9.

If offspring produced between time t and time $t + 1$ are still dependent on their parents at time $t + 1$, we do not include these offspring when calculating the direct contribution to reproductive success. Instead the offspring are included as a component of the parent's state at time $t + 1$ (see Sections 2.2 and 9.5 for examples of this).

(ii) *Survival probability.* The action taken determines the probability that the focal individual survives until time $t + 1$. The survival probability given that an organism in state $\boldsymbol{x}$ at time t takes action u is denoted by $S(\boldsymbol{x}, t; u)$.

There may be an interaction between the action taken and the organism's state in their effect on survival probability. To illustrate this, consider an organism whose action is its reproductive effort during a bout of reproduction. If the organism is in good condition then the level of reproductive effort may have little effect on the probability that the organism survives to breed again. In contrast, if the organism is in poor condition then survival probability may decrease strongly with effort. These sort of effects are discussed by Clutton-Brock (1991).

(iii) *Change in state.* The action taken determines the organism's state at time $t + 1$, given that it survives until this time, and hence determines the organism's reproductive value at time $t + 1$. In many cases a realistic model involves a probabilistic relationship between the action chosen and the organism's subsequent state. For example, when an organism searches for food it is not necessarily certain to find any. In general the probability distribution of the state at time $t + 1$ is determined by both the state at time t and the action chosen.

The next state (at time $t + 1$) is denoted by x'; if it is a random variable it is denoted by X'.

2.2 Optimal decisions

We have just described a framework within which we can model the relationship between an organism's behaviour and its reproductive success. The organism's state plays a central role in our approach. State is important in that the consequences of an action in terms of survival and reproduction may depend on the organism's state. This makes it advantageous to make state-dependent behavioural decisions. Indeed, as we illustrate in Section 3.6 and discuss in Section 5.8, organisms that do not make state-dependent choices will be at a considerable disadvantage. As a simple example, consider an animal outside its reproductive season that is seeking food to prevent starvation. We would expect that food is worth more in terms of avoiding starvation when the animal has low rather than high reserves of energy. Thus we would expect an animal's behaviour to depend on reserves, so that it is prepared to accept a greater risk of predation in its search for food when its reserves are low. The framework that we present allows these intuitive ideas to be given a firm formal basis.

We are concerned with the optimal choice of action at time t. In this chapter we assume that we already know how the reproductive value of the organism at time $t + 1$ depends on its state at this time, i.e. we know how $V(x', t + 1)$ depends on x'. Given this function we can now analyse the consequences of the actions performed by an organism in state x at time t. We define $H(x, t; u)$ to be the reproductive value of an organism in state x at time t that performs action u. In the deterministic case $H(x, t; u)$ is given by the following equation:

$$H(x, t; u) = B_{\text{off}}(x, t; u) + S(x, t; u)V(x', t + 1).$$

This reproductive value is a measure of the organism's expected future reproductive success and is decomposed into two components, the contribution to reproductive success $B_{\text{off}}(x, t; u)$ between times t and $t + 1$, and the contribution $V(x', t+1)$ to reproductive success from time $t+1$ onwards, given the organism survives until time $t + 1$. In the more general stochastic case we have

$$H(x, t; u) = B_{\text{off}}(x, t; u) + S(x, t; u)\mathbb{E}_u[V(X', t+1)], \qquad (2.1)$$

where $\mathbb{E}_u$ denotes expectation (i.e. the mean) taken over the possible states resulting from action u being chosen, conditional on survival until time $t+1$.

The H's provide a common currency for evaluating different actions (McNamara & Houston 1986). The best action u^* is the one with the highest H, i.e.

$$H(x, t; u^*) = \max_u H(x, t; u), \qquad (2.2)$$

where the maximization is taken over all possible actions in state x at time t. We now consider various examples to illustrate this approach.

Food versus danger of predation

There is considerable evidence that the danger of predation influences the foraging behaviour of animals (e.g. Lima & Dill 1990). When an animal eats food it gains energy or nutrients that may enhance future reproduction. In looking for food the animal may be killed by a predator and thus have no future reproductive success. Consider an animal in a refuge where it is safe from predators. Should the animal leave the refuge to feed in an area where it may be killed? We assume that food gives the animal energy. The state variable x might be energy reserves (appropriate for a small bird) or size (appropriate for analysing the growth of a fish).

Let the animal's state at time t be x. We compare two actions, u_1 and u_2.

u_1: Feed for unit time.
    The animal is killed by a predator with probability z.
    If the animal survives until time $t + 1$ its state at this time is $x + c$.

u_2: Remain in the refuge until $t + 1$.
    The animal is certain to survive until $t + 1$.
    The state at time $t + 1$ is x.

From our definitions we have

$$H(x, t; u_1) = (1 - z)V(x + e, t + 1)$$

and

$$H(x, t; u_2) = V(x, t + 1).$$

It is optimal to perform action u_1 if and only if $H(x, t; u_1) > H(x, t; u_2)$, i.e. if

$$V(x + e, t + 1) - V(x, t + 1) > zV(x + e, t + 1).$$

The left-hand side of this inequality is the increase in reproductive value that results from feeding. The right-hand side is the loss in reproductive value as a result of predation, since z is the probability of death and $V(x + e, t + 1)$ is the loss given that the animal is killed.

The condition for u_1 to be optimal can also be expressed in terms of the critical level of danger, z_c. It follows from the inequality that u_1 is optimal if $z < z_c$, where

$$z_c = \frac{V(x + e, t + 1) - V(x, t + 1)}{V(x + e, t + 1)}.$$

This ratio plays a central role in the analysis of the energy–predation trade-off given in Chapter 6.

Typically z_c will have a strong dependence on x, and may also depend on the time of day or time of year. In many circumstances the value of the animal's life, $V(x + e, t + 1)$, increases as its reserves, x, increase. The value of a food item, $V(x + e, t + 1) - V(x, t + 1)$, depends on the biological circumstances, e.g. whether energy is being used for survival or reproduction. When the animal is attempting to avoid starvation, the value of a food item typically decreases with increasing reserves x (Chapter 6). Under these circumstances, the maximum acceptable predation risk, z_c, decreases as the reserves increase.

Although this example could be developed in various ways (e.g. by including the energy spent while in the refuge or by specifying a distribution of possible gains while foraging) we have kept it simple to emphasise the logic of the procedure. Basically all we are doing is keeping track of the fitness consequences of behaviour by keeping track of the state changes produced by the behaviour.

Life-history trade-offs: current versus future reproduction

Williams (1966) pointed out that there can be a trade-off between reproductive success in the current breeding season and reproductive success in future years. He represented an organism's current reproductive value as the sum of its reproductive success in the current breeding attempt plus its residual reproductive value. By 'residual reproductive value' Williams meant the expected reproductive value of the organism in one year's time. Thus his decomposition of current reproductive value is precisely that given by equation (2.1) if the times t and $t + 1$ are one year apart. Since we are measuring reproductive value at the same time each year, and since expected future reproductive success is independent of the calendar year t, we can write reproductive value at the focal time of year as a function, $V(\boldsymbol{x})$, of state alone. Equation (2.1) then becomes

$$H(\boldsymbol{x}; u) = B_{\text{off}}(\boldsymbol{x}; u) + S(\boldsymbol{x}; u)\mathbb{E}_u\{V(\boldsymbol{X}')\}. \tag{2.3}$$

Let the action u be the effort that the organism puts into the current breeding attempt. If u is taken to be a proportion of the maximum possible effort we can suppose $0 \le u \le 1$.

The analysis of a sequence of state-dependent life-history decisions is given in Chapter 8. Here we illustrate some single reproductive decisions.

Current reproduction versus survival. We start by considering a simple model with no state variable other than whether the organism is alive at the start of the breeding attempt. In particular, the effects of age on breeding success and survival are ignored. As a result, we can write reproductive value as a constant, V. The organism's effort determines

$B_{\text{off}}(u)$,   the expected number of offspring that survive to maturity

and

$S(u)$,   the probability that the organism survives to the next breeding attempt.

We expect $B_{\text{off}}(u)$ to be an increasing function of u and $S(u)$ to be a decreasing function of u. For example, for a bird feeding its young, u might be the proportion of the day spent collecting food for the young, in which case as u increases the young get more food and so $B_{\text{off}}(u)$ should increase. Increasing u increases the time for which the parents are exposed to predators, so $S(u)$ is likely to decrease (see also Lima 1987a). Another example is provided by a male displaying to attract a female. In this case u is a measure of the strength of the signal. A stronger signal is more likely to attract a female, but it may also increase the male's probability of being killed. Calling by frogs attracts predatory bats (Tuttle & Ryan 1981, Ryan *et al.* 1982) and calling in crickets increases the risk of attack by the parasitic fly *Euphasiopteryx ochracea* (Cade 1975). In the case of courtship in the guppy, the signal is visual rather than auditory and the action u is now a physiological response that determines the male's colour pattern (Endler 1995).

In this case, where there is no state variable, equation (2.3) reduces to

$$H(u) = B_{\text{off}}(u) + S(u)V.$$

The optimal effort u^* maximises $H(u)$. Provided that suitable further assumptions about the functions B_{off} and S are made, it can be shown that u^* decreases as V increases (see Charnov & Krebs 1974, Schaffer 1974a). This is intuitively reasonable: the greater the expected future reproductive success, the lower the risks that the organism should take in the current reproductive bout.

Reproduction versus survival and future state. If the organism survives to breed in the next season, its state in that season may depend on its current state and reproductive effort. As a simple example we consider an organism that can allocate resources either to growth or to reproduction.

Let the current size of the organism be x, and assume that there are no age effects, so that reproductive value is a function of just the organism's size. A proportion u of resources is allocated to reproduction and a proportion $1 - u$ to growth. The choice of u determines

$$B_{\text{off}}(x; u), \quad \text{the current reproductive success}$$
$$S(x; u), \quad \text{the probability of surviving to the next year}$$

and

$$x' = x'(x; u), \quad \text{the size in the next breeding season.}$$

Then

$$H(x; u) = B_{\text{off}}(x; u) + S(x; u)V(x').$$

The optimal allocation to reproduction at size x is the value of u that maximises $H(x; u)$. This can be found by keeping x fixed and differentiating H partially with respect to u to obtain $\partial H / \partial u$. The optimal value of u is that for which $\partial H / \partial u = 0$. From the equation for H,

$$\frac{\partial H}{\partial u} = \frac{\partial B_{\text{off}}}{\partial u} + \frac{\partial S}{\partial u} V + S \frac{dV}{dx} \frac{\partial x'}{\partial u}.$$

The term $\partial B_{\text{off}} / \partial u$ is the marginal increase in current reproductive success with increased effort u. Effort is likely to decrease survival, so $\partial S / \partial u$ will be negative. Thus $-(\partial S / \partial u)V$ is the marginal cost, in terms of survival, of increased effort. Finally, increased effort is likely to reduce future reproductive value because growth is reduced. Thus $(dV/dx)(\partial x'/\partial u)$ is negative and $-S(dV/dx)(\partial x'/\partial u)$ is the marginal cost, in terms of future reproduction, of increased effort. At the optimal value of u, the marginal benefit equals the sum of the marginal costs.

Models like this can be used to explore patterns of growth and reproduction. In particular, predictions can be made about when there should be a period of intense growth $(u = 0)$ followed by a period of intense reproduction $(u = 1)$ as opposed to a life history that always involves both growth and reproduction at all ages (see, for example, León 1976, Sibly *et al.* 1985, Perrin & Sibly 1993 and reviews in Roff 1992, Stearns 1992).

The trade-off between energy and time

Many foraging decisions involve gains in energy at the expense of time. For example, when a foraging animal encounters a poor-quality item it can eat it and thus increase its energy reserves, but the associated handling time might have been better spent looking for further items. This is the prey choice problem (e.g. Charnov 1976a, Stephens & Krebs 1986). The same sort of trade-off arises when an organism is foraging in a patch and its rate of energetic gain is decreasing. Should the animal continue foraging in the patch or should it leave and look for a new patch?

Although it is not always made clear in discussions of foraging theory, the best decision in such cases depends on the current feeding options and on future expectations, i.e. on V. The worse the future expectations, the more worthwhile it is to spend time exploiting a current food source. We now show how our approach can be used to analyse such decisions. We will work with a particular form of V that is used (often implicitly) in models based on the maximization of the rate of energetic gain.

Let an animal be characterised by its level of energy reserves. To model the reproductive value, $V(x',t')$, of an animal with reserves x' at time t', we suppose that the animal can forage from time t' to some time T at mean rate γ. The mean level of reserves at time T is thus $x'+\gamma(T-t')$. We assume that reproductive value at T is proportional to reserves at this time. Thus the reproductive value of the animal with reserves x' at time t' is proportional to $x' + \gamma(T - t')$. Since the value of the constant of proportionality does not influence the optimality analyses given below we take it to be unity and set

$$V(x',t') = x' + \gamma(T - t'). \tag{2.4}$$

Now consider an animal with reserves x at time t. Suppose that the animal must choose a foraging option from an available range. An action u thus corresponds to a choice of foraging option. Foraging options differ in terms of energetic gain and the time taken to achieve this energetic gain, which we refer to as the handling time. We denote the net energetic gain under action u by $e(u)$ and the handling time by $h(u)$. Thus if an animal chooses action u, then its reserves will be

$$x' = x + e(u)$$

at time

$$t' = t + h(u).$$

The reproductive value of the animal if it chooses action u is thus $H(x, t; u) = V(x', t')$, where x' and t' are as above. Assuming that $V(x', t')$ is given by equation (2.4), we thus have

$$
\begin{aligned}
H(x, t; u) &= x + e(u) + \gamma\{T - [t + h(u)]\} \\
&= x + \gamma(T - t) + e(u) - \gamma h(u).
\end{aligned}
$$

The first two terms (i.e. $x + \gamma(T - t)$) are independent of u and hence the best option is the one that maximizes

$$\tilde{H}(u) = e(u) - \gamma h(u). \tag{2.5}$$

$\tilde{H}$ is composed of two terms, both of which are energies. The first term is the energetic gain that results from performing action u. The term $\gamma h(u)$ is the energy that could have been obtained if, instead of performing action u, the animal had spent its time foraging at the mean net rate of gain γ for the environment. Thus $\gamma h(u)$ represents an energy cost in terms of lost opportunity. We now illustrate how $\tilde{H}$ can be used to derive the optimal strategy in two standard foraging paradigms, prey choice and patch use.

Prey choice.   Assume that the animal has just encountered an item with energetic content e and handling time h. Should it accept it or reject it? The two possible actions are u_1 and u_2, a follows.

u_1:   Take the item.
             $e(u_1) = e, \ h(u_1) = h$
u_2:   Reject the item.
             $e(u_2) = 0, \ h(u_2) = 0.$

Thus we have

$$\tilde{H}(u_1) = e - \gamma h$$

and

$$\tilde{H}(u_2) = 0,$$

and so it is optimal to take the item if

$$e - \gamma h > 0$$

i.e. if

$$e/h > \gamma. \tag{2.6}$$

The ratio e/h is known as the profitability of the item. The argument shows that all items with a profitability greater than γ should be accepted, and all items with a profitability less than γ should be rejected (see also Charnov 1976a, Stephens & Krebs 1986).

Patch use. Consider an animal that encounters a patch of food. The intake rate after the animal has spent time v in the patch is $r(v)$. It is assumed that $r(v)$ decreases with time in the patch v. This might, for example, be the result of the animal's consumption of food in the patch. How long should the animal remain on the patch before moving on to another food source?

For this example the action u corresponds to the choice of total time on the patch. For given u the energetic gain to the animal is

$$e(u) = \int_0^u r(v)dv$$

and the handling time is

$$h(u) = u.$$

By equation (2.5) the optimal time to spend in the patch maximises

$$\tilde{H}(h) = e(u) - \gamma u.$$

The condition $d\tilde{H}(u)/du = 0$ gives us this best time u^*. It follows that since $e'(u) = r(u)$ we have

$$r(u^*) = \gamma,$$

i.e. it is optimal to leave the patch when the current rate drops to the overall rate for the environment. This is the marginal value theorem (e.g. Charnov 1976b).

The maximisation of energy obtained from foraging is discussed further in Chapter 4.

How long should a parent stay away from its young in response to a predator?

To illustrate the approach in a rather more complicated context than the previous cases, we analyse the question of how long a parent bird should spend away from its nest following the arrival of a predator. The purpose of this example is to show how dependent young can be included as part of their parent's state. For simplicity, we assume that the nest contains a single offspring. The predator is dangerous to the parent, but may also constitute a threat to the young. On sighting the predator, the parent flies away from the nest and then has to choose the time u at which it returns. The longer that the parent stays away, the greater is its chance of survival. There are, however, two costs of being away. The probability that the offspring is still alive decreases with u and, given that it is still alive, its condition decreases with u. We assume that if the parent dies then the offspring also dies.

The offspring is characterised by its energy reserves, x, and size, y. We ignore changes in the state of the parent, and simply characterise it as being alive or dead. Let $V(x, y, t)$ be the reproductive value of the parent given that it is alive and has an offspring with reserves x and size y at a time t in the breeding season. V is a measure of the descendants left by the parent far into the future. These descendants could be descendants of the current offspring or descendants of offspring produced by the parent in future breeding attempts. Thus V takes into account the probability that the current offspring survives until independence, its state at independence and the expected reproductive success of the parent in future breeding attempts.

The probability that the parent survives the current breeding attempt will depend on its behaviour, which in turn will depend on the current state of its offspring. Similarly, since the offspring is still dependent on its parent, the probability that the offspring survives to independence depends on future parental behaviour. Including the offspring's state as a component of the state of the parent allows all these possibilities to be taken into account. If the offspring is dead at time t in the breeding season, then we denote the reproductive value of the parent by $V_p(t)$. If the parent stays away for time u and then returns, then the probability that the parent survives is $S_p(u)$ and the probability that the offspring survives is $S_{off}(u)$. Typically S_p will increase with u and S_{off} will decrease. If the offspring is still alive when the parent returns then its reserves will have decreased from x to $x' = x - \alpha(y)u$. We assume that the size will not have changed, so that $y' = y$. Suppose that the predator appears at time t. If the offspring has reserves x and size y at this time and the parent is absent for time u, then the reproductive value of the parent is

$$H(x, y, t; u) = S_p(u)\{S_{off}(u)V(x', y', t + u) + [1 - S_{off}(u)]V_p(t + u)\}.$$

The optimal value of u is the one that maximizes H.

The model can be used to make predictions about how offspring reserves and size should influence the time for which the parent stays away. This example illustrates our general approach to the modelling of trade-offs. Behaviour is seen as influencing both immediate reproductive success and future state. Having found optimal behaviour, we can (if we wish) identify terms that correspond to benefits and costs of the behaviour. In other words, costs and benefits emerge from the analysis rather than being a starting point for the analysis. Taking costs and benefits as the starting point does not require an evaluation of the biological consequences of behaviour. As a result, such an approach may lead to an unrealistic model of the situation that is being analysed.

Further examples of trade-offs

In this chapter we have presented an approach based on states, actions and trade-offs. Even within foraging theory, trade-offs can be more complicated than the simple time versus energy trade-offs that we have discussed. For example, consider the foraging decisions of an ectotherm such as a lizard. The lizard's body temperature will influence many aspects of its performance (see Huey & Kingsolver 1989 for a review) and will be influenced by the lizard's behaviour. It is therefore necessary to take body temperature to be a component of state. Intake rate and digestive efficiency may improve as body temperature rises. This means that if the animal is foraging in a relatively cool habitat, by basking to raise its body temperature it can increase its foraging efficiency, but at the cost of decreasing the time available for foraging. A more detailed model would go beyond energetic considerations by including the dependence on body temperature of the ability to escape predation.

To model the behaviour of the mason bee *Chalicodoma sicula*, as described by Willmer (1986), several state variables might be required. This bee constructs its nests from sand. Although nests occur in colonies, each female constructs her own cells and provisions them with nectar and pollen. Willmer studied these bees in a dry and windy region of Israel, and concluded that choice of flower was largely determined by the need to obtain water. Bees preferred dilute nectar to concentrated nectar, even though this preference resulted in longer journeys. Water is important not only for regulating the concentration of the blood but also for controlling body temperature. In modelling the behaviour of this bee the state might include variables that characterize the nest (e.g. state of construction, amount of nectar, amount of pollen) and variables that characterize the bee (size, energy balance, blood concentration, body temperature).

The male smooth newt *Triturus vulgaris* courts the female under water while holding his breath (Halliday & Sweatman 1976). In analysing whether a male should continue courtship or should go to the surface to breathe, we need to establish the consequences of these actions for the male's state, which can be taken to be his oxygen balance and the receptiveness of the female. Courting is likely to increase the receptiveness of the female, and hence result in mating, but will also use up oxygen. Breathing will improve the male's oxygen balance, but the female may be lost to a rival.

In the context of avian migration, a bird on its wintering ground may have the choice between staying there for a while to increase its fat reserves or starting its journey to the breeding ground. Staying on the wintering ground to build up reserves may increase the reserves that the bird has on arrival at the breeding ground but it may also mean that the bird arrives later in the breeding season. Given that breeding success typically increases with reserves but decreases with arrival date, there is a trade-off between reserves at arrival and date of arrival. This could be analyzed in a model in which the state variables are reserves and location (e.g. Weber *et al.* 1998). Such an approach ignores the importance of weather conditions for migrating birds. Wind speed can influence flight costs and the distance that can be travelled on a given amount of fat. In order to understand the importance of these effects, the wind conditions could be included in the state, and the success of model birds that base their decisions on all the state variables could be compared with that of model birds that base their decisions just on reserves and location.

In all examples presented in this chapter, we have looked at optimal decisions at time t, assuming that we know how reproductive value at time $t + 1$ depends on state. Reproductive value at time $t + 1$ provides the link between current decisions and the future life history of the organism. In the next chapter we explicitly analyse the relationship between the value of current decisions and the future decisions of an organism.

3

Dynamic optimisation

3.1 Introduction

In Chapter 2 we looked at the choice between various actions at some time t. An action made a direct contribution to reproductive success between times t and $t+1$. It also determined the state of the organism at time $t+1$. In Chapter 2 we assumed that reproductive value at time $t+1$ was known. In this chapter we show how it can be derived by consideration of behaviour from time $t+1$ onwards.

The reproductive value of an organism at time $t+1$ is a measure of the reproductive success over the entire life from time $t+1$ onwards. It is often impractical, however, to analyse the whole of an organism's future life history. We would not want to follow a great tit to the end of its life in order to find out whether it should eat a particular prey item. Fortunately, it is often reasonable to carry out an analysis over a much shorter period of time. One approach is to choose a time interval such that reproductive value at the end of the interval is known from biological considerations. Alternatively it is sometimes possible to choose the time until the end of the interval to be sufficiently long that reproductive value at the end of the interval is irrelevant to behaviour at the time of interest. Once the time interval has been chosen and reproductive value has been assigned to states at the end of the interval, optimal behaviour over the interval is found by working backwards from the end. In this chapter we show how such an analysis can be carried out.

3.2 Modelling decisions over a finite time interval

We consider a time interval that starts at time 0 and ends at time T. The interval is referred to as $[0, T]$, and T is referred to as the final time. In principle T can be any time but, as indicated above and as we shall see later

25

in the chapter, there are practical reasons for choosing a final time that is biologically meaningful. The organism chooses between actions at each time $t = 0, 1, 2, \ldots, T - 1$. The set of possible actions available to the organism may depend on both the time and the organism's state.

The action performed by the organism determines the immediate contribution to reproductive success, which may of course be zero, and either the next state (if the model is deterministic) or the distribution of next states (if the model is stochastic) one time step later. Thus, given that action u is performed by an organism in state x at time t, the immediate contribution to reproductive success $B_{\text{off}}(x, t; u)$ and the next state x' are as defined in Chapter 2.

A strategy is a rule for choosing actions at each time t in $[0, T]$. Under such a rule the choice of action can depend on both the organism's state and on time. If the strategy is deterministic, then for a given state at a given time, the strategy specifies that the organism performs one particular action. In contrast, a stochastic strategy will specify the probabilities of performing various actions in a given state at a given time. To compare the performance of various strategies, we must calculate the fitness consequences of following each strategy.

If an organism survives until time T its reproductive value is some function $R(x)$ of its state at this time. R is called the terminal reward function. In Section 3.4 we discuss how R might be found, but for the moment we assume that this function is known. We now show how it links behaviour during the period $[0, T]$ to the rest of the organism's life-history.

An organism's reproductive value is given by the following equation:

$$\left\{ \begin{array}{c} \text{reproductive} \\ \text{value at} \\ \text{time } 0 \end{array} \right\} = \left\{ \begin{array}{c} \text{direct contribution to} \\ \text{reproductive success} \\ \text{during } [0, T] \end{array} \right\} + S_{\text{total}} \times \left\{ \begin{array}{c} \text{reproductive} \\ \text{value at} \\ \text{time } T. \end{array} \right\}$$

where S_{total} is the probability of survival from time 0 until time T.

A strategy that an organism adopts during $[0, T]$ will determine all three components of the right-hand side of this equation. There may often be a trade-off between the components, such that a high value of one component can only be achieved at the cost of a low value of another component. For example, an animal that devotes all its energy to reproduction during $[0, T]$ is likely to be in poor condition at T and hence have a low reproductive value at time T. The equation is the direct analogue of the trade-off equations that we gave in Chapter 2, but in that chapter we were concerned with a single action, whereas now we are choosing a sequence of actions, i.e. a strategy.

The optimal strategy over $[0, T]$ maximises the organism's reproductive value at time 0.

Finding the optimal strategy

The fitness consequences of an action depend on future actions. For example, in Chapter 2 we looked at the trade-off between food and predation risk. The level of predation risk that an animal should be prepared to accept at some time t depends on reproductive value at time $t + 1$. This in turn depends on the food that it can expect to obtain between $t + 1$ and T, which depends on both the availability of food and the actions taken between $t + 1$ and T. Similarly, if we are studying a parent bird that can mob a predator which threatens the bird's young, the bird's optimal decision depends on the probability that the young survive to independence, which in turn depends on how the parent feeds them and defends them in the future. It is therefore not possible completely to isolate a single action from future actions (McNamara & Houston 1986). This makes it natural to find an optimal strategy over an interval $[0, T]$ by working backwards from time t. In other words, once optimal behaviour from time $t + 1$ is known, and hence reproductive value at time $t + 1$ is known, we can use the method of Chapter 2 to find the optimal behaviour at time t.

Before we describe the technical details of the procedure of working backwards from T, we introduce some notation. The optimal strategy over the time interval $[0, T]$ is denoted by π^*. Under the optimal strategy, the action chosen when in state x at time t is denoted by $u^*(x, t)$. $V(x, t)$ is the reproductive value of an individual, in state x at time t, that follows the optimal strategy.

The technique that we now present is known as dynamic programming. To apply this technique, we start with a terminal reward $R(x)$ that gives the reproductive value at final time T, i.e.

$$V(x, T) = R(x).$$

We then work backwards finding V at time t in terms of V at time $t + 1$ as follows. For each state x set

$$H(x, t; u) = B_{\text{off}}(x, t; u) + S(x, t; u)\mathbb{E}_u[V(X', t + 1)], \qquad (3.1)$$

where, as in Chapter 2, B_{off} is the immediate contribution to reproductive success between times t and $t + 1$, S is the probability of survival until time $t + 1$ and X' is the state at time $t + 1$ given that the organism survives. $H(x, t; u)$ is thus the reproductive value of an organism, in state x at time t, that performs action u at time t and then behaves optimally from time $t + 1$ onwards. The optimal action $\pi^*(\boldsymbol{x}, t)$ for this organism is then the value of u that maximizes $H(\boldsymbol{x}, t; u)$, i.e.

$$H(\boldsymbol{x}, t; \pi^*(x, t)) = \max_u H(\boldsymbol{x}, t; u). \qquad (3.2)$$

The reproductive value of the organism given that it behaves optimally from time t onwards is thus

$$V(\boldsymbol{x}, t) = H(\boldsymbol{x}, t; \pi^*(\boldsymbol{x}, t)), \qquad (3.3\text{a})$$

or equivalently, by equation (3.2),

$$V(\boldsymbol{x}, t) = \max_u H(\boldsymbol{x}, t; u). \qquad (3.3\text{b})$$

We emphasise that this procedure finds $V(\boldsymbol{x}, t)$ for all possible states $\boldsymbol{x}$ given a knowledge of $V(\boldsymbol{x}', t + 1)$ for all possible states $\boldsymbol{x}'$. Starting with $V(\boldsymbol{x}, T) = R(\boldsymbol{x})$ for all $\boldsymbol{x}$, the first application of equations (3.1)-(3.3) gives us $V(\boldsymbol{x}, T - 1)$ for all $\boldsymbol{x}$. Repeating the procedure gives us $V(\boldsymbol{x}, T - 2)$, $V(\boldsymbol{x}, T - 3)$ and so on. In this way we can find $V(\boldsymbol{x}, t)$ for all states $\boldsymbol{x}$ and times t between 0 and T and $\pi^*(\boldsymbol{x}, t)$ for all $\boldsymbol{x}$ and t.

An illustrative model: Surviving a single day

To illustrate the dynamic programming procedure we describe a simple model based on an animal's foraging behaviour over a single day. The animal starts foraging at time 0 (dawn) and stops at time T (dusk). There is a single state variable x ($0 \leq x \leq L$) that characterises the animal's level of energy reserves. If x reaches 0 then the animal dies of starvation. There is an upper limit L to the reserves x such that any energy that would take x above L is lost. (For reasons explained below, we take L to be an integer.) In addition to death as a result of starvation, the animal may also be killed by predators. If the organism survives until final time T with reserves x then its reproductive value is $R(x)$.

At each of the times $t = 0, 1, 2, \ldots, T - 1$ the animal must choose between two feeding options u_1 and u_2. If it chooses option u_i ($i = 1, 2$) at time t then

z_i = probability that it is killed by a predator between t and $t + 1$.

If the animal is not killed then

p_i = probability that it finds a food item

$$1 - p_i = \text{probability that no item is found.}$$

The energetic value of food items is variable. We assume that it has the following distribution:

$$
\begin{array}{lll}
1 & \text{with probability} & 1/4 \\
2 & \text{with probability} & 1/2 \\
3 & \text{with probability} & 1/4
\end{array}
$$

regardless of which option is chosen. Thus the mean value is always 2. The animal always uses one unit of energy per unit time in metabolic expenditure. We assume that $p_1 < p_2$ and $z_1 < z_2$ so that option 1 offers less food but a lower probability of predation.

The energetic value of the food items, the upper limit L and the animal's metabolic expenditure are such that if the reserves x take integer values at some time, then they will take integer values at all subsequent times. It is therefore possible (and convenient) to model x as taking non-negative integer values.

The dynamic programming equation. An animal with reserves x at time t can be in one of four possible states at time $t+1$. We denote these possible states by x_0', x_1', x_2' and x_3', where

$$
\begin{aligned}
x_0' &= x - 1 \\
x_1' &= x \\
x_2' &= \text{the minimum of } x + 1 \text{ and } L \\
x_3' &= \text{the minimum of } x + 2 \text{ and } L.
\end{aligned}
$$

The animal gets to state x_0' if it is not killed and it does not find a food item, so that its reserves decrease by one unit as a result of metabolism. This occurs with probability $(1 - z_i)(1 - p_i)$. The animal gets to state x_1' if it is not killed and it finds food that has an energetic value of one unit. This occurs with probability $\frac{1}{4}(1 - z_i)p_i$. If the animal finds an item worth two or three units of energy then it is necessary to ensure that reserves cannot exceed the upper limit L. The minimisation incorporated in the definitions of x_2' and x_3' guarantees this. The probabilities of x_2' and x_3' are $\frac{1}{2}(1 - z_i)p_i$ and $\frac{1}{4}(1 - z_i)p_i$ respectively.

It follows from equation (3.1) that

$$
\begin{aligned}
H(x, t; u_i) = \;&(1 - z_i)[(1 - p_i)V(x_0', t + 1) + \tfrac{1}{4}p_i V(x_1', t + 1) \\
&+ \tfrac{1}{2}p_i V(x_2', t + 1) + \tfrac{1}{4}p_i V(x_3', t + 1)]
\end{aligned}
\tag{3.4}
$$

and from equation (3.3b) that

$$
V(x, t) = \max[H(x, t; u_1), \; H(x, t; u_2)].
\tag{3.5}
$$

Because the organism starves if $x = 0$, the lower boundary condition is

$$V(0, t) = 0.$$

At the final time T, V is given by the terminal reward, i.e.

$$V(x, T) = R(x).$$

To illustrate the calculation of the H's, we will use the model *Surviving a single day* with options described by the parameters

$$p_1 = 0.5 \qquad z_1 = 0.0 \tag{3.6a}$$

$$p_2 = 0.6 \qquad z_2 = 0.01 \tag{3.6b}$$

The maximum level of the reserves x is taken to be $L = 20$ and the terminal reward at dusk is given by

$$R_1(x) = \begin{cases} 1 & \text{if} \quad x \ge 10 \\ 0 & \text{if} \quad x < 10. \end{cases} \tag{3.7}$$

Recalling that $V(x, T) = R_1(x)$, we are able to calculate $H(x, T - 1; u_1)$ and $H(x, T - 1; u_2)$ and hence find $V(x, T - 1)$. To illustrate the calculation of $V(10, T - 1)$ note that by equation (3.4)

$$H(10, T - 1; u_1) = (1 - p_1)R_1(9) + p_1 \left[\tfrac{1}{4}R_1(10) + \tfrac{1}{2}R_1(11) + \tfrac{1}{4}R_1(12) \right].$$

Thus, since $R_1(9) = 0$ and $R_1(10) = R_1(11) = R_1(12) = 1$, it follows that

$$H(10, T - 1; u_1) = p_1 = 0.5.$$

Similarly

$$H(10, T - 1; u_2)$$
$$= (1 - z_2) \left\{ (1 - p_2)R_1(9) + p_2 \left[\tfrac{1}{4}R_1(10) + \tfrac{1}{2}R_1(11) + \tfrac{1}{4}R_1(12) \right] \right\}$$
$$= 0.99 \times 0.6 = 0.594.$$

Thus $H(10, T - 1; u_2) > H(10, T - 1; u_1)$ and so option u_2 should be chosen when $x = 10$ and $t = T - 1$. By equation (3.5),

$$V(10, T - 1) = H(10, T - 1; u_2) = 0.594.$$

Having found $V(x, T-1)$ for all states x one can then calculate $V(x, T-2)$ for all states x and so on. Examples of the two H's for various values of x at times $t = T - 1$, $t = T - 2$ and $t = T - 3$ are given in Table 3.1.

Table 3.1. *Surviving a single day*: the optimal strategy. $H(x, t; u_1)$ (top entry) and $H(x, t; u_2)$ (bottom entry) calculated using the procedure described in the text. The parameters of the foraging options are given by equation (3.6). The terminal reward is R_1 given by equation (3.7). This function defines $V(x, T)$. For $t \leq T - 1$, $V(x, t) = \max[H(x, t; u_1), H(x, t; u_2)]$ and is shown in bold.

	$H(x, t; u_i)$ for various t-values			
x	$T - 3$	$T - 2$	$T - 1$	$R_1(x)$
12	0.9146	**1.0000**	**1.0000**	1
	0.9224	0.9900	0.9900	
11	0.8337	0.7970	**1.0000**	1
	0.8499	**0.8292**	0.9900	
10	0.6460	0.6720	0.5000	1
	0.6784	**0.7101**	**0.5940**	
9	0.4587	0.4034	0.3750	0
	0.4969	**0.4499**	**0.4455**	
8	0.2867	0.2042	0.1250	0
	0.3188	**0.2426**	**0.1485**	

The optimal strategy. In Figure 3.1 we look at the optimal strategy for three forms of the terminal reward, $R_1(x)$, $R_2(x)$, $R_3(x)$, and for the values of p_i and z_i given by equation (3.6). The first terminal reward is given by equation (3.7). This reward function might be appropriate if the animal uses 9 units of reserves overnight. Under this interpretation the optimal strategy for this terminal reward maximises the probability of survival until the following dawn (see Section 9.3). The second terminal reward $R_2(x)$ is given by

$$R_2(x) = \begin{cases} 1 & \text{if } x \geq 1 \\ 0 & \text{if } x = 0. \end{cases}$$

The optimal strategy for this terminal reward maximises the probability of survival until dusk. Finally, our third terminal reward is appropriate if reproductive value at dusk is proportional to energy reserves at this time, i.e.

$$R_3(x) = kx \quad \text{if } x \geq 0,$$

where k is a positive constant.

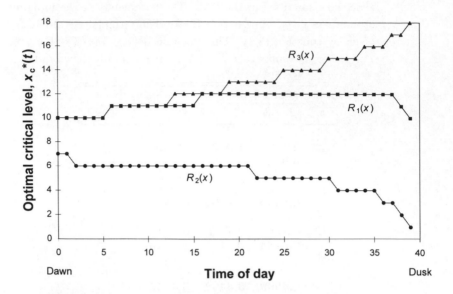

Figure 3.1. Optimal strategies for the model *Surviving a single day*. For given terminal reward the optimal strategy is to choose option u_2 at time t if reserves x are less than or equal to the critical level $x_c^*(t)$ and to choose option u_1 if reserves are above $x_c^*(t)$. The figure shows the function $x_c^*(t)$ for each of the terminal rewards $R_1(x)$, $R_2(x)$ and $R_3(x)$ discussed in the text. Parameter values for the options: $p_1 = 0.5$, $z_1 = 0.0$; $p_2 = 0.6$, $z_2 = 0.01$. Upper boundary $L = 20$. Daylength $T = 40$.

For each of the terminal rewards there is a critical level of reserves $x_c^*(t)$ such that the optimal strategy is to choose option u_1 at time t if $x(t)$ is above $x_c^*(t)$ and to choose option u_2 if $x(t)$ is equal to or less than $x_c^*(t)$. The general pattern of the results shown in Figure 3.1 is that when reserves are low it can be advantageous to accept an increase in the danger of predation in order to achieve a higher mean net gain and when reserves are high it is advantageous to avoid predation even though this results in a lower mean net gain. We will discuss this trade-off between predation and energy in detail in Chapter 6.

It can be seen from the figure that when t is close to T, the critical value of x is strongly dependent on t and on the form of the terminal reward. As t gets further away from T, the critical value settles down to a constant

x_c^* that is independent of t. When there is a long time to go, an animal should virtually ignore the terminal reward and concentrate on avoiding starvation and predation. The critical level reflects the best compromise between these conflicting pressures. As the final time T approaches, the form of the terminal reward becomes important. We first consider the terminal reward $R_2(x)$, under which the animal gets the same reward, 1, in all states as long as it is alive at T. There is then no advantage in having high reserves at time T, and the animal reduces the danger of predation by reducing the critical level $x_c^*(t)$ as T approaches. When the terminal reward is $R_3(x)$, the optimal strategy is independent of the value of the constant k. For this terminal reward it pays to have high reserves at T, and the animal is prepared to accept the dangerous option (option u_2) to raise reserves. This is reflected in the rise in $x_c^*(t)$ as T approaches. When the terminal reward is $R_1(x)$, the animal must attempt to maximize the probability that its reserves are at least 10 at T. Because the food supply is stochastic, the animal needs to start building up its reserves well before final time. Since there is metabolic expenditure, the level to which the animal attempts to build up its reserves is greater than the critical level of 10. When it is close to T an animal with reserves above 10 is prepared to let reserves drift downwards, in order to avoid predation. The combination of these effects results in a critical level $x_c^*(t)$ that first rises and then falls. We will return to this phenomenon when discussing daily routines of foraging and daily routines of singing in Section 9.3.

An illustrative model: Putting resources into growth or reproduction

As a second illustration of the dynamic programming procedure, we consider an organism of indeterminate growth such as a fish or a tree. The organism is characterised by its size, y, and its age, n. The organism reproduces annually. At reproduction it must decide how much of its resources to devote to reproduction and how much to devote to growth. We assume that total available resources are a function $r_{total}(y, n)$ of size and age. If an organism of size y and age n devotes u resources to reproduction and $r_{growth} = r_{total} - u$ resources to growth then it produces $B_{off}(y, n; u)$ offspring that survive to age 1; the focal organism survives to breed in the following year with probability $S(y, n, r_{growth})$, and if it does survive its size is $\hat{y}(y, n, r_{growth})$. We look at the strategy that maximises the expected number of offspring (surviving to age 1) produced over an organism's lifetime.

Consider an organism that behaves optimally. Then we can define $V(y, n)$ to be the expected future number of surviving offspring produced by this organism given that it has size y and age n and is about to reproduce. To derive the dynamic programming equation satisfied by V, suppose that the organism has size y and age n. If the organism allocates resource u to reproduction and then behaves optimally from age $n + 1$ onwards, its expected future number of surviving offspring is

$$H(y, n; u) = B_{\text{off}}(y, n; u) + S(y, n, r_{\text{growth}}) V(\hat{y}(y, n, r_{\text{growth}}), n + 1) \quad (3.8)$$

where $r_{\text{growth}} = r_{\text{total}}(y, n) - u$. Then

$$V(y, n) = \max_u H(y, n; u), \quad (3.9)$$

where the maximum is taken over all u in the range $0 \le u \le r_{\text{total}}(y, n)$.

Assume that there is senescence, so that an organism must die before age T. Then the terminal condition on V is

$$V(y, T) = 0 \qquad \text{for all } y. \quad (3.10)$$

Equations (3.8), (3.9) and (3.10) allow us to find $V(y, n)$ for all y and $n \le T - 1$ by working backwards in n. The optimal strategy for the organism is to allocate resources $\pi^*(y, n)$ to reproduction when its size is y and its age is n, where

$$H(y, n; \pi^*(y, n)) = \max_u H(y, n; u).$$

In order to compute an optimal life-history strategy using the above scheme some discrete grid of size values must be assumed. Suppose an organism has size y at age n, where y lies on the grid. Then, whatever grid is chosen, there will be choices of u for which the size $\hat{y}$ at age $n+1$ does not lie on the grid. One simple way to overcome this difficulty is to introduce some stochasticity, as follows. Suppose that $\hat{y}$ lies between two adjacent grid sizes y_1 and y_2, so that $y_1 \le \hat{y} < y_2$. Define

$$p_1 = \frac{y_2 - \hat{y}}{y_2 - y_1}$$

and set $p_2 = 1 - p_1$. Then, instead of assuming that the size at age $n + 1$ is exactly $\hat{y}$, assume that it is y_1 with probability p_1 and y_2 with probability p_2. In this way, the actual size at age $n + 1$ always lies on the grid and the mean size is $\hat{y}$. Details of the implementation of this type of scheme are given in Appendix 3.1.

There is also a slight difficulty in finding an optimal life-history strategy if there is no upper limit, T, to lifespan. For example, if there is no senescence there may be a non-zero but decreasing probability of survival to any age T. Under these circumstances we still assume that equation (3.10) holds for some T, but now the value of T chosen is successively increased until it has a negligible effect on behaviour in the age range of interest.

Including adequate stochasticity in models

In modelling a biological situation it is often important to include sufficient stochasticity in the model. Failure to do so can have the result that the model makes unrealistic predictions. This can occur for a variety of reasons.

Grid effects. In the model *Surviving a single day* the energetic value of a food item can take the three possible values 1, 2 or 3. Suppose instead that it had been assumed that all items have an energetic value of 2. Since metabolic expenditure per time interval is one unit, an organism's reserves would then change by ± 1 per time interval. Thus, for example, if reserves were 20 at time t, then there is a positive probability that reserves are still 20 at a time T such that $T - t$ is even, but it is impossible that reserves are still 20 at a time T such that $T - t$ is odd. This type of 'grid effect' in models can lead to an unrealistic form for the resultant optimal strategy. For example, the best action to perform may then depend on whether $T - t$ is odd or even and on whether reserves r are odd or even. This grid effect is an artefact of the way changes in reserves have been modelled. In reality an animal attempting to avoid starvation does not make successive foraging decisions at exactly one unit of time apart. Instead it finds food at variable times, and there is no precise time at which dusk falls and the animal must cease foraging. It is important to avoid grid effects in optimisation models since predictions that arise as a consequence of their presence are usually not biologically reasonable. To avoid such artefacts, adequate stochasticity must be incorporated into a model. It is for this reason that we have taken the value of a food item to be variable in the model *Surviving a single day*.

Starvation and variability in the food supply. Animals starve because of adverse fluctuations in their food supply. Hence the degree and type of fluctuation is important in any model in which starvation can occur. Clark & Ekman (1995) model the fat levels of a willow tit in winter. For some of the parameter values that they use, there is so little stochasticity that the bird can guarantee that its net energy gain over each day is non-negative and hence that it does not starve. Consequently the optimal fat level is just above zero regardless of the capacity of the bird to carry fat. If, however, the stochasticity is increased so that the bird cannot be absolutely sure of achieving a non-negative net gain on each day, the picture is completely altered. In the absence of any costs of carrying fat, the optimal level of reserves for surviving an extended period of winter is close to the maximum amount of fat that can be carried (Houston *et al.* 1997).

McNamara *et al.* (1994a) show that not only the amount but also the type of stochasticity is important. In one class of models they allowed the amounts of food found in successive time intervals to be positively correlated. The resulting optimal daily routines of feeding and resting were very different from the routines obtained when food found in successive time intervals was independent.

Variation in observed behaviour. Suppose that all members of a population are in the same state as one another at time 0 and all use the same behavioural strategy. If there were no stochasticity, so that the state changed in a deterministic manner over time, then all population members would be in the same state as one another at a subsequent time t, and hence all population members would behave identically at t. Most natural populations exhibit variation in behaviour. Such variations imply that organisms either have different initial states, follow different strategies, have state changes that are stochastic or have a stochastic relationship between state and behaviour.

One of the fundamental questions in life-history theory is when an animal should start to reproduce. In the absence of frequency-dependent effects, an optimal life-history strategy is both unique and deterministic. Such a strategy specifies those states for which it is optimal to reproduce and those states for which it is not. If age is the only state variable in the model, then because age increases deterministically, all animals will start reproduction at the same age. In many species, this is not the case (see McNamara & Houston 1996). As we argue in Chapter 8, aspects of state other than age are important in life-history decisions. Incorporating a state variable, such as condition, that changes stochastically can account for the variation in age at the time of first reproduction.

There are various ways in which stochasticity can be introduced. For a foraging animal, we can assume that the number of items found in a given time is stochastic, the energy content of an item is stochastic or the time to capture or handle an item is stochastic. In addition, even if processes such as metabolic expenditure and growth can be viewed as deterministic, it may be advisable to model them as being stochastic. Modelling continuous variables on a discrete grid means that the next state takes two possible values instead of one, and hence introduces stochasticity (see Appendix 3.1). We can allow the next state to take more than two possible values; Appendix 3.1 gives a simple algorithm that assigns probabilities to four possible outcomes.

3.3 Expected behaviour

We have seen that the optimal strategy can be calculated using dynamic programming. This strategy does not by itself specify the action that we can expect to observe at a given time. This is so because a strategy specifies what action to perform *if* an organism is in a particular state, but does not specify the likelihood that it is in this state. Most studies of organisms record their behaviour but not their state, so some further work is necessary if we are to use a strategy to make predictions about behaviour. What we can do is to consider a large group of individuals where each follows the strategy. We can then follow this cohort forward in time and find the proportion of cohort members in the various possible states at each time. We assume that the states of cohort members have some given distribution at time 0. The optimal strategy at this time tells us the action that should be taken in each possible state, and the consequences of the action then give us the distribution of states of cohort members at time 1. By repeating this procedure we are able to calculate the distribution of states at any time t $(0 < t \leq T)$ and hence find the proportion of cohort members that choose each of the possible actions at this time. We refer to this proportion as the expected behaviour under the optimal strategy. We can, of course, apply the procedure to find the expected behaviour that results from following not just the optimal strategy but any strategy.

If organisms are dying during the period $[0, T]$ then the mortality in each time interval can be calculated. Since observations are based on organisms still alive, to compare our results to data it is necessary in the model to calculate the proportion of those organisms that are still alive that perform each action at a given time t.

The procedure is now illustrated using the model *Surviving a single day* given in Section 3.2 above. Assume that we have a strategy and that we wish to find the resulting expected behaviour. We start with some distribution of the reserves of cohort members at time 0. Let $P(x, 0)$ be the proportion of these animals that have reserves x at this time. For example, if we let

$$P(x, 0) = \begin{cases} 1 & \text{if } x = x_0 \\ 0 & \text{otherwise,} \end{cases}$$

then all animals start the day with reserves x_0.

To find the distribution of reserves through time we use the following iterative procedure for finding the distribution of reserves at time $t + 1$ from the distribution of reserves at time t. The procedure also finds the proportion of cohort members dying in any time period, together with the cause of death.

Define $P(x,t)$ as the proportion of the original cohort of animals that have reserves x at time t ($x > 0$). To look at the numbers dying, we define $P(0, t+1)$ to be the proportion of the original cohort that dies of starvation between times t and $t+1$, and $Pred(t+1)$ to be the proportion of the original cohort that is killed by a predator between times t and $t+1$. Then the iterative procedure is as follows.

1. Initially, set $P(y, t+1) = 0$ for all y, $1 \leq y \leq L$, and set $P(0, t+1) = 0$ and $Pred(t+1) = 0$.

2. For each $x = 1, 2, \ldots, L$

   (a) use the strategy to find the action taken when the reserves are x. Let z be the probability of being killed by a predator and p be the probability of finding a food item under this action. Then

   (b) add $(1-z)(1-p)P(x,t)$ to $P(x_0', t+1)$,
   add $\frac{1}{4}(1-z)pP(x,t)$ to $P(x_1', t+1)$,
   add $\frac{1}{2}(1-z)pP(x,t)$ to $P(x_2', t+1)$,
   add $\frac{1}{4}(1-z)pP(x,t)$ to $P(x_3', t+1)$,
   add $zP(x,t)$ to $Pred(t+1)$,

   where, as before, $x_0' = x - 1$, $x_1' = x$, $x_2' = \min(x+1, L)$ and $x_3' = \min(x+2, L)$.

To motivate the calculation in 2(b) consider the first line. A proportion $P(x,t)$ of the original cohort members are in state x at time t. Of these a proportion $(1-z)(1-p)$ survive until time $t+1$ and do not find a food item. Their state at time $t+1$ is then x_0'. Thus individuals in state x at time t contribute $(1-z)(1-p)P(x,t)$ to the proportion $P(x_0', t+1)$ of original cohort members in state x_0' at $t+1$. Other terms in 2(b) have a similar interpretation. Some results of applying the above computational procedure to the model *Surviving a single day* are shown in Table 3.2.

Given the distribution of states, we can calculate the proportion, $Alive(t)$, of original cohort members that are still alive at time t:

$$Alive(t) = \sum_{x=1}^{L} P(x,t).$$

Let $\tilde{\rho}_u(t)$ be the proportion of the original cohort members that perform action u at time t. Then

$$\tilde{\rho}_u(t) = \sum_{x:\pi(x,t)=u} P(x,t),$$

Table 3.2. *Surviving a single day:* iterating forward. We assume that all cohort members start with reserves $x = 8$ at time 0. Subsequently they follow the optimal strategy for terminal reward R_1 (see Figure 3.1). For the range of t considered in this table this strategy is to choose option u_1 when x is 11 or more and option u_2 when x is 10 or below. (*a*) gives the proportion $P(x,t)$ of the original cohort in each possible state at times $t = 0, 1, 2, 3$. (*b*) gives the proportion $\tilde{\rho}_2(t)$ choosing option u_2, the proportion $Alive(t)$ that are still alive, the proportion $\rho_2(t) = \tilde{\rho}_2(t)/Alive(t)$ of those still alive that choose option u_2, the proportion $Pred(t+1)$ of the original cohort that are killed by a predator between t and $t + 1$, and the proportion $Pred(t+1)/Alive(t)$ of cohort members alive at t that are killed by a predator between t and $t + 1$.

(*a*)

x	$t = 0$	1	2	3
		$P(x,t)$		
14	0	0	0	0.0028
13	0	0	0	0.0165
12	0	0	0.0221	0.0445
11	0	0	0.0882	0.0919
10	0	0.1485	0.1323	0.1631
9	0	0.2970	0.2058	0.1768
8	1	0.1485	0.2573	0.1779
7	0	0.3960	0.1176	0.1659
6	0	0	0.1568	0.0699
5	0	0	0	0.0621

(*b*)

	$t = 0$	1	2	3
$\tilde{\rho}_2(t)$	1	0.99	0.8698	0.8157
$Alive(t)$	1	0.99	0.9801	0.9714
$\rho_2(t)$	1	1	0.8875	0.8398
$Pred(t+1)$	0.01	0.0099	0.0087	0.0082
$Pred(t+1)/Alive(t)$	0.01	0.01	0.0089	0.0084

where the summation is taken over all states x for which the strategy specifies that action u should be performed. What really interests us is the proportion $\rho_u(t)$ of those cohort members still alive at time t that perform action u.

This proportion is given by

$$\rho_u(t) = \frac{\tilde{\rho}_u(t)}{Alive(t)}.$$

The results of calculating $\rho_u(t)$ in this manner are presented in Table 3.2 and Figure 3.2. The mortalities $P(0, t + 1)$ and $Pred(t + 1)$ give the proportions of the original cohort present at $t = 0$ that die of starvation and predation, respectively, between times t and $t + 1$. It may, however, be of more interest to focus on the proportion of individuals alive at time t that die between times t and $t + 1$. $P(0, t + 1)/Alive(t)$ and $Pred(t + 1)/Alive(t)$ are the relevant proportions dying of starvation and predation respectively.

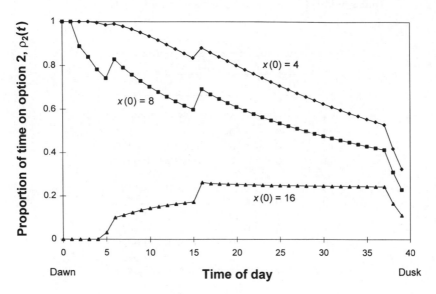

Figure 3.2. Expected behaviour for the model *Surviving a single day*, given that the optimal strategy for terminal reward R_1 is followed. (This strategy is illustrated in Figure 3.1). The figure shows the proportion of time spent on option u_2 as a function of time of day for the three initial states at dawn $x_0 = 4$, $x_0 = 8$ and $x_0 = 16$. Parameter values are as in Figure 3.1.

Behavioural sequences

The proportions $\rho_{u_1}(t), \rho_{u_2}(t), \dots$ give the distribution of behaviours at a time t. However, many studies are concerned with observing sequences of decisions performed by individuals over time. For example, it has been observed that grey-headed albatrosses do not breed every year, but tend to skip a year between each breeding attempt. In particular there is a negative correlation between whether a grey-headed albatross had a successful breeding attempt in one year and whether it attempted to breed in the following year (Prince *et al.* 1981).

Once an optimal strategy has been calculated using dynamic programming, we can analyse the sequences of behaviour that are predicted to emerge if an organism follows the strategy. One method is to perform a large number of simulations in which an organism follows the strategy, and analyse the resulting model 'data' just as if it were empirical data. When, however, a strategy is used to predict how behaviour at time t correlates with behaviour at $t - 1$, simulation may not be the most efficient computational method. Instead, it may be better to introduce a dummy state variable and calculate the expected behaviour by working forwards. This state variable, a, keeps track of the previous activity of the organism. The state at time t now comprises a pair (x, a) where x is the usual state variable describing the organism and a is the action taken by the organism at $t - 1$. We then run forward to find, for each x, a and t, the proportion $P(x, a, t)$ of the organisms that are in state (x, a) at time t. This proportion is the proportion of the original cohort that took action a at $t - 1$ and are in state x at t. Since the action chosen at t depends only on the state x we can thus determine how the action at t correlates with the action at $t - 1$.

3.4 The terminal reward

We have shown how dynamic programming can be used to find the reproductive value $V(x, t)$ given a terminal reward $R(x)$ at T. This still leaves us with the problem of specifying the terminal reward. This reward is defined as the reproductive value of the organism at time T, so it clearly depends on the organism's life-history after time T. We now discuss various ways in which we could find a suitable terminal reward.

(i) If T is beyond the maximum age to which the organism can survive, then the terminal reward is irrelevant because T is never reached. In this case, strategies will differ in their resulting expected contribution to reproductive success before time T, and the optimal strategy will maximize expected reproductive success from time 0 until the death of the organism.

The situation is similar if T is such that the organism cannot reproduce beyond time T and cannot make a contribution to its reproductive success by caring for kin after this time. In this case, the organism may survive until time T, but if it does so it obtains a reward of zero. Examples are provided by studies of the egg-laying behaviour of insects (e.g. Mangel 1987, 1989).

(ii) We could model the whole of the organism's life history from T until the organism dies and find the optimal behaviour over this period. We could then take $R(x)$ to be the reproductive value of an organism in state x at T given that its future behaviour is optimal. This is an example of what Mangel & Clark (1988) call sequential coupling. Using this approach, the optimal behaviour found by dynamic programming over the period $[0, T]$ is the same as would be found over this period if the whole life history from birth to death were analyzed.

Approaches (i) and (ii) entail considerable effort so, except in the case of short-lived organisms, some simplification is desirable.

(iii) We might base R on data concerning reproductive success after time T. For example, in an animal whose reproductive success depends only on its size we could take T to be the start of the breeding season and use data on the correlation between size and reproductive success to determine R. In this way we would avoid the need to model the details of how size influences reproductive success. When we find the optimal behaviour up to time T in such cases, we are finding an optimum subject to the constraint that performance after T follows the observed pattern.

This method of determining R gives a constrained optimum, as opposed to the global optimum that is found by taking behaviour after T to be optimal (cf. Houston & McNamara 1985a, McNamara & Houston 1986). If the whole life-history is optimal, then the constrained optimisation must correctly predict behaviour over $[0, T]$. The fact that behaviour over $[0, T]$ follows the constrained optimum does not, however, imply that the whole life-history is optimal. In other words, following the constrained optimum is a necessary but not sufficient condition for the whole life history to be optimal.

(iv) Although finding a terminal reward for some arbitrary time T may be difficult, it is possible that a reasonable guess about the form of R can be made if T corresponds to some biologically relevant time. A sensitivity analysis can then explore how changes in the terminal reward change the model's predictions. If the predictions are not very sensitive to changes in the terminal reward, the model is said to be robust. We can then have some confidence in the model's predictions, even though we do not know the terminal reward exactly. In contrast, if the model's predictions are very sensitive to changes in the terminal reward, then we have established that a detailed knowledge of the terminal reward is required if accurate predictions are to be made about behaviour during the period $[0, T]$. This is a general fact about the biology of the system, not a feature that is unique to our approach. Any attempt to model the relationship between behaviour and reproductive success in such a system will also need this detailed information. We can use the sensitivity analysis to investigate which aspects of the terminal reward are the most important, so that further empirical work can be focused upon them.

3.5 Convergence

In the previous section we discussed the sensitivity of predictions to the form of the terminal reward. There is a general feature of dynamic models that is useful in this context. The optimal strategy near final time T is likely to depend on the terminal reward but, as the time to go, $T - t$, increases, the strategy at time t becomes virtually independent of the terminal reward. This can be seen from the model *Surviving a single day*. As Figure 3.1 shows, the critical levels associated with the three terminal rewards are all very different when $T - t$ is small, but are similar at the beginning of the day when $T - t = 40$. Had the day been longer, the three critical levels would have been exactly the same for $T - t \geq 64$. There are two forms of backward convergence that can occur.

(i) Weak backward convergence: when $T - t$ is large, the strategy is independent of the terminal reward but depends on t.

(ii) Strong backward convergence: when $T - t$ is large, the strategy is independent of both the terminal reward and time.

When environmental conditions do not depend explicitly on time, strong backward convergence will usually hold. McNamara (1990a) gives sufficient conditions for it to hold in a particular class of models, and his results prove that strong backward convergence holds for the model *Surviving a single day*. When environmental conditions depend explicitly on time, usually only weak backward convergence would hold.

As we have seen, a strategy must be followed forward in time in order to find the resulting expected behaviour. To start the procedure, we have to choose a state (or distribution of states) at time 0. How does the expected behaviour under the strategy depend on this choice? In general, the probability that an organism is alive at time t will depend on the initial state, but given that an organism is alive, its probability of performing a given action becomes independent of the initial state as t increases. In the model *Surviving a single day*, the day is not long enough for the effect of the initial reserves at dawn to have completely disappeared by dusk (Figure 3.2). Had the day been significantly longer, behaviour at dusk would have been totally independent of initial reserves. As in the case of the optimal strategy, two forms of convergence can be distinguished.

(i) Weak forward convergence: the expected behaviour of organisms alive at time t depends on t but does not depend on the initial state for large t.

(ii) Strong forward convergence: the expected behaviour is independent of initial state and t for large t.

When backward convergence applies, to predict the behavioural strategy employed well before final time it is not necessary to know the terminal reward. When forward convergence applies, to predict expected behaviour under a strategy a long time after the initial time it is not necessary to know the initial distribution of states. In general, the time required for either backward or forward convergence to occur decreases with increasing stochasticity in the state changes and increases with the maximum time required to get from one state to another. We illustrate the use of convergence in two cases.

Overwinter survival of a small animal.   Suppose the only state variable relevant to a small animal in winter is its energy reserves. For models based on animals with a limited capacity to carry fat and a high metabolic rate, such as shrews or tit species, backward and forward convergence is rapid, occurring in a few days. Suppose behaviour is modelled over winter, with initial time 0 taken in autumn and final time T taken just before breeding in spring. During this time interval, environmental conditions might have seasonal trends and so need not be constant. The form of the terminal reward at T will depend on how reserves contribute to the ability of the animal to breed and survive in the breeding season. But because of weak backward convergence we do not need to know this terminal reward to find the optimal strategy in the middle of winter: in this case, when the breeding season is a long way off, the animal is just trying to survive. The behaviours observed in a large group of animals following the optimal strategy depend on the reserves of group members. Thus expected behaviour at the start of winter depends on how well fed the animals are at this time. Weak forward convergence ensures that later on in winter the distribution of reserves amongst those group members still alive is independent of the initial distribution at the beginning of winter. Thus expected behaviour is independent of the initial distribution.

If there are underlying condition variables such as the state of the immune system which change on a much slower time scale than reserves, then winter may not be long enough for convergence to occur.

Life histories for structured populations. Suppose we are concerned with the life-history strategy employed by members of a population that is structured by a state variable such as age or size (or both). Suppose that the population is observed annually and that environmental conditions are the same each year (although there may be seasonal effects within a year). First, assume that population members are following some given life-history strategy. Then the distribution of states amongst population members settles down over time to a stable state distribution that is independent of the initial distribution of states within the population (Caswell 1989; see also Section 8.3). Consequently the distribution of behaviours also settles down to a stable distribution. This is an example of strong forward convergence.

Under an optimal life-history strategy each individual maximises the expected number of descendants that it leaves far into the future. We can find this optimal strategy by dynamic programming (Section 8.6). A target year T is chosen far into the future and all descendants present in this year are assigned a reward. We then work backwards to find the strategy that maximises the expected total reward received by all descendants alive at T. Under appropriate technical conditions strong backward convergence applies, and the strategy employed a long time before T is independent of T and the exact rewards assigned at T. This limiting strategy is the optimal life-history strategy.

3.6 Robustness and selective advantage

In order to estimate how likely it is that behaviour will follow the predictions of a particular model, various things must be considered. One is the extent to which predictions depend on the model's parameters. For example, the model *Surviving a single day* involves the probability of being killed by a predator. In many cases the exact value of this probability will not be known. In the absence of this knowledge, the model can be used to explore the extent to which behaviour depends on the danger of predation. Another consideration is whether features that are not currently included in the model are likely to have a substantial effect. For example, if a model of foraging does not include the danger of being killed by a predator, to what extent are the model's predictions changed if this factor is included?

Both of these issues are concerned with the sensitivity of the model's predictions, i.e. with robustness. It is also of interest to establish the advantage, in terms of reproductive success, that results from following the optimal strategy. If various suboptimal strategies achieve a reproductive success close to that of the optimal strategy, then the optimal strategy will take a long time to spread through the population by natural selection. What this means for observed behaviour depends on how close the behaviour produced by 'good' suboptimal strategies is to the behaviour of the optimal strategy. It is logically possible for it to be quite different (i.e. strategies that are similar in terms of resulting reproductive success may be different in terms of resulting behaviour) in which case the model's predictions are unlikely to be accurate.

The selective advantage of state-dependent behaviour

This book is largely concerned with state-dependent behaviour, and an obvious question is whether state-dependent strategies give a significant advantage over non-state-dependent strategies. To illustrate the selective advantage of state-dependent strategies in a particular case, consider an animal that is attempting to survive an extended period of time such as winter. Suppose that at each time the animal has a choice between the same two feeding options as in the model *Surviving a single day*. Recall that option u_1 is safe from predators but gives a low net rate of energy gain. Option u_2 involves a predation risk but gives a higher net rate of energy gain. These options do not change over time. Since the period that we are considering is long and conditions are constant we will restrict attention to stationary strategies, i.e. strategies that are independent of time. Each stationary strategy results in a characteristic mortality rate per unit time, μ. This rate is the proportion of individuals alive at time t that are dead by time $t + 1$ (at equilibrium; see Section 3.5).

Consider first strategies that do not depend on state. For q in the range $0 \leq q \leq 1$ consider a strategy that, at each time, chooses option u_1 with probability $1 - q$ and option u_2 with probability q. The choices of option in successive time intervals are independent. Figure 3.3 illustrates how the mortality rate μ depends on the parameter q. For the particular parameters used to characterise options in this figure, the best non-state-dependent strategy is to use option u_1 the whole time. This strategy results in a mortality rate of $\mu = 0.00363$.

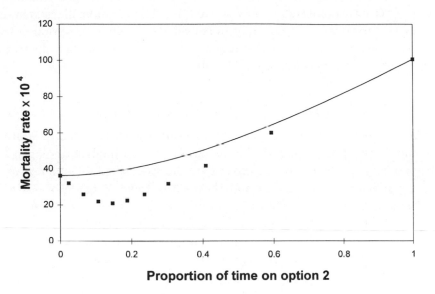

Proportion of time on option 2

Figure 3.3. Equilibrium mortality rates. The solid line gives the mortality rate μ as a function of the proportion q of time spent on option u_2 when behaviour does not depend on state. When behaviour is state dependent with critical level x_c, mortality is $\mu(x_c)$ and the proportion of time on option u_2 is $q(x_c)$: individual points plot the pairs $(\mu(x_c), q(x_c))$ for $x_c = 0, 2, 4, 6, \ldots, 20$. (As x_c increases so does $q(x_c)$.) Parameters describing the options are as for Figure 3.1. $L = 20$.

A stationary state-dependent strategy is characterised by a critical level of reserves x_c. Under the strategy, option u_1 is chosen when reserves are above x_c and option u_2 is chosen when reserves are equal to or below x_c. Each choice of x_c results in a characteristic level of mortality, $\mu(x_c)$. For the particular parameters used in Figure 3.3 mortality is minimised when $x_c = 8$.

In following a strategy with critical level of reserves x_c an animal spends (at equilibrium) a proportion of time $1 - q(x_c)$ choosing option u_1 and a proportion of time $q(x_c)$ choosing option u_2. The solid points in Figure 3.3 show the relationship between the mortality $\mu(x_c)$ and the proportion of time $q(x_c)$ as x_c varies. It can be seen that for any given proportion of time q spent on option u_2, the state-dependent strategy achieving this q does better than the non-state-dependent strategy. It can also be seen that the optimal state-dependent strategy spends a different proportion of time using option u_2 than the best non-state-dependent strategy and results in a much lower level of mortality: $\mu(8) = 0.00210$. This example illustrates that it can be very important not only to consider how often an option is used but also when (i.e. in what state) it is used. An example further illustrating this point is given by McNamara (1990b).

The canonical cost of an action

Although there may be very strong selection pressure to adopt a state-dependent strategy, small deviations from the optimal state-dependent strategy may not be costly. To investigate the strength of selection pressure on deviations from optimality and the related topic of model robustness, McNamara & Houston (1986) introduced the idea of canonical costs. The canonical cost $c(\boldsymbol{x}, t; u)$ of performing action u when in state $\boldsymbol{x}$ at time t is the loss in reproductive value that results from performing u rather than the optimal action u^*. In calculating the cost, it is assumed that behaviour at times $t + 1, t + 2, \dots, T - 1$ is given by the optimal strategy. Thus

$$c(\boldsymbol{x}, t; u) = H(\boldsymbol{x}, t; u^*) - H(\boldsymbol{x}, t; u),$$

where the H's are given by equation (3.1). Table 3.3 gives some canonical costs from the model *Surviving a single day* presented in this chapter.

It can be seen that the canonical costs tend to be small near the critical level of reserves $x_c^*(t)$. What this means is that in this region the loss in terms of reproductive value that results from a suboptimal decision is not great (Mangel & Ludwig 1992, Houston *et al.* 1992). Thus selection pressure on behaviour is low and one might not expect this model, even if it were realistic, to be a good predictor of behaviour when reserves are near the critical level. When reserves are high it is optimal to avoid predators by choosing option u_1. When reserves are low, the advantage of food outweighs the predation risk and option u_2 is optimal. In both these cases the canonical cost of taking the wrong decision is large. The only exception to this occurs at very low reserves near dusk, since the animal will not survive overnight whatever it does. Large canonical costs do not, however, translate directly into large selection pressure. The reason is that if an animal rarely finds itself in a particular region of state space there is little opportunity for selection to act on the behaviour adopted in this region. To take this effect into account, Houston & McNamara (1986a) defined the selection pressure associated with a suboptimal choice of option u in state x at time t to be the product of the canonical cost $c(x, t; u)$ and the probability $P(x, t)$ of being in state x at time t.

Table 3.3. *Surviving a single day:* canonical costs. The canonical costs $c(x, t; u_1)$ (upper entry) and $c(x, t; u_2)$ (lower entry) of options u_1 and u_2 respectively. When the cost of an option is 0 it is optimal to choose that option. The costs are multiplied by 10^4 and rounded to the nearest integer. The parameters are as for Figure 3.1. The terminal reward is R_1 given by equation (3.7).

x	$c(x, t; u_i)$ for various t-values					
	$t = 0$	8	16	24	30	36
20	0	0	0	0	0	0
	87	90	93	97	100	100
14	0	0	0	0	0	0
	29	26	24	24	31	100
13	0	0	0	0	0	0
	20	16	11	5	2	19
12	0	0	1	15	32	38
	11	6	0	0	0	0
11	0	2	14	37	71	163
	4	0	0	0	0	0
10	3	12	28	60	111	259
	0	0	0	0	0	0
9	11	21	42	82	143	304
	0	0	0	0	0	0
8	20	32	55	100	163	297
	0	0	0	0	0	0
2	169	159	140	103	56	1
	0	0	0	0	0	0

Canonical costs are also relevant to the issue of model robustness. If choosing one option results in a reproductive value similar to that resulting from the choice of another option, we might expect the optimal choice to be sensitive to the exact values of the model parameters. To illustrate this in the model *Surviving a single day* we alter the predation risk under option u_2, z_2, taking the value used to compute Table 3.3 as a baseline. The resulting change in the optimal strategy is shown in Figure 3.4. It can be seen from Table 3.3 and the figure that at any given time, for those states where the canonical costs of the suboptimal action is large the optimal choice of option is robust to changes in z_2. In contrast, near the original critical level, where canonical costs are small, optimal choice is sensitive to the choice of z_2.

Small canonical costs also mean that an organism's choice may be determined by factors that are not included in the model. For example, in the model *Surviving a single day* such a factor might be the time and energy that are required to switch between options. (These could be significant if the two options correspond to microhabitats that are not adjacent.) In a model based just on energetic gains the factor could be the danger of predation associated with the different options or the possibility of social interactions.

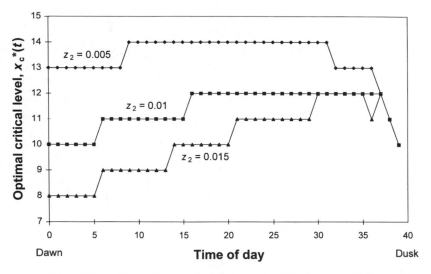

Figure 3.4. Dependence of the optimal critical level $x_c^*(t)$ on the time of day for three values of the predation risk z_2 under option u_2 in the model *Surviving a single day*. Other parameters are as for Figure 3.1.

3.7 Further topics

Dependent offspring. In all the examples considered in this chapter, offspring produced at time t are independent by time $t + 1$. The framework that we have described can also be applied to cases in which young are dependent on their parents for several time units (provided that the young themselves are not making decisions whilst they are dependent). To do so, dependent young are included in the parents' state (Section 2.2). Thus the state of a focal organism includes information about whether the organism is currently caring for young and also about the state of any young that are being cared for. The parent does not obtain a contribution to reproduction B_{off} immediately the young are produced; it obtains a contribution later, when they become independent. The contribution depends on the number and state of the young, and the time at which they become independent. An example illustrating this is given in Section 9.5.

McNamara *et al.* (1994b) analyse a simple model that can be interpreted as either a parent allocating food between two dependent offspring or a dominant offspring allocating food between itself and a subordinate sib. Young are assumed to be independent at some fixed time T, and the immediate contribution to reproductive success is thus included in the terminal reward to the controlling animal. In both this example and that in Section 9.5, the optimal strategy for the controlling animal can be found by working backwards from independence by dynamic programming. Having found the optimal strategy, it is possible to identify quantities that can be interpreted as costs and benefits and characterize decisions made in terms of Hamilton's rule (Hamilton 1964) based on these costs and benefits.

Fluctuating environments. We refer to a fluctuating environment as one in which large-scale fluctuations are experienced by all organisms in the environment. Such fluctuations might be caused by changes in weather conditions from year to year or changes in population density. When there are such fluctuations, fitness measures based on the lifetime reproductive success of an individual are no longer appropriate; measures based on geometric mean fitness or some generalisation of it must be used. This topic is discussed in detail in Chapter 10. In fluctuating environments, there is no individual optimisation and an optimal dynamic strategy cannot be found by simply using dynamic programming. Attempts to do so are flawed. In particular, the example presented by Mangel & Clark (1988, Appendix 8.1) would not be appropriate if the stochasticity in the example were the result of genuine environmental fluctuations. McNamara *et al.* (1995) illustrate the correct procedure for dealing with sequential decisions when there are environmental fluctuations. Their analysis and generalisations to more complex cases are discussed in Chapter 10.

Appendix 3.1 Interpolating between grid points

Here we show how the dynamic programming equations for the model *Putting resources into growth or reproduction* (see Section 3.2) can be solved on a discrete grid of size values.

Suppose that the size, y, of an organism lies in range $0 \leq y \leq y_{\max}$. Take a discrete grid with positions labelled $0, 1, \ldots, k_{\max}$. We take grid position k, where $0 \leq k \leq k_{\max}$, to correspond to the size

$$y = \left(\frac{y_{\max}}{k_{\max}} \right) k. \qquad (A3.1.1)$$

As in the main text, let $V(y, n)$ be the reproductive value of an individual of size y and age n. Then we can define reproductive values $\tilde{V}$ on the grid of size (and age) by setting $\tilde{V}(k, n) = V(y, n)$ where y is given in terms of k by equation (A3.1.1).

Let k be a point of the grid with corresponding size y given by equation (A3.1.1). Suppose an individual of this size and age n allocates resources u for reproduction. Then, given that it survives to age $n + 1$, its size at this time is $\hat{y} = \hat{y}(y, n, r_{\text{growth}})$, where $r_{\text{growth}} = r_{\text{total}}(y, n) - u$. Let

$$\hat{k} = \left(\frac{k_{\max}}{y_{\max}} \right) \hat{y}.$$

In general this $\hat{k}$ will not be an integer, i.e. the size at age $n + 1$ does not fall on our grid of sizes. Set $\hat{k}_-$ to be the greatest integer that is less than or equal to $\hat{k}$ (i.e. it is the integer part of $\hat{k}$). Set $p_2 = \hat{k} - \hat{k}_-$ and set $p_1 = 1 - p_2$. Finally, to keep within the range of the grid, set $\hat{k}_1 = \max[0, \min(k_{\max}, \hat{k}_-)]$ and $\hat{k}_2 = \max[0, \min(k_{\max}, \hat{k}_- + 1)]$. Then we can interpret the new size at age $n + 1$ as lying at grid point $\hat{k}_1$ with probability p_1 and $\hat{k}_2$ with probability p_2. The dynamic programming equations for $\tilde{V}$ are then as follows. Given k and n, define y by equation (A3.1.1). For each u in the range $0 \leq u \leq r_{\text{total}}(y, n)$ set

$$\tilde{H}(k, n; u) = B_{\text{off}}(y, n; u) + S(y, n, r_{\text{growth}})[p_1 \tilde{V}(\hat{k}_1, n + 1) + p_2 \tilde{V}(\hat{k}_2, n + 1)]. \qquad (A3.1.2)$$

Then

$$\tilde{V}(k, n) = \max_u \tilde{H}(k, n; u). \qquad (A3.1.3)$$

Equations (A3.1.2) and (A3.1.3) determine $\tilde{V}$ at age n in terms of $\tilde{V}$ at age $n + 1$. The terminal condition (3.10) becomes $\tilde{V}(k, T) = 0$ for all k.

In the above we have assumed that the state at the next time epoch can take one of two values, $\hat{k}_1$ and $\hat{k}_2$. Sometimes it is convenient to introduce further stochasticity. One simple way is to allow the state to take four, rather than two, values, as follows. Set

$$\hat{k}_0 = \max(0, \hat{k}_1 - 1)$$
$$\hat{k}_3 = \min(k_{\max}, \hat{k}_2 + 1).$$

Let α be a parameter such that $0 < \alpha < \frac{1}{3}$ and set

$$\hat{p}_0 = \alpha p_1$$
$$\hat{p}_1 = (1 - 2\alpha)p_1 + \alpha p_2$$
$$\hat{p}_2 = \alpha p_1 + (1 - 2\alpha)p_2$$
$$\hat{p}_3 = \alpha p_2.$$

Then we can assume that the state at the next time epoch takes the values $\hat{k}_0$, $\hat{k}_1$, $\hat{k}_2$ or $\hat{k}_3$ with probabilities $\hat{p}_0$, $\hat{p}_1$, $\hat{p}_2$ and $\hat{p}_3$ respectively. With this specification of probabilities then, in the absence of the effects of the boundaries at 0 and $k_{\max}$, the mean value of the state is still $\hat{k}$. The parameter α can be used to control the degree of stochasticity.

4

Maximising the energy gained from foraging

4.1 Introduction

In the previous chapter we described a general method for finding optimal sequences of behaviour over a given period of time $[0, T]$. We now look at a special case in which the analysis is relatively simple. We assume that reproductive success after time T is an increasing function of the amount of energy that the animal has at time T. For most of the chapter we also ignore other components of state. We ignore the possibility of death from starvation and any stochastic effects associated with foraging (these are discussed in Chapter 5) and we also ignore any effects of predation (these are discussed in Chapter 6). Under these circumstances, fitness is maximised by maximizing the net amount of energy obtained by time T.

An animal's foraging options will typically differ in both the mean rate b at which the animal gains energy and the rate c at which it expends energy. If the animal is able to forage for the whole time interval and the time interval is long enough that we can perform a suitable averaging (see Section 4.2), then it is clear that the net amount of energy obtained is maximised by maximising the mean net rate of gain $\gamma = b - c$. The mean net rate of gain γ has often been used as a currency for evaluating foraging options (see Stephens & Krebs 1986 for a review).

Although we do not explicitly model it in this chapter, an animal may perform activities other than foraging during the interval $[0, T]$. If we assume that, for a given level of energy reserves at the end of the interval, an animal's reproductive success is maximised by maximising the time spent on these activities, then the animal should achieve a given level of energy by devoting the shortest possible time to foraging. This is the argument for a time-minimising forager given by Schoener (1971). If the time available for foraging is long, then time minimisation and energy maximisation are both achieved by maximising the mean net rate of energetic gain.

We start this chapter with a description of how to find the strategy that maximises the mean net rate of energetic gain. We then apply this approach to problems of patch use and prey choice. Next we examine the effects of time constraints on an animal that is attempting to maximise its net energy intake. It has been suggested that animals may be limited in the amount of energy that they can expend (e.g. Drent & Daan 1980) or acquire (e.g. Kirkwood 1983) during a given time period. If either or both of these constraints is operative, then the animal may not be able to maximise its net rate of energy gain for the whole time interval. We give a general graphical analysis of these constraints. Each of the above scenarios assumes that the net energy gain is all that matters to an animal. At the end of the chapter, we discuss some implications of relaxing this assumption to include the possible costs of expending energy. As we point out, these costs may be the result of adverse changes in some condition variable.

4.2 Rate maximisation

Consider an animal exploiting patchily distributed food in its environment. The animal's behaviour will typically follow a regular sequence or cycle. First the animal searches for a food patch. On finding a patch it consumes prey items. This foraging activity depletes the patch and as a result the animal's intake rate falls. The animal then leaves the patch and starts searching for a new patch. The cycle here consists of: search; exploit food; leave patch. If the animal faces the same distribution of patch types and has the same knowledge about its environment at the beginning of each cycle we can think of the system 'renewing' each time the animal leaves a patch. The whole cycle is thus referred to as a renewal cycle and the time at which the animal leaves a patch is called the renewal time.

In the standard prey choice model (Stephens & Krebs 1986, see also Section 2.2) the animal goes through the following cycle. It searches for a prey item. On finding an item it either eats it or rejects it. The animal then searches for a new item. If the renewal time is taken to be the time at which the animal recommences searching, then we can regard the period of search and subsequent time spent dealing with an item as a renewal cycle.

A parent bird delivering food to young in its nest will also exhibit cyclic behaviour. Here the cycle is: leave the nest; find food; return to the nest; feed the young. This is a renewal cycle, the renewal time being the time of departure from the nest. Orians & Peason (1979) refer to the bird as a central place forager; the nest is the central place.

We need to define the mean net rate of energy gain for an animal foraging in a cyclically renewing system. Let G denote the net energy gain to the animal over a single cycle and let D denote the duration of the cycle. These quantities will typically differ from cycle to cycle because of stochastic effects. Thus G and D are random variables. Their distribution is determined by the behavioural strategy adopted by the animal. For example, in the standard model of prey choice G is the energetic value of the item minus the metabolic expenditure on the cycle if the item found is accepted and G is just minus the metabolic expenditure if the item is rejected. D is the search time plus any handling time. We assume that the animal adopts the same strategy on each cycle. The net gains on successive cycles are then independent random variables each having the same distribution as G. Similarly, the durations of successive cycles are independent random variables having the same distribution as D.

Consider an animal that starts foraging at time 0. Set

$$\gamma = \lim_{t \to \infty} \frac{1}{t} \text{Total net energy gained by time } t.$$

It can easily be shown that this limit exists. We call γ the mean net rate of energy gain under the animal's behavioural strategy. There are other ways to express γ. For example, renewal theory arguments show that γ is given by

$$\gamma = \lim_{t \to \infty} \frac{1}{t} \mathbb{E}(\text{Total net energy gained by time } t)$$

where $\mathbb{E}$ denotes the mean or average value.

Neither of the above formulae is very useful for computing γ. Fortunately, γ is also given by

$$\gamma = \frac{\mathbb{E}(G)}{\mathbb{E}(D)}, \tag{4.1}$$

as is shown by Johns & Miller (1963). An example showing how to calculate γ from this formula is presented in Box 4.1.

Templeton and Lawlor (1981) have questioned this formula for the rate, and have suggested $\mathbb{E}(G/D)$ as an alternative. Gilliam *et al.* (1982), Turelli *et al.* (1982), Stephens & Charnov (1982) and Houston *et al.* (1982) pointed out errors in Templeton and Lawlor's paper. One way to see that $\mathbb{E}(G/D)$ does not give a sensible long-term average is to consider an infinite sequence of tosses of a fair coin and look at the mean rate at which heads are obtained

Box 4.1. Calculations of the mean rate: an example based on patch use. Suppose that food in the environment is distributed in well-defined patches. There are three patch types of which a proportion p_i are of type i. Food is obtained as a continuous flow on each patch. The rate of gain when an animal has been on a patch of type i for time t is $r_i(t)$. Parameter values are as shown in part (a) of the table. The interpatch travel time is $\tau = 1$. Metabolic costs are ignored.

(a)				(b)		
Type	p_i	$r_i(t)$		Type	t_i	g_i
1	0.5	e^{-t}		1	0.5108	0.4
2	0.4	$2e^{-2t}$		2	0.6020	0.7
3	0.1	$2e^{-t}$		3	1.2040	1.4

Let an animal foraging in this environment use the following rule: leave a patch when the instantaneous rate falls to 0.6. We will find the mean gross rate of energy gain under this strategy.

Let t_i be the time spent on a patch of type i and let g_i be the total food gained in this time. Then, for example, t_1 can be found from the equations $r_1(t_1) \equiv e^{-t_1} = 0.6$ and g_1 is given by

$$g_1(t_1) = \int_0^{t_1} r_1(t)dt = 1 - e^{-t_1}.$$

Three such easy computations give the values shown in part (b) of the table.

We are now able to compute the expected gain over a cycle, $\mathbb{E}(G)$, and the expected duration of a cycle, $\mathbb{E}(D)$. We have

$$\mathbb{E}(G) = p_1 g_1 + p_2 g_2 + p_3 g_3 = 0.62$$

and

$$\mathbb{E}(D) = \tau + p_1 t_1 + p_2 t_2 + p_3 t_3 = 1.6166.$$

Thus the mean food intake rate under this strategy is

$$\gamma = \frac{\mathbb{E}(G)}{\mathbb{E}(D)} = 0.3835.$$

per throw. Define a renewal to occur after each occurrence of a head. Let G be the number of heads per renewal cycle and D the number of throws in a renewal cycle. Then $G = 1$ and D has a geometric distribution with mean 2. Thus $\gamma = \mathbb{E}(G)/\mathbb{E}/D) = 1/2$. That is, the mean rate of occurrence of heads per throw is $1/2$ as one would expect. In contrast, $\mathbb{E}(G/D) = \mathbb{E}(1/D)$ which can be shown to be equal to 0.6931!

Maximising mean rate

Now that the concept of the mean net rate of gain γ has been established, we are able to consider how an animal's behaviour influences γ. We will be particularly interested in finding the behavioural strategy that maximizes γ. In the context of patch use, a strategy will determine when the animal leaves each sort of patch; in the model of prey choice a strategy determines which items are eaten and which are rejected. In general, different strategies will result in different mean net rates of gain. We define γ^* to be the maximum possible value of γ. We will refer to γ^* as the optimal rate and the strategy that achieves γ^* will be called the optimal strategy.

When the strategy can be given by a single number, it is usually easy to write down an expression for γ as a function of the strategy used and hence find γ^*. When problems are more complex, it is possible to solve them by working backwards using dynamic programming to maximise the net energy obtained over a long interval of time. There is, however, an alternative approach that makes use of the renewal cycle. This approach is both computationally more efficient and conceptually instructive. The basis of the approach is to use γ^* to compare the advantage of energy gained on the cycle with the disadvantage of the time spent on the cycle. Consider behaviour on a single cycle, and assume that after the cycle finishes the animal behaves optimally and so achieves a mean net rate of energetic gain γ^*. A strategy over the cycle gives a gain in energy G but also takes time D. During this time an animal following the optimal strategy could have been foraging at rate γ^* and so could have gained an amount of energy $\gamma^* D$. Thus if the cycle has duration D, the animal pays an opportunity cost $\gamma^* D$, and hence γ^* acts as an opportunity cost per unit time. Viewed in this way, it is not surprising that the optimal strategy maximises the net energy gained minus the opportunity cost. This result was introduced into behavioural ecology by McNamara (1982) and can be stated formally as follows.

Theorem 4.1.   The optimal strategy maximises $\mathbb{E}(G - \gamma^* D)$ on each cycle, where G is the net gain over a cycle and D is the duration of a cycle.

This result allows us to transform a long-term problem into a problem during a single cycle. The theorem can be extended to look at behaviour within the cycle.

Corollary.   At any stage in a cycle, the optimal decision is the one that maximises

$$\mathbb{E}(\text{net gain from remainder of cycle} - \gamma^* \times \text{time left}).$$

The theorem can be used to derive some general results concerning the form of the optimal strategy, e.g. increasing γ^* decreases the time in cycle. In the context of patch use, increasing the travel time τ decreases γ^* and hence increases the optimal stay time. In the case of smoothly depleting patches this result follows from the marginal value theorem, but the theorem shows that it is in fact a very general result (see McNamara 1982 for further discussion).

Stephens & Krebs (1986, p. 44) mention the theorem but say that it suffers from the disadvantage of requiring a knowledge of γ^* before it can be used to find optimal behaviour. This is a reference to the circularity at the heart of the theorem. It enables us to find the optimal strategy from γ^*, but we do not know γ^* until we find the optimal strategy. This is not really a problem: it is relatively easy to compute γ^* and the optimal strategy by iteration.

The computation of γ^*.   For any rate γ we can find the strategy that maximises $\mathbb{E}(G - \gamma D)$. We can regard this strategy as specifying the optimal behaviour on a patch given that the animal forages at mean rate γ after it leaves the patch. Let $f(\gamma)$ be the mean rate achieved if this strategy is used on every patch that the animal encounters. The optimal rate is the unique solution to the equation

$$f(\gamma^*) = \gamma^*. \tag{4.2}$$

In other words, a strategy is optimal if the behaviour that it specifies on a patch is the best, given that the strategy is adopted on all future patches. To find γ^* and hence the optimal strategy, we perform the following iteration. We start off by taking any strategy. Let γ_0 be the mean rate achieved if this strategy is used on all patches. We then find the strategy that maximises $\mathbb{E}(G - \gamma_0 D)$, and the resulting rate $f(\gamma_0)$ under this strategy. Now let $\gamma_1 = f(\gamma_0)$. We then find the strategy that maximises $\mathbb{E}(G - \gamma_1 D)$ and the resulting rate $\gamma_2 = f(\gamma_1)$, and so on. In this way we achieve a sequence of rates $\gamma_0, \gamma_1, \gamma_2, \ldots$ where $\gamma_n = f(\gamma_{n-1})$. It can then be shown that

$$\gamma_0 \leq \gamma_1 \leq \gamma_2 \leq \cdots \leq \gamma^*$$

and $\gamma_n \to \gamma^*$ as $n \to \infty$ (McNamara 1985). Furthermore if $\gamma_n = \gamma_{n+1}$ for any n then by equation (4.2) $\gamma_n = \gamma^*$. An example is given in Box 4.2 and Figure 4.1.

If the rate of energetic expenditure is the same for all activities performed by the animal then the mean rate of energy expenditure is independent of the strategy adopted. It follows that the strategy that maximises the mean net rate of energy gain also maximises the mean rate of energy intake from food, i.e. the mean gross rate of gain. Under these circumstances, it is convenient to ignore metabolic expenditure altogether, defining γ to be the mean gross rate of energy gain and working directly with this quantity.

Box 4.2. Rate maximisation with three patches.   To maximise γ in the three-patch environment described in Box 4.1, we look at the consequences of leaving a patch when the intake rate r falls to γ. Under this strategy a patch of type i is left after time $t_i(\gamma)$, where $r_i(t_i(\gamma)) = \gamma$. From the expressions for $r_i(t)$ in part (a) of the table in Box 4.1,

$$
\begin{aligned}
t_1(\gamma) &= -\log \gamma \\
t_2(\gamma) &= -\tfrac{1}{2}\log(\tfrac{1}{2}\gamma) \\
t_3(\gamma) &= -\log(\tfrac{1}{2}\gamma)
\end{aligned}
$$

for $0 < \gamma \leq 1$. The strategy achieves a mean intake rate

$$
f(\gamma) = \frac{1.1 - 0.8\gamma}{0.3\log 2 - 0.8\log \gamma + 1}.
$$

This function is shown in Figure 4.1. The optimal strategy can now be found by iterating to find the fixed point

$$
f(\gamma^*) = \gamma^*,
$$

as is shown in the figure. To start the iteration we can use any strategy. Here we use the strategy of leaving each patch when the intake rate falls to 0.05. The mean intake rate under this strategy is $\gamma_0 = 0.2941$. Then

$$
\begin{aligned}
\gamma_1 &= f(\gamma_0) &= 0.39539 \\
\gamma_2 &= f(\gamma_1) &= 0.40184 \\
\gamma_3 &= f(\gamma_2) &= 0.40186 \\
\gamma_4 &= f(\gamma_3) &= 0.40186.
\end{aligned}
$$

Thus $\gamma^* = 0.40186$.

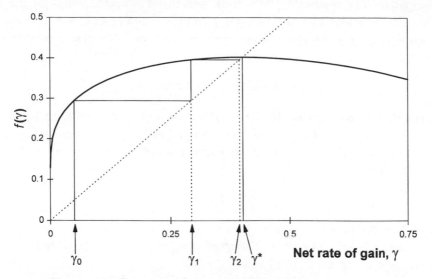

Figure 4.1. Iterative scheme to find the maximum mean rate of gain, γ^*. For given γ, $f(\gamma)$ is the rate achieved by behaving as if γ were the true value of γ^*. Since $f(\gamma^*) = \gamma^*$, the maximum occurs where the curve $f(\gamma)$ intersects the 45° line. The sequence $\gamma_0, \gamma_1, \gamma_2 \ldots$ satisfies $\gamma_n = f(\gamma_{n-1})$ and converges to γ^*.

4.3 Patch use and the marginal value theorem

Consider an environment in which there are n patch types. When feeding on a patch an animal obtains a continuous flow of energy. This assumption would be appropriate for an animal feeding on nectar. It might also be a reasonable assumption for an animal that obtains a frequent stream of small items while feeding. On patch type i, food is found at a rate $r_i(t)$, where r_i decreases with the time t spent on the patch. The proportion of type i patches is p_i and the mean travel time between patches is τ. Patches are encountered at random. For this scenario, a strategy is a rule specifying how long to spend on each patch before travelling to the next patch. Since there are n patch types, a strategy $(t_1, t_2, \ldots, t_n)$ is given in terms of times $t_1, t_2, \ldots, t_n$, where t_i is the time spent on each patch of type i encountered. Let $\gamma(t_1, t_2, \ldots, t_n)$ be the mean net rate of energy gain under the strategy $(t_1, t_2, \ldots, t_n)$. An optimal strategy $(t_1^*, t_2^*, \ldots, t_n^*)$ maximises this rate. Thus

$$\gamma^* = \gamma(t_1^*, t_2^*, \ldots, t_n^*) = \max_{t_1, t_2, \ldots, t_n} \gamma(t_1, t_2, \ldots, t_n).$$

For this problem we can define a renewal cycle by taking the renewal time to be the time at which an animal leaves a patch and starts to search for a new patch. The renewal cycle is thus: search for a patch; exploit a patch; leave. If an animal encounters a patch of type i and spends a total time t_i on this patch, the duration of the cycle is $\tau + t_i$ and the energetic gain on the cycle is

$$g_i(t_i) = \int_0^{t_i} r_i(s)\, ds.$$

This gain increases with t_i ($g_i'(t_i) > 0$), but because r_i is decreasing the gain decelerates ($g_i''(t_i) < 0$). Consider an animal that has just arrived on a patch of type i. Then the gain from the remainder of the cycle minus γ^* times the time left is

$$g_i(t_i) - \gamma^* t_i = \int_0^{t_i} r_i(s)\, ds - \gamma^* t_i.$$

From the corollary to Theorem 4.1 the optimal stay time t_i^* on patch type i must maximise this quantity. Differentiating with respect to t_i and setting the derivative equal to zero gives

$$g_i'(t_i^*) = \gamma^*,$$

which means that $r_i(t_i^*) = \gamma^*$. Since this holds for each i we can write

$$r_i(t_i^*) = \gamma^*, \qquad i = 1, 2, \ldots, n \tag{4.3}$$

which is the marginal value theorem (MVT) (Charnov 1976b).

The usual derivation of the MVT, as given by Stephens & Krebs (1986), involves writing an explicit expression for γ as a function of $t_1, t_2, \ldots, t_n$ and then using the n equations

$$\frac{\partial \gamma}{\partial t_i}(t_1^*, t_2^*, \ldots, t_n^*) = 0, \qquad i = 1, 2, \ldots, n. \tag{4.4}$$

Equation (4.3) is useful conceptually, but does not immediately identify the precise values of $t_1^*, t_2^*, \ldots, t_n^*$. The problem is that the equation gives each t_i^* in terms of γ^*, but γ^* is the rate achieved under the optimal strategy, i.e. $\gamma^* = \gamma(t_1^*, t_2^*, \ldots, t_n^*)$, and so is not known until $t_1^*, t_2^*, \ldots, t_n^*$ have been found. Thus equation (4.3) appears circular. It is not circular, and uniquely determines both the times $t_1^*, t_2^*, \ldots, t_n^*$ and the rate γ^*, but the numerical evaluation of these quantities requires the application of a suitable computational algorithm. One method is to solve the n equations (4.4) for the n unknowns $t_1^*, t_2^*, \ldots, t_n^*$ directly. But given that the solution has to be found numerically, we feel that the approach based on iteration of γ given in Section 4.2 is much less clumsy than this direct method. Iteration on γ works directly with the renewal process and has several desirable features:

it highlights the essential structure of the problem; it is robust; it is fast; it always works. An example is given in Box 4.2.

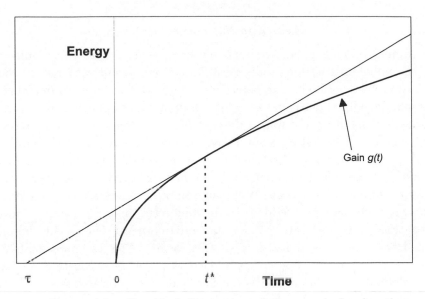

Figure 4.2. Graphical illustration of the marginal value theorem when all patches are the same. A straight line through the point $(-\tau, 0)$ that intersects the gain curve g at time t has slope $g(t)/(t+\tau) = \gamma(t)$. Thus $\gamma(t)$ is maximised by choosing the line intersecting the gain curve to have maximum slope. This is achieved when the line is a tangent to the gain curve. The time t^* at which the line and the curve touch is then the optimal patch resident time. Since the curve and the line have the same slope at t^*, $r(t^*) = g'(t^*) = \gamma(t^*) = \gamma^*$.

The MVT is usually described in terms of the graphical construction shown in Figure 4.2. The tangent drawn from $-\tau$ to touch the cumulative gain curve gives the optimal time t^* to spend in the patch. This construction can only be used when there is a single patch type. When there is more than one patch type, constructing a tangent to the mean gain curve produces the wrong answer. For example, if there are two equally common patch types, with rates $r_1(t) = e^{-t}$ and $r_2(t) = 3e^{-t}$, then $g_1(t) = 1-e^{-t}$ and $g_2(t) = 3(1 - e^{-t})$. Applying the MVT to the mean gain $g(t) = 2(1 - e^{-t})$ does not find the optimal strategy. This approach can be thought of as finding the best behaviour given that the animal spends the same amount of time in all patch types. This is clearly not a desirable feature of a good strategy. The true optimum involves spending longer in the good patches than in the poor patches. Using the average gain involves ignoring the difference between patches and hence always underestimates the value of γ^*.

Learning the value of γ^*

An animal foraging in an environment composed of smoothly depleting patches maximises its long-term reward rate by leaving each patch when the instantaneous rate on the patch falls to γ^*. But in order to implement this rule the animal must first know γ^*. If an animal's behaviour on a patch is based on the average reward rate that it has experienced in an environment, an animal exploiting a novel environment faces a potentially circular problem. To experience γ^* it must first behave optimally, while to behave optimally it must first experience γ^*. McNamara (1985) presented a rule by which an animal learns about any environment where the patches give a smoothly decreasing flow of food. In the long term this rule always achieves the optimal mean net rate, γ^*, for that environment. Under this rule an animal has a prior estimate $\gamma_0 = G_0/D_0$ for γ^*. This estimate might be based on the animal's previous experience, or might be a constant selected by evolution. The animal leaves the first patch encountered when the reward rate on this patch falls to γ_0. It then updates its estimate of γ_0 to $\gamma_1 = (G_0 + G_1)/(D_0 + D_1)$, where G_1 is the gain on the first patch and D_1 is the sum of the travel time to this patch and the time on the patch. The second patch encountered is left when the reward rate on this patch falls to γ_1, and so on. In general after n patches have been visited the animal's estimate of γ^* is

$$\gamma_n = \frac{G_0 + G_1 + \cdots + G_n}{D_0 + D_1 + \cdots + D_n}, \tag{4.5}$$

where G_i is the gain on the ith patch encountered and D_i is the sum of the travel time to this patch and the time on this patch. The animal then leaves the $(n + 1)$th patch when the instantaneous reward rate on this patch falls to γ_n.

McNamara (1985) showed that the animal's estimate γ_n tends to the maximum achievable rate γ^* as the number of patches visited tends to infinity. This is true whatever the initial estimate γ_0 and whatever the distribution of patch types in the environment. McNamara & Houston (1985a) illustrated the convergence of γ_n to γ^*, and discussed a modification to the rule given by formula (4.5) designed to cope with patchy environments that are changing over time.

Learning about current patch type

When patches give a smoothly decreasing flow of rewards then, as we have seen above, the animal can potentially learn the value of the optimal mean net rate of gain γ^*. Once it knows γ^*, it does not need to know anything about the distribution of patch types or the type of patch that it is currently on. It just needs to leave each patch when the instantaneous rate on the patch falls to γ^*. The situation is different when the food on a patch consists of discrete items found at irregular and unpredictable times. In an environment of this sort then, even if an animal has learnt γ^* and the types of patches and their frequency of occurrence, it will not know the current patch type on finding a patch. It can obtain information about the patch type from the times at which items are found. The distribution of patch types in the environment together with experience on the current patch give information about the food that the animal is likely to find if it continues to exploit the patch. This information, together with γ^*, determines whether it is optimal to leave the patch or to continue searching.

In some cases, patches either contain no prey items or just a single item. For example, in laboratory studies of the detection of prey by the blue jay (*Cyanocitta cristata*) Kamil *et al.* (1985) showed the jays a series of slides. Some slides contained a cryptic moth against its typical background, other slides just consisted of the background. Each slide can be thought of as a patch that can have at most one prey item. A predator that is aware of the nature of such an environment will always leave when it finds an item. The strategy that maximises γ may also involve leaving a patch after a period t of unsuccessful search. Let $F(t)$ be the probability that an item is found by time t on a randomly chosen patch. It might seem that the best value of t can be found by taking the cumulative gain $g(t)$ on a patch to be $F(t)$ and applying the graphical approach shown in Figure 4.2. Although this graphical approach seems to be a version of the MVT, it does not find the optimal strategy. The reason is that the MVT is based on the fact that the slope of the line that is tangent to the gain curve is the gain on a cycle divided by the time on a cycle. When the cumulative gain is taken to be $F(t)$ this is no longer the case. The expected time in the patch is less than t, because the animal may find the item before t and hence leave after a time that is less than t (McNamara & Houston 1985b). To look at this case in more detail, let $f(t) = F'(t)$ so that f is the probability density function of the time taken to find an item. This function gives the unconditional probability per unit time of finding the item at time t. Using f as a reward rate on the patch, and hence leaving a patch when f falls to γ^*, does not give the optimal strategy. The function f ignores the information that has been obtained by unsuccessful search until time t. The quantity

$$r(t) = f(t)/[1 - F(t)]$$

takes this into account and is the conditional probability per unit time of finding the item at time t given that the item has not been previously found. Since $r(t)$ is based on future expectations given current circumstances it is the reward rate appropriate for this stochastic problem. As long as $r(t)$ is a monotone decreasing function of t then it is optimal to leave a patch when $r(t)$ falls to γ^* (Oaten 1977, McNamara 1982). It can be seen from the above equation that $r(t) > f(t)$ and hence the true optimum involves staying for a longer time on a patch than would be predicted by the incorrect procedure described above.

4.4 Prey choice

The standard model of prey choice assumes that prey items can be classified into types. Prey of type i have an energetic content e_i and a handling time h_i. Encounters with items of type i follow a Poisson process with rate λ_i while searching. All the Poisson processes are independent of each other (for a review see Stephens & Krebs 1986).

For this model a strategy specifies which types of item are included in the diet and which are rejected. In Section 2.2 we found the optimal strategy using an approach that takes energy reserves as an animal's state variable. Here we derive the optimal strategy using renewal cycles. The two derivations are in essence the same. We assume for simplicity that metabolic expenditure while searching equals that while handling prey. This allows us to ignore metabolic expenditure and work with gross rates. In the following analysis, γ^* is the maximum mean gross rate of energy gain rather than the maximum mean net rate.

We take the renewal time to be the time at which an animal recommences searching, having dealt with an item. Suppose that the animal encounters a type-i item. If it eats the item the gross energy gain during the remainder of the cycle is e_i and the opportunity cost during the remainder of the cycle is $\gamma^* h_i$. If the animal rejects the item then the cycle renews immediately, so that future gains and opportunity costs in the current cycle are both zero. By the corollary to Theorem 4.1 it is optimal to accept the item if and only if $e_i - \gamma^* h_i > 0$, i.e. if and only if

$$e_i/h_i > \gamma^*,$$

which is condition (2.6) of Section 2.2.

The ratio e/h is known as the profitability. It is clear from the above inequality that all items above a given profitability should be included in the diet. This critical profitability γ^* depends only on the encounter rate with items that are accepted.

Box 4.3. Computing the optimal choice of prey type. Let there be 100 prey types, each with energetic content $e_i = 1$, handling time $h_i = i$ and encounter rate $\lambda_i = 10^{-3}$. A strategy of the form 'take all items with handling time $\leq k$' has a mean intake rate

$$g_k = \left(\frac{1000}{k} + \sum_{i=1}^{k} \frac{i}{k} \right)^{-1}$$

$$= \left(\frac{1000}{k} + \frac{k+1}{2} \right)^{-1}.$$

Let the initial strategy be to take all items whose handling time h is 60 or less. (The initial choice of strategy is arbitrary; any choice would suffice.) The resultant gain rate under this strategy is $\gamma_0 = 0.021201$. The best strategy taking γ_0 as the opportunity cost per unit time is to accept an item if and only if $e/h \geq \gamma_0$, i.e. if and only if $h \leq \gamma_0^{-1} = 47.17$, i.e. if and only if $h \leq 47$. The iteration then proceeds as follows:

$$\gamma_1 = f(\gamma_0) = g_{47} = (45.28)^{-1}$$

$$\gamma_2 = f(\gamma_1) = g_{45} = (45.22)^{-1}.$$

Because of the discrete nature of the options in this case, we have reached exact convergence, with $\gamma_2 = \gamma_3$, i.e. $f(\gamma_2) = \gamma_2$, which means that $\gamma_2 = \gamma^*$ and so $\gamma^* = 0.022113$ and the optimal strategy is to take an item if and only if its handling time is less than or equal to 45.

The standard algorithm for finding the optimal choice of prey items is described by Stephens & Krebs (1986). It involves ranking the items in terms of profitability e_i/h_i and adding types to the diet in decreasing order of profitability until the profitability of the next type to be considered is less than the rate that results from taking all types with higher profitability. We show in Box 4.3 that the iterative procedure of Section 4.2 can be a more powerful method.

Simultaneous choice

We now consider a renewal cycle that starts when the animal travels to a foraging area. On arrival at the foraging area the animal has a simultaneous choice between a number of prey items that differ in their energetic content and handling time. After an item has been chosen and handled, the system renews. Suppose that there are n prey types, where type i has energetic content e_i and handling time h_i. By the corollary to Theorem 4.1, the optimal choice of item maximises

$$e_i - \gamma^* h_i,$$

where γ^* is the maximum achievable long-term rate (see Engen & Stenseth 1984 for a version of this result).

To illustrate this, consider a parent bird feeding its young. The renewal cycle starts when the parent leaves the nest to go to the foraging area. The handling time is the time required by the parent to deal with the item in the foraging area, return with it to the nest and transfer it to the young. We assume that the parent bird is attempting to maximise the rate at which it brings energy to the young. Suppose that there is a choice between two prey types on the foraging area. Type 1 has the higher energetic content ($e_1 > e_2$) but also has the longer handling time ($h_1 > h_2$). If the time to fly back to the nest is independent of the type of item, then maximisation of $e_i - \gamma^* h_i$ is equivalent to maximisation of $e_i - \gamma^* \tilde{h}_i$, where $\tilde{h}_i$ is the sum of the time required to deal with the item in the foraging area and the time required to transfer the item to the young at the nest. Under these circumstances, the type-1 item should be preferred to the type-2 item if and only if

$$e_1 - e_2 > \gamma^*(\tilde{h}_2 - \tilde{h}_1).$$

As the round-trip travel time increases, γ^* decreases and there can be a switch from choosing the small item with shorter handling time to choosing the large item with longer handling time (e.g. Orians & Pearson 1979). This sort of switch has been observed by Carlson (1983) in a field experiment on the red-backed shrike (*Lanius collurio*).

In all the prey-choice models that we have discussed in this section, there are simple rules by which an animal that initially knows nothing about the environment can asymptotically achieve a rate of γ^*. One such rule is the prey-choice analogue of equation (4.5) for patch use. The animal forms a sequence of estimates $\gamma_0, \gamma_1, \gamma_2, \ldots$ for γ^*. On the nth cycle it behaves as if the true rate for the environment were γ_{n-1}. Having made its decision on this cycle it updates γ_{n-1} to γ_n.

4.5 Exploitation of prey distributed in patches

McNamara *et al.* (1993) investigated the optimal behaviour of a foraging animal that encounters prey items sequentially in well-defined patches. Encounter rates on a patch may decrease with time on a patch as a result of depletion by the forager. The forager knows the number of each type of item on a patch. It can choose which items to accept and which to reject while on a patch and can also decide, on the basis of the items remaining, when to leave the patch. McNamara *et al.* argue that an item should only be taken if its profitability e/h is greater than γ^*. This means that, in contrast to the erroneous conclusions of Heller (1980) and Lucas & Schmid-Hempel (1988), the optimal prey choice does not change with time on the patch.

A worked example based on diving

In the model of McNamara *et al.* (1993), the time that the animal has spent on a patch is irrelevant. All that matters is the numbers of each type of item remaining in the patch. We now give a worked example of a foraging problem in which the time spent on the patch is important. The example serves to illustrate both the use of the corollary to Theorem 4.1 and the iterative procedure for calculating γ^* and the optimal strategy.

Several species of birds and mammals search for food underwater. In doing so, they are constrained by the need to return to the surface to breathe. There is evidence that the time spent on the surface after a dive is an increasing and accelerating function of the duration of the dive. Houston & McNamara (1985b) investigated the effect of such a cost on the foraging behaviour of a diver that can only take one prey item to the surface to be eaten (a single-prey loader). They assumed that the diver was maximising its rate of energetic gain. For such a diver a renewal cycle starts when the animal dives and consists of the time spent underwater and the subsequent time spent on the surface before the next dive. They showed that

(i) There is a time $t_{\max}$ such that if the diver has been searching for this length of time it will return to the surface even if it has not found an item.

(ii) The diver becomes less selective about the sort of item that it will accept as the dive proceeds. In the context of a diver that can encounter two prey types with equal handling times, the diver should either always take both types of item or should initially accept only the item with the greater energetic content.

We now describe how to formulate a model of diving for a single-prey loader and compute the optimal strategy.

A model of the diving behaviour of a single-prey loader.   For the diver, a dive cycle consists of three phases.

(i) *The round-trip travel time τ.* This is the time spent getting from the surface to the foraging area at the start of the dive plus the time getting from the foraging area back to the surface at the end of the dive.

(ii) *The time t spent in the foraging area.*

(iii) *The time spent on the surface recovering.* We assume that the diver's oxygen debt $d(t, \tau) = t d_{\text{search}} + \tau d_{\text{travel}}$. The resulting recovery time is $J(d)$. For simplicity, we assume that it does not depend on whether an item was found (see Houston & McNamara 1985b for an example in which recovery time depends on what is found).

The search process is modelled as follows. The time at which the animal reaches the foraging area is taken to be time 0. The time spent searching is divided into time intervals $0, 1, 2, \ldots, T$ of equal duration. The time T is the maximum time that can be spent in the foraging area when there is no travel time ($\tau = 0$). When the animal must dive down to the foraging area, the maximum time in the foraging area will be less than T; for example, when $d_{\text{search}} = d_{\text{travel}}$, then the maximum time in the foraging area is $T - \tau$. Typically it will be optimal to leave the foraging area before the maximum time. While searching, the diver may encounter a prey item. Items are of two types, labelled type 1 and type 2, with energetic contents e_1 and e_2 respectively ($e_1 > e_2$). If the diver is searching at time t then, by time $t + 1$, it finds a type-1 item with probability p_1 or finds a type-2 item with probability p_2 or finds no item with probability $1 - (p_1 + p_2)$.

At the start of any interval, if the diver has accepted an item then it must return to the surface. If it has not accepted an item then it chooses between returning to the surface or staying in the foraging area to search for one more time interval (after which it decides again). We seek the pattern of choice that maximises the diver's gross rate of energetic gain. (It is easy to extend the model to find the behaviour that maximises the net rate of energetic gain.) Given that the time to recover does not depend on the type of item taken, it is clear that type-1 items should always be accepted. Thus the forager has to make two sorts of decision: if it finds a type-2 item should it accept this, and if no item is found should it continue to search?

The opportunity cost of lost time. In addition to the direct time lost while travelling and searching, the longer the time spent in these activities the greater the time needed to recover on the surface. This means that the opportunity cost of lost time while performing an activity is no longer γ^* per unit time. Instead, the opportunity cost has also to take into account the extent to which the performance of the activity increases recovery time.

Regardless of the time spent searching, the animal takes a time τ travelling to and from the foraging area. If the animal returned immediately on reaching the foraging area then its recovery time would be $J(d(0, \tau))$. Thus the lost opportunity cost incurred if the animal just travels to the foraging area and returns is

$$c_{\text{dive}} = \gamma^*[\tau + J(d(0, \tau))]. \tag{4.6}$$

Spending time searching on the bottom incurs additional lost opportunity costs. Suppose that an animal has already spent time t searching on the bottom. If it decides to continue searching until time $t + 1$ rather than to surface immediately, it will require one unit of time to perform the actual search and an extra time $J(d(t + 1, \tau)) - J(d(t, \tau))$ to recover. The opportunity cost of searching between times t and $t + 1$ is thus

$$c_t = \gamma^*[1 + J(d(t + 1, \tau)) - J(d(t, \tau))]. \tag{4.7}$$

Computational procedure. To compute the optimal strategy, we start off with an initial strategy and compute the resulting rate γ_0. We then compute the sequence $\gamma_1, \gamma_2, \ldots$ of successive approximations to γ^* as follows. Suppose that γ_n has been calculated.

Step 1. Set

$$c_t = \gamma_n[1 + J(d(t + 1, \tau)) - J(d(t, \tau))]$$

for $t = 0, \ldots, T - 1$ (cf. equation (4.7)).

Step 2. Use dynamic programming to calculate the strategy that maximises the expected energetic gain on a dive minus the total opportunity cost paid. Let

$$W(t) \quad = \quad \max[\mathbb{E}(\text{energy gain after } t)$$
$$-\mathbb{E}(\text{expected cost paid after } t)],$$

where the maximisation is over all strategies for which the diver searches for at least one more time step. Then for $t \leq T - 1$

$$W(t) \quad = \quad -c_t + p_1 e_1 + p_2 \max[W(t+1), e_2]$$
$$+ [1 - (p_1 + p_2)] \max[W(t+1), 0],$$

where we set $W(T) = 0$ by convention.

Thus $W(t)$ can be found for all $t \leq T$ by working backwards from T. Having found $W(t)$, we also know the optimal strategy. A type-1 item is always accepted, and a type-2 item is accepted at time t if and only if $W(t) \leq e_2$. If $W(t) \leq 0$ then a diver that is still in the foraging area at time t should return to the surface whether or not it has found an item.

Step 3. Find the expected energetic gain on a dive under the strategy found in step 2. One way to do this is to define

$$H(t) \quad = \quad \mathbb{E}(\text{energy gain after time } t$$
$$\text{given no item found at time } t).$$

Note that finding no item is equivalent to finding an item and rejecting it; thus $H(t)$ is also the expected future gain given that an item has been rejected at time t. If it is optimal to leave without an item at time t then $H(t) = 0$, in particular $H(T) = 0$. If it is not optimal to leave without an item at time t, then

$$H(t) = p_1 e_1 + (1 - p_1) H(t+1)$$

if it is optimal to specialise on type-1 items at time $t + 1$ and

$$H(t) = p_1 e_1 + p_2 e_2 + [1 - (p_1 + p_2)] H(t+1)$$

if it is optimal to take either type of item at time $t + 1$. These equations can be used to find $H(t)$ for all t by working backwards from T. Note that $H(0)$ is the expected energy gain on a dive.

Step 4. Find the expected duration of a dive cycle under the strategy found in step 2. One way to do this is to note that

$$\mathbb{E}(\text{total search cost in the foraging area}) = H(0) - W(0).$$

Thus if we set $c_{\text{dive}} = \gamma_n[\tau + J(d(0, \tau))]$ (cf. equation (4.6)), then

$$c_{\text{dive}} + H(0) - W(0) = \gamma_n \mathbb{E}(\text{cycle duration}).$$

Step 5. Calculate γ_{n+1} as

$$\gamma_{n+1} = \frac{\mathbb{E}(\text{energy gain})}{\mathbb{E}(\text{cycle duration})}.$$

Go to step 1.

Results. The optimal strategy for this problem has the following form. Items of energetic content e_1 are always taken. Before a time t_{gen} items of energetic content e_2 are rejected. At time t_{gen} or later the animal becomes a generalist, taking the first item of either type it encounters. Finally, if no item has been encountered by a time t_{leave}, the animal surfaces without an item. Figure 4.3(a) illustrates the dependence of the switch times t_{gen} and t_{leave} on the travel time τ. As can be seen, when τ is small the animal can afford to return to the surface often and it is optimal to generalise ($t_{\text{gen}} = 0$). As τ increases it pays the animal to be more discerning and more persistent in its search. Finally, as τ increases further, and oxygen needs severely constrain the maximum time at the bottom, the time t_{leave} is close to this maximum and the animal is again a generalist while searching.

By following behaviour forward under the optimal strategy we can obtain a number of measures. For instance, we can record the proportion of each prey type taken, and the proportion of trips where no item is taken (Figure 4.3(b)). Another obvious measure is the correlation between time underwater and the type of item found.

4.6 Effects of finite time

We have assumed so far that the period of time available for foraging is long enough for us to ignore the possibility that the animal may run out of time before it can complete the behaviour associated with a foraging option. For example, it might be optimal for an animal exploiting patchily distributed food to remain on a patch even when its rate is very low, because there is insufficient time to travel to a new patch before dusk, when the animal is no longer able to forage.

We illustrate the effect of a time constraint in a simple deterministic context. There is a total time T available for foraging. The animal can exploit patches of food. The time to travel between patches is τ, and at the

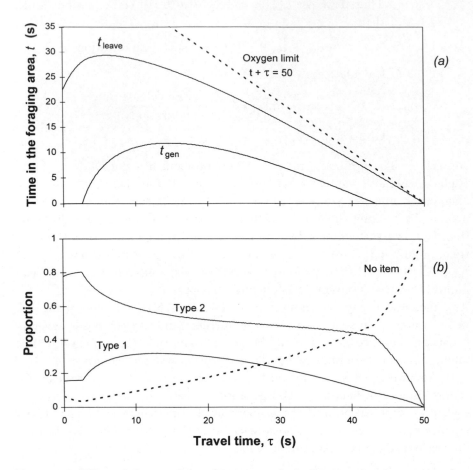

Figure 4.3. Effect of the travel time (time to reach the bottom plus time to resurface from the bottom) in the diving model of Section 4.5. (*a*) The time t_{gen} spent in the foraging area after which it is optimal to generalise by accepting both types of item; and the foraging time t_{leave} at which it is optimal to surface if no item has been accepted. (*b*) The proportions of times when a surfacing animal has a type-1 item, a type-2 item and no item. To compute the figure we have taken $T = 5000$ and 100 decision times during each second. Thus there is a maximum time of 50 seconds under water. The rate of oxygen consumption while searching is taken as equal to that while travelling and is 1 per second ($d_{search} = d_{travel} = 0.01$). The recovery time from an oxygen debt d is $5 + 0.0005 \log[50/(50 - d)]$ seconds, so $J(d)$ is 100 times this quantity. Prey energetic contents are $e_1 = 38$ kJ and $e_2 = 3.1$ kJ. Encounter rates are 0.02 and 0.1 per second respectively, giving $p_1 \cong 0.0002$ and $p_2 \cong 0.001$.

start of the period the animal has to travel to get to a patch. This means that if a forager visits n patches, it will spend a time τn travelling. It follows that the time available for foraging is $T - \tau n$. If the forager spends a time t in a patch, then it obtains an amount of energy $g(t)$, where $g'(t) > 0$ and $g''(t) < 0$.

Suppose that the forager visits more than one patch. Then it can be shown that the energy obtained from a given foraging time will be maximised when the time is divided equally between patches. Thus if n patches are visited, the time in each of them is $T/n - \tau$. The total amount of energy obtained is $ng(T/n - \tau)$. We can now find the value n^* of n that maximises this quantity and the associated optimal time $t^* = T/n^* - \tau$ spent in each patch. Some examples are given in Table 4.1, from which it can be seen that the optimal time in a patch can either increase or decrease as τ is increased. In contrast, when there are no time constraints it is always optimal to increase time in a patch as τ increases.

Table 4.1. Effect of finite time on patch use. All patches are the same, with $g(t) = t^{0.5}$. The optimal number n^* of patches to exploit and the optimal time in patch t^* are given for various values of the travel time τ. For this form of $g(t)$, the long-term rate is maximised when $t = \tau$.

τ	$T = 20$		$T = 10$	
	n^*	t^*	n^*	t^*
0.5	20	0.5	10	0.5
1.0	10	1.0	5	1.0
1.5	7	1.357	3	1.833
2.0	5	2.0	$\begin{cases} 2 \\ 3 \end{cases}$	$\begin{cases} 3.0 \\ 1.333 \end{cases}$
3.0	3	3.667	2	2.0
4.0	$\begin{cases} 2 \\ 3 \end{cases}$	$\begin{cases} 6.0 \\ 2.667 \end{cases}$	1	6.0
5.0	2	5.0	1	5.0
6.0	2	4.0	1	4.0
7.0	1	13.0	1	3.0

Table 4.1 shows that when the travel time τ is small, the optimal time is close to or equal to the time that maximises long-term rate in the absence of a time constraint. In contrast, when τ becomes relatively large, the time constraint means that the best constrained time can be very different from

the long-term optimum. As τ increases, the optimal number n^* of patches to visit decreases. When n^* falls to 1, once an animal reaches a patch it stays there for the remainder of the time available.

Table 4.1 shows that the optimal solution to the constrained problem is not necessarily unique. For example, when $T = 20$ and $\tau = 4$, visiting two patches and spending 6 time units in each results in the same amount of energy as visiting three patches and spending $2\frac{2}{3}$ time units in each.

It is instructive to see how the strategy that maximises long-term rate performs when time is limited. Assume that $T = 10$ and $\tau = 2$. It can be seen from Table 4.1 that one of the two best strategies is to visit two patches, spending a time $t^* = 3$ in each one. Thus the animal's behaviour involves travelling for 2 time units, spending 3 time units in a patch, then again travelling for 2 time units and spending 3 time units in a patch. The energy gained is $3^{0.5} + 3^{0.5} = 3.464$. The strategy of spending 2 units of time on each patch maximises the long-term rate. If the animal adopts this patch time, then it will travel for 2 time units, spend 2 time units in a patch, travel for 2 time units and spend 2 time units in a patch. At this point, it has used 8 time units and so has 2 time units left. If it travels, then it will have no time for a subsequent patch, and the energy gained will be $2^{0.5} + 2^{0.5} = 2.828$. If it spends the remaining 2 units in the current patch, then the total time in this patch will be 4 units and the energy gained will be $2^{0.5} + 4^{0.5} = 3.414$.

4.7 The effect of constraints on b and c

Hammond & Diamond (1997) review evidence for a limit to the sustainable rate of energy expenditure. (An energy budget is sustainable if it can be maintained over a long enough period for average body mass to be constant because average energy intake equals average energy expenditure.) The limit could be the result of a limit to the rate of intake (e.g. Kirkwood 1983) or of a limit to the rate of expenditure (e.g. Drent & Daan 1980). In this section (as in Sections 4.2 – 4.5) we assume a long time interval, so that a suitable averaging may be carried out, and we explore the consequences of the above constraints.

We introduce the effect of constraints with the following simple example. We consider an animal's behaviour over a total time T. There is an upper limit K on the animal's total energy intake during this period. This upper limit might, for example, result from digestive constraints. The animal chooses a foraging option, which it adopts until it reaches its maximum energy intake K. During the remainder of the period it performs an activity other than foraging, during which it expends energy at a rate c_r.

If the foraging option chosen has a gross rate of gain b, then the total time spent foraging is t_f, where

$$t_f = K/b; \tag{4.8}$$

hence the total energy expended while foraging is ct_f, where c is the rate of energy expenditure under the chosen foraging option. Since during the remaining time $T - t_f$ energy is spent at a rate c_r, the animal's total energy expenditure during T is thus

$$ct_f + c_r(T - t_f). \tag{4.9}$$

Because the gross energy gained is K regardless of the option that is chosen, the net energy gained is maximised by minimising expression (4.9). Using equation (4.8) to substitute for t_f, it follows that the net energy gained is maximised by minimising

$$c_r T + K(c - c_r)/b,$$

i.e. by choosing the foraging option that maximises

$$\frac{b}{c - c_r}. \tag{4.10}$$

(cf. Hedenström & Alerstam 1995, Houston 1995). The ratio b/c is known as the efficiency of a foraging option, so the ratio $b/(c - c_r)$ can be thought of as a modified form of the efficiency. Ydenberg *et al.* (1994) claim to have justified efficiency as a currency, but their argument ignores the energy that the animal must spend while it cannot forage (see Houston 1995 for further discussion). We now introduce a graphical approach that not only makes the criterion based on expression (4.10) clear but also leads to several novel conclusions.

We consider a reasonably long period of foraging and assume that there is a constraint on either the total energy intake during the interval or the total energy expenditure during the interval. The animal has a range of basic foraging options. It can switch freely between options without loss of time or energy. At any time during the interval, the animal can choose any foraging option. The animal's problem is to allocate its time to the foraging options in such a way that the net energetic gain from the period is maximised, subject to the constraints on total intake or expenditure. Many graphical analyses of foraging plot the gross rate b and the rate of expenditure c against some aspect of the animal's behaviour (e.g. the time spent in a patch). Our

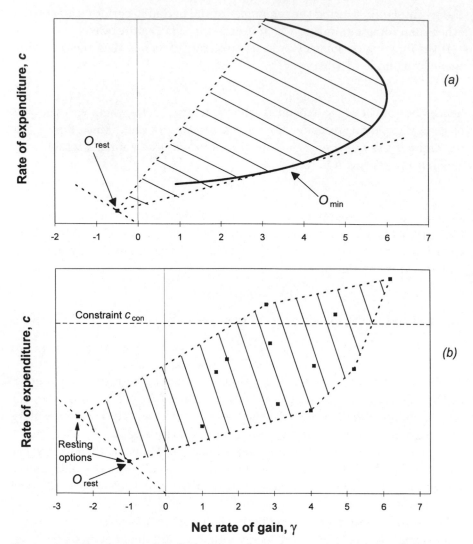

Figure 4.4. Characterisation of foraging options in terms of their mean net rate of energy gain, γ and mean rate of energy expenditure c. In (a) the basic foraging options are the resting option together with the options of the curved line; in (b) the basic options are the filled squares. In both cases the animal can generate new options by dividing its time between several basic options: the resultant generalised options are shown by the shaded region. Figure (b) also shows the effect of a constraint on mean energy expenditure. When mean expenditure must not exceed c_{con} only those generalised options below the given line can be used.

approach (McNamara & Houston 1997) involves plotting c against the net rate $\gamma = b - c$ for each possible foraging option. The options could form a continuum, as shown in Figure 4.4(a), or could be a discrete set, as is shown in Figure 4.4(b). The animal can generate new options from a mixture of these basic options. For example, suppose that the animal has available option O_1 with gross rate b_1 and rate of expenditure c_1, and option O_2 with gross rate b_2 and rate of expenditure c_2. The animal can then choose option O_1 a proportion p of the time and option O_2 a proportion $1 - p$ of the time, to create an option with mean gross rate $= pb_1 + (1 - p)b_2$ and mean rate of expenditure $= pc_1 + (1 - p)c_2$. This option then has mean net rate of gain $\gamma = p\gamma_1 + (1 - p)\gamma_2$, where γ_1 and γ_2 are the net rates under options 1 and 2 respectively. By creating new options in this way, the animal's effective range of foraging options is given by the shaded areas in Figures 4.4(a), (b). Thus, whatever basic set of options is available, the set of effective options forms a convex subset of the γc-plane. It is clear, however, that for a given value of c, the animal will obtain more energy if it has a high rather than a low value of γ. This means that only options on the right-hand boundary of the option set should be used.

So far, we have been concerned with options that involve foraging. The animal may also have the option of resting. When the animal rests, $b = 0$ and so

$$\gamma = -c.$$

Thus any resting option will be on the straight line given by this equation (see Figure 4.4). If there is more than one resting option let O_{rest} be the one with the lowest value of c, and let this value of c be c_r. In terms of energy expenditure, this resting option is the only one that should be used.

Because of the way that we have defined effective options, an animal that chooses a particular allocation of time to various basic foraging options can be thought of as choosing a single effective option for the whole time period. A constraint on how much energy can be acquired or spent during the time period can be translated into a constraint on the effective options that can be used for the whole time period.

We now identify various foraging options that are important in the analysis of the optimal choice of option. These options, together with the resulting values of γ and c, are illustrated in Figure 4.5 and summarised in Table 4.2.

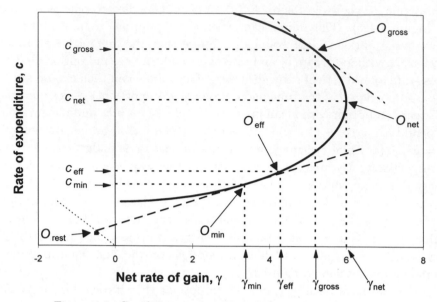

Figure 4.5. Graphic representation of the options O_{min}, O_{eff}, O_{net} and O_{gross} (see text).

By definition, the gross rate of gain $b = \gamma + c$, so lines of constant b are straight lines with slope $= -1$ in a space with axes γ and c. Thus when there is a unique option O_{gross} that maximises the gross rate of gain, this option will correspond to the point where a straight line of slope -1 is tangent to the option set.

O_{net} is the option that maximises the net rate of gain γ. This is obviously the best option when there are no constraints. The resulting maximum value of γ is denoted by γ_{net}.

O_{eff} is the option that minimises c/γ, which is equivalent to maximising the efficiency b/c. This option can be found graphically by constructing a tangent to the option set which passes through the origin, although this tangent is not shown in Figure 4.5.

O_{min} is the option for which a line from the option O_{rest} is tangent to the right-hand edge of the option set. It follows that at O_{min}

$$\frac{c - c_{\text{r}}}{\gamma + c_{\text{r}}} \tag{4.11}$$

is minimised, which means that expression (4.10) is maximised.

We note from Figure 4.5 that as we move from O_{gross} to O_{net} to O_{eff} to

O_{min}, both the gross rate of intake b and the rate of energy expenditure c decrease, i.e.

$$b_{gross} > b_{net} > b_{eff} > b_{min}$$

and

$$c_{gross} > c_{net} > c_{eff} > c_{min}.$$

Table 4.2. *Classification of some key foraging options.*

Option	Quantity maximised	γ	c
O_{gross}	gross rate b	γ_{gross}	c_{gross}
O_{net}	net rate γ	γ_{net}	c_{net}
O_{eff}	efficiency b/c	γ_{eff}	c_{eff}
O_{min}	$b/(c - c_r)$	γ_{min}	c_{min}
O_{rest}	metabolic rate while resting	$-c_r$	c_r

It is now relatively straightforward to analyse the effect of a constraint. We illustrate this by considering the effect of a constraint on the total energy spent during the period. A constraint on total intake can be handled in a similar way. Suppose that the total energetic expenditure during a period T cannot exceed E_{con}. Then this constrains the effective foraging options to have a rate of expenditure $\leq c_{con}$, where

$$c_{con} = E_{con}/T.$$

The effective foraging options that satisfy this constraint are illustrated in Figure 4.4b. We emphasise that the constraint does not prevent an animal from devoting some time to a basic foraging option with a rate of expenditure greater than c_{con}. All that matters is that the average rate of expenditure of the entire time interval does not exceed c_{con}.

We distinguish three cases (McNamara & Houston 1997).

(i) When $c_{con} \geq c_{net}$, the constraint has no effect; the best option is still O_{net}.

(ii) When $c_{net} > c_{con} \geq c_{min}$, O_{net} can no longer be used exclusively, and so the resulting net rate under the effective option will be less than γ_{net}. The best option may involve a mixture of two basic options, one with $c > c_{con}$ and the other with $c < c_{con}$. This has interesting implications for prey choice (McNamara & Houston 1997).

(iii) When $c_{con} < c_{min}$, the animal should divide its time between the options O_{min} and O_{rest}. No foraging option with a net rate less than γ_{min} should be used, since a suitable mixture of O_{min} and O_{rest} will give the same value of c but a higher value of γ (McNamara & Houston 1997). Thus when $c_{con} < c_{min}$, all the time devoted to foraging (as opposed to resting) is spent on the option that minimises expression (4.11). The same conclusion holds if the constraint is on total energetic intake rather than total expenditure, as we concluded from the algebraic argument.

4.8 Other components of state

In this chapter, we have concentrated on the amount of energy gained by some time T as giving a measure of the expected future reproductive success after this time. Although this is a simple and appealing approach to foraging, it is clearly not realistic. One major limitation is that it completely ignores the probability that the forager is killed by a predator. We analyse this case in Chapter 6. A perhaps less obvious drawback of energy as a currency is that there may be various components of an animal's state that are influenced by the animal's foraging behaviour and will have consequences for the animal's reproductive success. We can distinguish two broad categories.

(i) *Damage.* The particular foraging behaviour adopted may incur a risk of damage. For example, the prey taken by a crab may influence the wear to its claws (e.g. Juanes & Hartwick 1990) and the intensity with which an oystercatcher tries to open a shell may influence the probability that it damages its bill (e.g. Swennen *et al.* 1989).

(ii) *Condition.* There are general costs associated with expending energy. For example, a high level of energetic expenditure might have a detrimental effect on the immune system and thus make the animal more susceptible to disease (Deerenberg *et al.* 1997; see also Sheldon & Verhulst 1996, Apanius 1998).

In the approach that we have taken in this chapter, we have assumed that the basic option set can be used to generate options that are equivalent to mixtures of basic options. This assumes that, for example, average energy expenditure is all that matters and that spending half the day working very hard and half the day resting is equivalent to spending all the day working at an intermediate level that results in the same expenditure of energy over the day. In other words, all that is important is the effective foraging option and not the details of the basic options that make up the effective option. We now look at two cases. In the first we keep the basic assumption that an option can just be characterised in terms of average rates and describe how the approach of Section 4.7 can be modified to deal with cases in which adverse effects on condition depend only on the mean rate of energetic expenditure over an interval. We then go on to analyse circumstances where the effect of expenditure on condition cannot be characterised just in terms of mean expenditure but depends on the patterns of expenditure over time.

Cost depends only on mean rate of energy expenditure

As a simple example of a fitness cost of energy expenditure, consider a forager that has to achieve a given net rate of gain γ_{crit} over a certain period. If the achieved rate γ is less than γ_{crit} then the fitness is zero, whereas if γ is at least as great as γ_{crit} then the fitness depends only on the mean rate of energy expenditure c and decreases as c increases. In these circumstances, the best option is the one that minimises c subject to the constraint that $\gamma = \gamma_{crit}$. Lines of constant fitness are horizontal in the γc-plane. When the forager has two options, it can be optimal to use a mixture of them if the option with the lower value of c has a value of γ that is less than γ_{crit}. (The model is analogous to the model of Gilliam & Fraser 1987 in which predation is minimised subject to a constraint on γ.)

This criterion of minimising c subject to the constraint that $\gamma = \gamma_{crit}$ has been used in the special case $\gamma_{crit} = 0$ (e.g. Pyke 1979, Masman *et al.* 1988, Hedenström & Alerstam 1995). Pyke analysed the territorial behaviour of golden-winged sunbirds (*Nectarinia reichenowi*) during a period when they were not building up reserves. He found that predictions based on minimising energy expenditure subject to the energy balance requirement (i.e. $\gamma = 0$) gave a better fit to the data than predictions based on maximising the net rate of energetic gain.

In the case just considered, there is no advantage to increasing γ once a critical level has been reached. In general, however, an increase in γ will confer an advantage, and lines of constant fitness in the γc plane will have positive slope. As metabolic expenditure c increases, the marginal cost, in terms of fitness, of a further increase in c is likely to increase. The marginal benefit from increasing γ will depend on the biological circumstances. Provided the marginal benefit is either constant or decreasing, the effect of c means that lines of constant fitness have a decreasing slope as γ increases. It follows that (i) the option chosen need not be the one with the highest net rate, (ii) it may be optimal to adopt a mixture of two options, and (iii) adding an extra option that has lower gain rate and lower rate of energy expenditure than the existing options can increase fitness. All these effects are also found when c has no cost but there is a constraint on the maximum possible value of c.

It is important to note that, when c has a cost, its value at the optimum may be below any reasonable estimate of the upper limit on c. Masman *et al.* (1988) observed that kestrels (*Falco tinnunculus*) typically use two forms of hunting (high-yield high-cost flight hunting and low-yield low-cost perch hunting) even though exclusive use of flight hunting would maximise net rate of energetic gain and exclusive use of perch hunting would minimise the energetic expenditure required to balance the energy budget. The use of a mixture of options is predicted if the forager is up against a constraint on energy expenditure, but Masman *et al.* did not think this was likely for the kestrels that they were studying. Our analysis shows that a mixture of options with an average cost below the upper limit can be optimal when metabolic expenditure is costly.

Condition or damage as state variables

So far, we have assumed that condition is determined by the mean rate of energetic expenditure. The way in which an animal's condition changes may depend on the temporal details of its energetic expenditure. To understand such cases, condition can be introduced explicitly as a state variable. Within this approach, an animal's reproductive value at time t is $V(x, w, t)$, where x is the animal's level of energy reserves and w is the animal's condition. An animal's choice of basic foraging option can be characterised by two quantities. One is the net rate of energetic gain γ. The other is the rate $\dot{w} = dw/dt$ at which condition increases (note that $\dot{w}$ will be negative if condition decreases). The results in Appendix 4.1 show that, in the absence of predation, the optimal foraging option maximises

$$\gamma \frac{\partial V}{\partial x} + \dot{w} \frac{\partial V}{\partial \omega}. \tag{4.12}$$

Since $\partial V/\partial x$ is the rate at which reproductive value increases as reserves increase and γ is the rate at which reserves increase with time, the product $\gamma \partial V/\partial x$ is the rate at which reproductive value increases with time as a result of food intake. Similarly, $\partial V/\partial w$ is the rate at which reproductive value increases as a result of an increase in condition, and hence $\dot{w} \partial V/\partial w$ is the rate at which reproductive value changes with time as a result of a change in condition. Both $\partial V/\partial x$ and $\partial V/\partial \omega$ will usually be positive. Given a range of foraging options, options that are advantageous in having a high value of γ are likely to be disadvantageous overall, because of a high rate of expenditure. As expenditure increases, $\dot{\omega}$ will decrease and may be negative for large expenditures. The dependence of $V(x, \omega, t)$ on x and ω and the resulting optimal actions are illustrated in Section 9.5. (The analysis in Appendix 4.1 also takes predation risk into account. The trade-off between energetic gain and danger of predation is discussed in Chapter 6).

The above analysis has been given in terms of condition, but could have equally well been phrased in terms of a state variable that represents the state of repair of some structure, such as the bill of an oystercatcher or the claw of a crab. If w is now taken to be this state variable, then again the best foraging option maximises expression (4.12).

Although the effects of condition are perhaps less obvious than the effects of damage, we believe that the former are likely to prove very important. It is reasonable to argue on current evidence that condition underlies many fundamental life-history trade-offs (see McNamara & Houston 1996). In Chapters 8 and 9 we discuss the possible physiological basis of condition and explore some of the consequences of including condition in life-history models.

Future directions

In this section we have talked about condition as an abstract measure. To go beyond the simple schematic approach that we have taken here and will take in Chapter 9, detailed empirical work on the physiological basis of condition is required. In particular, we need to answer the following questions.

(i) *What aspects of an organism's physiology are influenced by work?* We have mentioned immuno-competence as a measure of condition, but it may also be necessary to include other variables such as muscle glycogen levels and lactic acid levels.

(ii) *What is the quantitative relationship between intensity and duration of work and changes in condition variables?* In particular we need to understand the extent to which we can average over any work rates. For example, for some species 5 minutes of hard work followed by 5 minutes of rest may be equivalent to 10 minutes of work at an intermediate level, but 5 hours of hard work followed by 5 hours of rest will not be equivalent to 10 hours of work at an intermediate level.

(iii) *What are the time scales of recovery for the various aspects of condition?* When the animal is not working hard, different components of condition may recover at different rates. It is tempting to suggest that rapidly recovering components of state, such as muscle glycogen levels, may be responsible for details of foraging-bout structure, whereas immuno-competence may be important in determining whether an organism breeds more than once in a year. This may, however, be too simple a view.

(iv) *What are the consequences of poor condition for survival and reproduction?*

If we knew the answers to questions (i) – (iv) above then we could construct an optimisation model taking the appropriate physiological measures as state variables. In this model, the answers to questions (ii) and (iii) would give us the dynamics, and the answer to question (iv) would link condition to consequences for fitness. The model would be able to predict not just overall time allocation but detailed bout structure. Out of such a model would emerge the currency that is maximised, but this currency would no longer depend on just mean net energetic gain or mean expenditure. Instead the currency would be some integrated measure of gains and expenditures over time.

Appendix 4.1 The trade-off between gaining energy, increasing condition and avoiding predation

Suppose that an animal has reserves x and condition w at time t. We will analyse the optimal behaviour of the animal between times t and $t+\delta$, where δ is small. It is assumed that behaviour from time $t + \delta$ onwards is given, and that if the animal has reserves x' and condition w' at time $t + \delta$ its reproductive value is $V(x', w', t + \delta)$.

Suppose the animal chooses option u between t and $t + \delta$. Under this option the animal has a net rate of energetic gain $\gamma(u)$, a rate of increase of condition $\dot{w}(u)$ and predation rate $M(u)$. The probability that the animal is not killed by a predator by time $t+\delta$ is thus $1 - M(u)\delta$, ignoring terms of order δ^2 or smaller. If it survives, its reserves at this time are $x' = x + \gamma(u)\delta$ and its condition is $w' = w + \dot{w}(u)\delta$. Thus the reproductive value of the animal at time t is

$$H(x, w, t; u) = [1 - M(u)\delta]V(x + \gamma(u)\delta, w + \dot{w}\delta, t + \delta)$$

to first order in δ. The optimal choice of u maximises the quantity. Expanding V as a Taylor series and ignoring terms of order δ^2 or higher gives

$$H(x, w, t; u) = V + \delta \frac{\partial V}{\partial t} + \delta \left[\gamma(u) \frac{\partial V}{\partial x} + \dot{w}(u) \frac{\partial V}{\partial w} - M(u)V \right] \quad \text{(A4.1.1)}$$

where V and its derivatives are evaluated at (x, w, t). Since V and $\partial V/\partial t$ do not depend on u, expression (A4.1.1) is maximised by choosing u to maximise

$$\gamma(u) \frac{\partial V}{\partial x} + \dot{w}(u) \frac{\partial V}{\partial w} - M(u)V. \quad \text{(A4.1.2)}$$

In particular, when $M(u)$ is the same for all options the optimal choice of option maximises expression (4.12).

5

Risk-sensitive foraging

5.1 Introduction

The basic logic of risk-sensitive foraging is illustrated by the following example, based on a single decision (cf. McNamara & Houston 1986). Imagine an animal that can choose between two foraging options. Option u_1 involves no uncertainty – the animal is sure to gain one unit of energy. In contrast, option u_2 results in the animal gaining either nothing or two units of energy; each alternative occurs with probability 0.5. We wish to compare the options in terms of the resulting contribution that they make to the animal's expected future reproductive success. Assume that one unit of energy results in a unit increase in reproductive value for an animal with a given current state (i.e. energy reserves) x, whereas two units of energy result in an increase of $1 + \delta$. Thus increasing the reserves from x to $x + 1$ increases the reproductive value by 1, whereas a further increase in reserves from $x+1$ to $x + 2$ increases the reproductive value by δ. This means that the increase in reproductive value that results from a given gain in energy depends on the animal's state. If $\delta < 1$, then the second unit of energy is worth less (in terms of reproductive value) than the first unit, i.e. a unit of energy is worth less when the reserves are $x + 1$ than when the reserves are x. In contrast, if $\delta > 1$ then the second unit of energy is worth more than the first unit, i.e. a unit of energy is worth more when the reserves are $x + 1$ than when the reserves are x. If $\delta = 1$, then the second unit of energy is worth the same as the first unit.

If the animal chooses the variable option, u_2, it only gets an increase in reproductive value if it obtains two units of energy – this occurs with probability 0.5. Thus the mean increase in reproductive value is 0.5 times the increase $1+\delta$ in reproductive value that results from obtaining two units of energy. It follows that if $\delta < 1$ then option u_1 has the higher associated reproductive value and should be preferred, whereas if $\delta > 1$ then option u_2 has the higher associated reproductive value and should be preferred. If $\delta = 1$ then the two options are equivalent in terms of reproductive value.

This simple example illustrates that both the mean energetic gain and its variability can be important in determining which action is best. Assume that an animal is given a choice between two alternatives, one of which gives a constant amount of energy and the other of which gives a variable amount of energy with the same mean. The animal is said to be risk-prone if it chooses the variable option and risk-averse if it chooses the constant option. Risk-sensitive foraging has been investigated in a wide range of animals, including birds, mammals, fish and spiders (see Kacelnik & Bateson 1996 for a review). There are various ways in which an animal can influence the variance in its energetic gain (Sutherland & Anderson 1987). Choice of foraging area is one obvious possibility. Another possibility is that the size of the group in which the animal forages determines the mean and variance of its intake rate (e.g. Clark & Mangel 1986).

In analysing risk-sensitive behaviour, the key question is how reproductive value depends on energy reserves. Given that this relationship is known, there are simple general principles that determine optimal behaviour. The form of this relationship depends on the biological circumstances of the animal under consideration. This means that, as emphasised by McNamara & Houston (1992a) and Kacelnik & Bateson (1996), there is no single model of risk-sensitive foraging. In this chapter we describe a range of models to illustrate how various factors influence the dependence of reproductive value on state.

5.2 The basic principle behind risk-sensitive foraging

Whether an animal should be risk-sensitive is determined by the dependence of $V(x,t)$ on the reserves x at a fixed time. $\partial V/\partial x$ is the rate of increase in reproductive value with reserves, i.e. the marginal value of energetic reserves. The fundamental issue is how this marginal value changes with reserves. If $\partial V/\partial x$ does not depend on x, then the second derivative $\partial^2 V/\partial x^2 = 0$ and $V(x,t)$ is a linear function of x. If $\partial V/\partial x$ increases with x, then $\partial^2 V/\partial x^2 > 0$ and $V(x,t)$ is a convex function of x. If $\partial V/\partial x$ decreases with x, then $\partial^2 V/\partial x^2 < 0$ and $V(x,t)$ is a concave function of x. The incremental value of a unit of food to an animal with reserves x is $V(x+1,t) - V(x,t)$. It can be shown that this incremental value is constant if V is linear, increases with reserves if V is convex and decreases with reserves if V is concave. We now show how deviations of V from linearity make it advantageous to be risk-sensitive.

A single decision

We consider an animal with reserves x at time t. The animal can choose between two feeding options. The mean net rate of energy gain is γ under both options. Under option u_1, the net change in reserves by time $t+1$ is a constant γ. Thus the level of reserves at time $t+1$ is $x' = x + \gamma$. Under option u_2, the net change in reserves is a random variable Y with mean γ. The reserves at time $t+1$ are thus the random variable $X' = x + Y$. If we assume that the reproductive value of being in state x' at time $t+1$ is $V(x', t+1)$, the reproductive value of an animal that chooses option u_1 at time t is

$$H(x, t; u_1) = V(x + \gamma, t + 1).$$

Similarly, averaging over possible values of Y we have

$$H(x, t; u_2) = \mathbb{E}\left[V(x + Y, t + 1)\right].$$

Thus it is optimal to choose option u_2 (i.e. be risk-prone) if and only if

$$V(x + \gamma, t + 1) < \mathbb{E}[V(x + Y, t + 1)]. \tag{5.1}$$

Whether this condition is satisfied depends on the form of V as a function of the reserves x at time $t+1$. Jensen's inequality (see Box 5.1 for details) tells us that if V is a convex function (i.e. the incremental value of energy is increasing with reserves), then inequality (5.1) is satisfied, and it is optimal to be risk-prone. If V is concave, the reverse inequality holds and it is optimal to be risk-averse. When V is linear, the value of a unit of energy is constant, and the animal should not be sensitive to variance. A Taylor series approximation (e.g. Real 1980) can be used to quantify the difference in payoff between an option with variable net energetic gain and a fixed option with the same mean. This difference is approximately equal to $\frac{1}{2} \times$ variance $\times \partial^2 V / \partial x^2$.

In the above we have emphasised that, for options with the same mean net energetic gain, the variance in gain can be important. Other aspects of the probability distribution of the net energetic gain can also be relevant. In particular, the skew may be important because it influences the probability that reserves will decrease in any time interval (Caraco & Chasin 1984, Houston & McNamara 1986a).

Box 5.1. Jensen's inequality.   Let X be a random variable and $V(x)$ be a function. We compare $\mathbb{E}[V(X)]$ with $V(\mathbb{E}(X))$.

If V is smooth, then the condition that V is convex is equivalent to $d^2V/dx^2 \geq 0$. Similarly V is concave if and only if $d^2V/dx^2 \leq 0$. Jensen's inequality states that

$$\mathbb{E}[V(X)] \geq V(\mathbb{E}(X)) \quad \text{if } V \text{ is convex} \qquad (B5.1.1)$$

and

$$\mathbb{E}[V(X)] \leq V(\mathbb{E}(X)) \quad \text{if } V \text{ is concave.} \qquad (B5.1.2)$$

To illustrate the first inequality, suppose that V is convex and that X takes two values, $\gamma-\sigma$ and $\gamma+\sigma$, each with probability 0.5. Thus $\mathbb{E}(X) = \gamma$, so that $V(\mathbb{E}(X)) = V(\gamma)$ and $\mathbb{E}[V(X)] = 0.5[V(\gamma-\sigma) + V(\gamma+\sigma)]$. Figure 5.1 illustrates that we then have

$$0.5[V(\gamma-\sigma) + V(\gamma+\sigma)] > V(\gamma), \qquad (B5.1.3)$$

so that inequality (B5.1.1) is satisfied. A similar argument applies when V is concave.

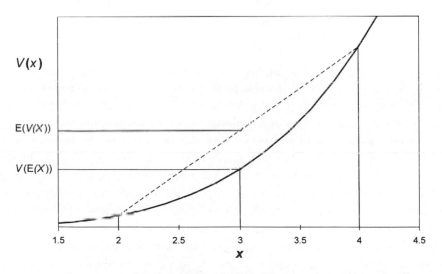

Figure 5.1. Graphical illustration of Jensen's inequality. To illustrate formula (B5.1.3), and hence formula (B5.1.1), we have taken V to be convex (solid line) and chosen the random variable X such that $P(X = 2) = P(X = 4) = 0.5$. Thus, in the notation of equation (B5.1.3), $\gamma = 3$ and $\sigma = 1$.

5.3 When is V linear and what causes deviations from linearity?

A risk-sensitive animal responds to both the mean and the variance of its foraging options. We have seen that when V is linear the animal should not be sensitive to the magnitude of the variance. Its best behaviour is to choose the option that gives the highest mean net gain. In this section we look at a particular case in which V is linear. (For a technical discussion, see Mangel 1992.) We then analyse effects that can cause deviations from linearity.

We consider an interval $[0, T]$ during which there is no reproduction. As usual, the state variable x is the animal's level of energy reserves. We assume that the amount of energy gained by the animal during any time interval is independent of x. The terminal reward $R(x)$ is taken to be a linearly increasing function of x, so that $R(x) = k_0 + k_1 x$.

Consider an animal with reserves x at time t, and suppose that the amount of energy gained by the animal between t and T is a random variable Y. Thus the reproductive value of the animal is

$$\begin{aligned} V(x, t) &= \mathbb{E}[R(x + Y)] \\ &= k_0 + k_1[x + \mathbb{E}(Y)]. \end{aligned}$$

Two points can be seen from this equation. One is that $V(x, t)$ depends only on the mean of Y, not on its variance. The other is that V is a linear function of x. It follows that if we give the animal a choice of options at any time, it should prefer the option with the biggest mean net gain, i.e. it should rate-maximise (see also Chapters 2 and 4).

One feature that is essential for this result is the linear terminal reward, but in fact there are good reasons for expecting the terminal reward to be non-linear in some circumstances. We discuss this topic later in the chapter. The other essential feature for the above result is that the amount of energy obtained by an animal in a given interval does not depend on its reserves at the start of the interval. This may not hold if an animal's state can influence the consequences of its behaviour. For example, the energy expended during movement may increase with mass, so that an animal with a high level of reserves expends more energy than an animal with a low level of reserves. Under these circumstances, the net amount of energy obtained over the interval under a given option will decrease with increasing initial level of reserves. Similarly, if the options available to an animal depend on its state then the energy obtained may depend on initial state. For example, if the animal's state is its size, the type of prey that the animal can capture is likely to depend on its size.

An upper limit to the amount of food that a foraging animal can store or the fact an animal starves if its energy reserves fall to zero can also mean that the amount of food obtained depends on initial reserves, as we now describe.

5.4 The effect of boundaries

Suppose that an animal has a single foraging option at each of the times $t = 0, 1, 2, \ldots, T - 1$. Under this option the net energetic gain in a unit time interval has mean γ and variance σ^2. Gains in successive intervals are independent. The terminal reward at final time T equals the energy reserves at this time, i.e. $R(x) = x$. Under this assumption an animal with reserves x at time t has reproductive value

$$V(x, t) = x + \mathbb{E}(\text{net energy gain between } t \text{ and } T). \qquad (5.2)$$

If the animal could always forage at mean net rate γ the expected net energy gain between t and T would be $\gamma(T - t)$. We would then have $V(x, t) = x + \gamma(T - t)$, so that $V(x, t)$ would be a linear function of x for each t. However, limits on energy reserves mean that the animal is not always able to forage at the mean net rate γ, and this causes $V(x, t)$ to deviate from linearity.

Effect of an upper limit on reserves. There is likely to be an upper limit on the amount of energy that an animal can store. An animal that reaches this upper limit will be unable to forage until reserves decrease below the limit. Thus the animal will gain less energy by time T than an animal that was able to forage for the whole time interval. If an animal has reserves x at time t then the expected time it spends at the upper limit between times t and T increases as x increases. Thus the expected net energetic gain between t and T decreases with x at time t and hence, by equation (5.2), $V(x, t)$ increases at a slower rate than x, for x close to its upper limit (Figure 5.2).

Effect of a lower lethal limit on reserves. An animal will starve if its energetic reserves reach some critical lower level, which we take to be zero. An animal that dies gains energy reserves at rate 0 for the rest of the time interval until T.

Suppose first that $\gamma > 0$. If an animal starves before time T it can no longer gain energy. Thus the possibility of starvation implies that an animal's mean net energy gain between t and T is less than $\gamma(T - t)$. Since the probability of starvation decreases as initial reserves increase, the mean net energy gained between t and T increases as reserves at time t increase.

Thus, by equation (5.2), $V(x,t)$ will be increasing faster than x for x close to 0 (Figure 5.2).

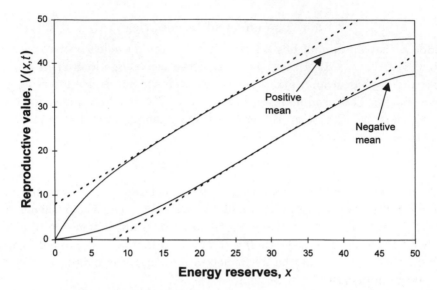

Figure 5.2. The effects of boundaries. An animal has a single foraging option with mean net rate of gain γ. The animal dies if its reserves x fall to the lower boundary $x = 0$. Reserves can never exceed the upper boundary $L = 50$. The figure shows reproductive value $V(x,t)$ as a function of x for $t = T - 80$. The terminal reward at T is taken to be $R(x) = x$. Two cases are shown: negative mean, $\gamma = -0.1$, and positive mean, $\gamma = 0.1$. For each case the broken line shows the value that $V(x,t)$ would have taken if there were no upper or lower boundaries.

Conversely, when $\gamma < 0$ an animal that forages for the whole time interval has a negative mean net energetic gain. By arguments analogous to those above, $V(x,t)$ increases more slowly than x for x close to 0 (Figure 5.2).

In the cases illustrated in Figure 5.2, the time to go until final time, $T - t$, is small, and away from the boundaries $V(x,t)$ is still an approximately linear function of x. Thus the value of a unit of food, $V(x+1,t) - V(x,t)$, is approximately constant in this range of x. As $T - t$ increases, $V(x,t)$ tends to zero for every state x. This is so because given enough time the animal is sure to die. If the mean net gain rate γ is high and the upper level of reserves is also high then $V(x,t)$ may tend to zero very slowly. Conversely, if γ is negative $V(x,t)$ may tend to zero rapidly. However, given sufficient stochasticity to ensure that there is a possibility that energy reserves can increase, $V(x,t)$ will never equal zero for large $T-t$ even when $\gamma < 0$, because there is always a possibility that the animal will survive until time T.

Given any choice between foraging options, the optimal choice at time t depends on how the reproductive value at time $t+1$ depends on the energy reserves. The absolute value of V is irrelevant to this choice: all that matters is the relative reproductive values of different levels of reserves. Now although for any two levels of energy reserves x and y we have $V(x,t) \to 0$ and $V(y,t) \to 0$, strong backward convergence (Section 3.5) ensures that the ratio $V(x,t)/V(y,t)$ tends to a limit as $T-t$ increases. This limit is independent of the terminal reward and is hence the same regardless of whether $R(x) = x$ for $x > 0$ or $R(x) = 1$ for $x > 0$. The limiting value of $V(x,t)/V(y,t)$ is equal to the ratio of the probabilities of surviving until T for initial states x and y respectively. Let L be the upper limit on x. Then the limiting value, as $T-t$ increases, of

$$\frac{V(x+1,t) - V(x,t)}{V(L,t)} \tag{5.3}$$

is the value, in terms of increased survival probability, of a unit of food to an animal with reserves x relative to the survival probability of an animal with maximum reserves L. Figure 5.3 illustrates this limiting relative value of food. As can be seen, when the mean net energetic gain γ is positive the value of food decreases strongly as reserves increase. This is so because at low reserves there is a high probability that the animal will starve before it can increase its reserves. This probability decreases rapidly as x increases, since the animal has a positive mean net rate of energetic gain. When γ is negative the relative probability of survival is low unless initial reserves are well above zero. Thus for low reserves an increase in reserves by one unit makes little difference to survival prospects: the animal is very likely to die whether it gains this unit of reserves or not. Consequently the relative value of a unit of food given by expression (5.3) is small for low reserves and initially increases as reserves increase. The existence of the upper boundary means that the relative value of food decreases as reserves approach this boundary, even when γ is negative.

Avoiding starvation while foraging

To look at the effect of boundaries on optimal choice we consider an extended period of time that ends at time T. It is assumed that the feeding options available and the amount of food that they deliver do not depend explicitly on time. In particular, we ignore the effect of the day–night cycle

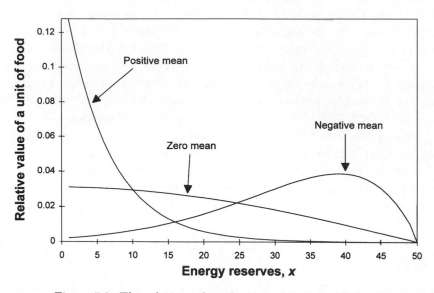

Figure 5.3. The relative value of a unit of food, given by equation (5.3), when the time to go until final time T is large. As in Figure 5.2 the animal has a single foraging option with mean net rate γ. Starvation occurs at 0 reserves and the reserves can never exceed $L = 50$. Three cases are shown: negative mean, $\gamma = -1$; zero mean, $\gamma = 0$; positive mean, $\gamma = 0.1$.

on food availability. One biological context would be an animal such as a shrew that does not have to stop foraging at night. For such an animal, the decision times might be of the order of minutes apart. An alternative biological context would be a relatively larger forager, for which each time period was a day. On this time scale any interruption to foraging imposed by darkness does not need to be represented explicitly. The problem of overnight starvation is considered in the next section.

Given a choice of foraging options the optimal choice depends on energy reserves. However, by strong backward convergence (Section 3.5) this optimal choice does not depend on time t for large values of the time to go, $T - t$. It is this limiting state-dependent strategy that we discuss here. The strategy can be thought of as maximising the animal's long-term survival probability (for more details see McNamara 1990a, McNamara & Houston 1992a). The form of this optimal strategy depends on the form of the stochasticity in the food supply.

Independent energetic gain.   Suppose that the amounts of food found in successive time intervals under an option are independent. This might apply if variations in food are the result of good or bad luck when searching.

Let all options have the same mean net rate of energetic gain, γ. Then the value of a unit of energy, $V(x+1,t) - V(x,t)$, can be expected to have the qualitative dependence on x already illustrated in Figure 5.3. Consider first the case $\gamma \geq 0$. Since the value of a unit of energy is then a decreasing function of reserves (Figure 5.3), options with low variance in energetic gain will be preferred to options with high variance. In other words, risk-averse behaviour is predicted (McNamara & Houston 1990a). In this case of positive mean net gain, the animal's most important consideration is not to let reserves drift down towards zero; the likelihood of this drift is minimised by playing safe.

In contrast when $\gamma < 0$ the value of a unit of energy first increases and then decreases as reserves increase (Figure 5.3). Thus risk-prone behaviour is predicted for low energetic reserves and risk-averse behaviour is predicted for high reserves (McNamara & Houston 1990a). There is a critical level of reserves at which behaviour switches (Figure 5.4(a)). In this negative-mean-gain case an animal with low reserves is very likely to starve in the near future and must take risks to try to increase its reserves by a large amount.

An intuitive justification of the form of the optimal strategy can be obtained by considering what happens in the absence of variability. When the mean is positive, the animal is sure not to starve, and so variability can only make things worse. When the mean is negative, the animal is sure to starve, and variability when reserves are low gives the animal the chance to increase its reserves. Once reserves are high, low variance is again advantageous because of the upper boundary.

Correlated energetic gain. The food supply may fluctuate at irregular unpredictable intervals as a result of, for example, bad weather or the presence of predators. Under these circumstances the amount of food found in successive time intervals is likely to be positively correlated; for example, if little food is found in one small time interval because of snow cover, the snow is still likely to be around and the food supply poor in the next time interval.

Barnard *et al.* (1985) and Merad (1991) considered the case where interruptions, during which no food can be found, alternate with periods of good food. They take the overall mean net gain (averaged over interrupted and good periods) to be positive. The details of the optimal strategy depend on the distribution of interruption times but, for a negative binomial distri-

bution, the general form of strategy is to be risk-averse when reserves are
very low, to be risk-prone when they are slightly higher and again to be
risk-averse when they are high. The region of risk-prone behaviour arises
because within this region an interruption is likely to lead to starvation,
whereas if the animal can increase its reserves somewhat it should be able
to survive an interruption.

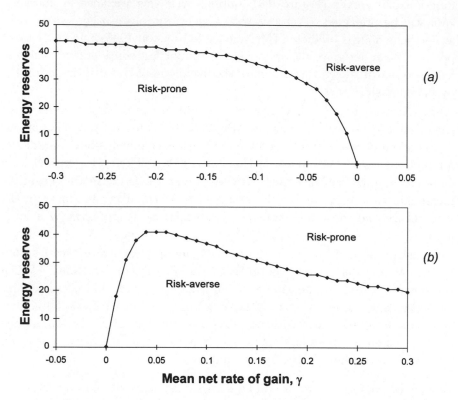

Figure 5.4. The optimal long-term strategy when the animal can
choose between two options with the same mean net gain rate γ.
In (a) the animal's strategy maximises its chances of long-term
survival, where mortality is due to starvation at the lower boundary. In (b) the animal reproduces on reaching the upper boundary at $L = 50$. Its strategy maximises lifetime reproduction success. Mortality is due to starvation and to a background mortality
$m = 0.0005$.

The models of Barnard *et al.* (1985) and Merad (1991) assume no upper
boundary and assume that the probability of escape from the region near
the lower lethal boundary is maximised. Similar results hold, however, when
there is an upper level on reserves, provided that this level is sufficiently high.

A model that includes reproduction

McNamara *et al.* (1991) and Merad & McNamara (1994) presented a model
that involves both starvation and reproduction. As in the previous models
of this chapter, the animal is characterised by its level of energy reserves x.
The animal starves if x falls to zero and reproduces when its reserves are
above L. In addition to starvation there is a probability m that the animal
dies in each time interval, independently of its reserves or behaviour. We
refer to this as the background mortality. It is assumed that the animal
continues to forage and reproduce until it dies. The optimal strategy is the
one that maximises the animal's lifetime expected reproductive success.

Merad & McNamara (1994) showed that when the feeding options have
the same negative mean, it is optimal to choose the higher variance for all
values of reserves, regardless of the value of m. When the options have the
same positive mean and there is no background mortality (i.e. $m = 0$), it
is optimal to choose the lower variance at all values of reserves. When the
common mean is positive but m is positive then it is optimal to choose the
low-variance option if reserves are low in order to avoid starvation. When
reserves are higher, the background mortality is more important than the
possibility of starvation, and the animal should choose the high-variance
option in order to maximise its probability of reaching L and reproducing
before it dies (Figure 5.4(b)).

For other models of risk-sensitivity in the context of reproduction, see
Bednekoff (1996).

5.5 Building up reserves to survive a period without food

Under some circumstances there is a predictable period during which an
animal cannot forage. For example, an animal may have to stop foraging
at dusk or when the tide comes in, or be unable to obtain much food after
a certain time of year if the food supply is seasonal. To cope with such
a period the animal must build up its reserves beforehand. To model the
build up of reserves we take the period of no food to start at time T and
look at optimal behaviour at the preceding times $t = 0, 1, 2, \ldots, T - 1$. At
each of these times the animal has a choice between two feeding options.
Under option u_i ($i = 1, 2$) the net energetic gain in a time unit has mean
γ_i and variance σ_i^2. It is assumed that $\sigma_1^2 < \sigma_2^2$, so that option u_2 is the
more variable option. The animal's state is again taken to be its energetic
reserves x. The terminal reward at T is $R(x)$ and the animal dies if reserves
fall to 0.

We start by taking the terminal reward to be a step function. This can be given the following biological interpretation. Assume that we are studying the foraging behaviour of a small bird during a winter's day. We take the final time T to be dusk, when the bird stops foraging and goes to roost for the night. This bird dies from starvation overnight if its reserves are less than x_{night} at dusk. A step function, with $R(x) = 1$ if $x \geq x_{\text{night}}$ and $R(x) = 0$ if $x < x_{\text{night}}$, can then represent the bird's probability of surviving the night as a function of its level of reserves at dusk. Using this terminal reward at dusk we can work backwards over the day using dynamic programming. The strategy during the day obtained by this procedure maximises the bird's probability of surviving until dawn on the next day. An example is given in Figure 5.5. The general picture that emerges is that the animal should be risk-prone in a wedge-shaped region near final time (e.g. Houston & McNamara 1985c, McNamara & Houston 1986). The upper boundary of this region is approximately a straight line. The slope of the line is given by the diffusion approximation described in Appendix 5.1. When the two options have the same mean, the slope is equal to the common mean. The wedge-shaped region arises because the animal is attempting both to avoid starvation during the day and ensure that its energy reserves at dusk are sufficient to survive the night. Within the region, by preferring variability the animal improves its chances of achieving a level of reserves of at least x_{night} at dusk and thus surviving the night. Above the region, only bad luck will result in reserves being below x_{night} at T, and so the probability of surviving the night is increased by preferring the less variable option. Below the region, the danger of immediate starvation is greater than that of overnight starvation and it is optimal to prefer the less variable option.

Houston & McNamara (1985c) investigate a forager that encounters prey of two types. Prey type i has an energetic content e_i and a handling time h_i (see also Sections 2.2 and 4.4). The forager's pattern of accepting or rejecting prey types determines the mean and variance of its energetic gain. Assuming that $e_1/h_1 > e_2/h_2$, the only options that need to be considered are option u_1, 'generalise', i.e. take both types of item as encountered and option u_2, 'specialise on type-1 items', i.e. take all type-1 items that are encountered and reject all type-2 items that are encountered. Suppose that encounter rates are such that if the animal were rate maximising (Section 4.4) it would take only type-1 items. Then option u_2 has the higher mean but also the higher variance, so that the choice of options involves a trade-off between mean and variance. Houston and McNamara found a wedge-shaped region near dusk in which option u_2 (i.e. high mean and high variance) is optimal. They also found that the diffusion approximation gives a slope very close to the slope of the upper boundary of this region.

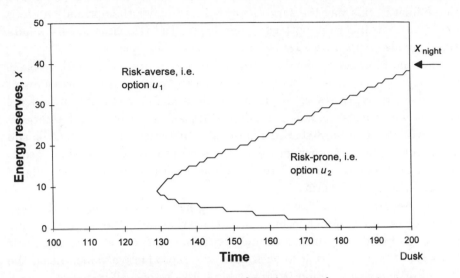

Figure 5.5. Optimal strategy when the animal must raise its reserves to $x_{night} = 40$ by dusk if it is to survive the night. Under option u_1, the mean net gain rate is $\gamma_1 = 0.5$ and the variance in gain is $\sigma_1^2 = 1.125$. Under option u_2, $\gamma_2 = 0.6$ and $\sigma_2^2 = 4.04$. The irregular nature of the boundary between the risk-prone and risk-averse regions is the result of the integer grid of reserves and time used in the computation.

When both foraging options give the same mean net energetic gain $\gamma(= \gamma_1 = \gamma_2)$, the diffusion approximation shows that the less variable options should be preferred if $x + \gamma(T - t) \geq x_{night}$. This strategy is known as the expected daily energy budget rule (Caraco 1980, Stephens 1981). What it says is that if an animal can expect to obtain enough energy during the day to survive the following night, then it should minimise the variance, whereas if it does not expect to obtain enough energy then it should maximise the variance. The expected daily energy budget is thus the net gain during the day minus the energy requirement during the night. It is not the mean net gain while foraging – this could be negative and yet the expected daily energy budget could still be positive if the initial reserves were sufficiently high.

In the above models an animal can make a sequence of choices before dusk. It can thus adapt its behaviour according to how well it is doing. But in the model proposed by Stephens (1981) the animal does not have this flexibility of behaviour. Stephens assumed that an animal in state x at time t must choose to spend all the remaining time before T exploiting either a high-variance source or a low-variance source. As a result of constraining the animal in this way the survival probability under the optimal choice is less than that under the optimal dynamic strategy (Houston & McNamara 1982, see also Appendix 5.1).

The step-function terminal reward assumes that an animal with reserves less than x_{night} at dusk is sure to die, whereas an animal with reserves at least equal to x_{night} is sure to survive. This is not very realistic; it is likely that various effects, including random fluctuations in overnight temperature, will result in an S-shaped relationship between reserves at dusk and overnight survival. Numerical calculations suggest that the optimal strategy is essentially unchanged if the terminal reward is taken to be S-shaped rather than a step function.

Furthermore, a model based on surviving just one night may not be very realistic. A small bird in winter has to survive until the start of spring if it is to breed and thus have any reproductive success. McNamara & Houston (1982) argue that in such cases the reproductive value during the period of non-reproduction is proportional to the probability of long-term survival. For our present purposes we note that in addition to surviving the coming night it may be advantageous to start the next day with relatively high reserves in order to improve subsequent survival. Although we could make *ad hoc* modifications to the terminal reward to incorporate such an effect, it is possible to represent it directly by maximising survival over a period of several days. To do so one uses dynamic programming to work backwards over this period, taking into account energy expenditure on each night. This procedure is described in more detail in Section 9.3. As explained in that section, as the period for which the animal must survive increases, the behaviour on any day a long time from the end of the period settles down to a limiting routine that maximises long-term survival. A terminal reward at dusk appropriate for long-term survival emerges from the analysis.

5.6 Risk-sensitive behaviour in a changing environment

In experimental procedures intended to test risk-sensitivity it is usual to keep the available foraging options constant throughout the experimental session. In particular, the probability that an option yields a reward and the size of this reward do not change. Models that attempt to predict the results of such experiments are usually based on an exact representation of the laboratory procedure and assume that

(i) the distribution of food obtained under an option does not change with time during a session.

Since experimental animals have usually been exposed to the experimental set-up previously, in a training session, models also assume that

(ii) the animal knows the distribution of food under each option (this is discussed further by Kacelnik & Bateson 1996).

However, animals have evolved in environments that are far more complicated than the simple experimental procedures in which they are tested (see Houston & McNamara 1989 and Section 1.3). In their natural environment, feeding options will vary over time, and an animal attempting to exploit these options will usually have incomplete information on how good are current options. Such an animal has a single, possibly complex, rule for making choices on the basis of its past experience and its present state. We can regard this rule as enabling the animal to both sample and exploit various possible food sources. This rule may involve the continual updating of information (McNamara & Houston 1980, 1985a, 1987a, Stephens & Krebs 1986; see also Section 4.3). The animal continues to use this rule in the laboratory and it is consequently not immediately clear whether an animal in the laboratory can be expected to show the risk-sensitive behaviour predicted by simple models based on assumptions (i) and (ii).

In an investigation of this issue, McNamara (1996) analysed a series of models that relax assumptions (i) and (ii). Here we describe one of these models, in which the environment can change over time but the animal has complete information on current environmental conditions. The consequences of incomplete information are discussed in the next section.

Example 5.1. Survival in a changing environment. An animal is attempting to maximise its probability of surviving a period of many days. There is a day–night cycle. During daylight the animal forages, at night it rests. Energy reserves are used up both during daylight and at night, and the animal starves if the reserves reach zero. There is an upper limit on the reserves. The day is divided into T time intervals. At each of the times $t = 0, 1, 2, \ldots, T - 1$ the animal chooses between option u_1 and option u_2. At any given time t these options have the same mean net gain $\gamma(t)$ but differ in their variability, option u_2 having the greater variance in energetic gain. The mean net gain $\gamma(t)$ changes stochastically over time. In all, $\gamma(t)$ can take one of 11 values, $\overline{\gamma} - 5\epsilon, \overline{\gamma} - 4\epsilon, \ldots, \overline{\gamma}, \ldots, \overline{\gamma} + 4\epsilon, \overline{\gamma} + 5\epsilon$ centred on an overall mean of $\overline{\gamma}$. Values close to $\overline{\gamma}$ are more likely than extreme values. A change occurs between successive times t and $t + 1$ with probability

p_{change}. If a change in $\gamma(t)$ occurs, it is to an adjacent value, but a crucial assumption is that changes that take $\gamma(t)$ closer to $\overline{\gamma}$ are more likely than changes that take $\gamma(t)$ further away from $\overline{\gamma}$. The animal is assumed to know the current value of $\gamma(t)$ when choosing its foraging option.

The optimal strategy for the above model specifies the choice of foraging option as a function of energy reserves, current value of $\gamma(t)$ and time of day. For very low reserves it is important to avoid daytime starvation, and the optimal choice is option u_1. For higher reserves the need to survive the night is the predominant influence on the optimal strategy, and it is this aspect of the optimal strategy on which we focus here. For each time of day and current value of $\gamma(t)$ there is a critical level x_c below which it is optimal to be risk-prone and choose option u_2 and above which it is optimal to be risk-averse and choose option u_1. When the parameter p_{change} describing the rate of environmental change is small, the critical level x_c, at time t, is approximately given by $x_c + \gamma(t)(T - t) = x_{\text{night}}$, where x_{night} is the mean overnight energetic expenditure. This is to be expected from the analysis of Section 5.5. When p_{change} is not small, however, the current mean net gain $\gamma(t)$ is likely to change before dusk. If $\gamma(t) < \overline{\gamma}$ the food supply is more likely to improve than to deteriorate and the mean net energetic gain before dusk is greater than $\gamma(t)(T - t)$. Thus $x_c < x_{\text{night}} - \gamma(t)(T - t)$. Conversely, if $\gamma(t) > \overline{\gamma}$ a deterioration in the food is more likely, and $x_c > x_{\text{night}} - \gamma(t)(T - t)$. These results are illustrated in Figure 5.6.

Most laboratory experiments that attempt to obtain risk-prone behaviour expose the study animal to low rates of food intake. This rate may well be lower than the animal experiences most of the time in the wild. Under these conditions, modelling the laboratory procedure as involving a fixed mean net gain rate $\gamma(t)$ may predict risk-prone behaviour, while making a more realistic model of the natural environment in which $\gamma(t)$ changes may not (Figure 5.6).

The model used to obtain Figure 5.6 makes certain assumptions. Option u_1 and option u_2 always have the same mean net rate of energetic gain. In practice, one might not expect the gains under different options to be either the same or perfectly correlated with each other. It is an empirical question whether the way in which McNamara (1996) models environmental changes is realistic. The key feature of the model presented here is that a change is more likely to be towards the overall mean, $\overline{\gamma}$, than away from it. This may not be so in the wild, or it may be that the change in $\gamma(t)$ is not described by a Markov process at all, and depends on both the current value of $\gamma(t)$ and its recent history. Finally, an animal may not have complete information on the mean net gain rates under current options. We now examine the issue of gaining information.

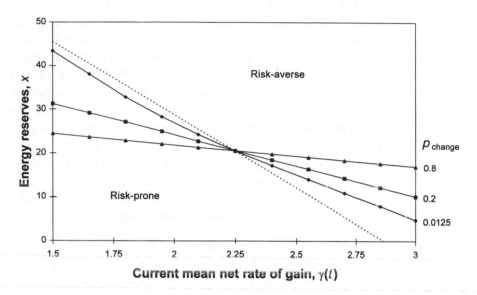

Figure 5.6. The optimal strategy when an animal is maximising its long-term survival probability in an environment that is changing over time (Example 5.1). There is a day–night cycle. The figure gives the optimal strategy at midday as a function of energy reserves x and current mean net gain rate $\gamma(t)$. Results are shown for a range of values of the probability that the environment will change p_{change}. The broken line gives results based on the expected energy budget rule in an environment that does not change. For given p_{change} it is optimal to be risk-prone below the curve and risk-averse above the curve. There is a small region at very low reserves where risk aversion is optimal. This region is not shown. (From McNamara 1996).

5.7 Gaining information on foraging options

Almost any food option exploited by an animal in its natural environment will change in unpredictable ways over time. An animal that had not exploited a particular option for a while would thus have incomplete knowledge of the option and would have to sample this food source to improve its knowledge.

Sampling may itself only provide partial information. Consider an option which in unit time delivered no food with probability $1 - p$ and one item of food of value e with probability p. Let $\gamma = ep$ be the mean gross gain per unit time under this option. For a given value of γ, the larger the value of e, and hence the smaller the value of p, the greater the variability in gain under the option. Assume that e is constant over time but that p and hence γ vary in an unpredictable way over time. An animal that was continually exploiting this food source would never have complete knowledge of the current value of γ. For example, if there was a run of fewer items than usual the animal would not know for sure whether this was the result of a decrease in γ or a run of bad luck (cf. McNamara & Houston 1985a). For given mean gain rate γ the greater the value of e, and hence the smaller the value of p, the less information gained about γ per unit time spent exploiting the option (McNamara 1996). Thus, given a choice of options, even if reproductive value is a linear function of energetic reserves one might expect to observe a preference for the less variable options (bigger p, smaller e) because these options give more information. McNamara (1996) presented the following rate-maximisation model which illustrates this preference.

Example 5.2. Risk-sensitivity under rate maximisation.   At each of the times $t = 0, 1, 2, \ldots$ an animal can choose between three options. Option u_0 always gives $\gamma_0 = 2.5$ units of energy in unit time. Under option u_i $(i = 1, 2)$ the animal gains a food item of energetic content e_i with probability p_i and no item with probability $1 - p_i$. Metabolic expenditure is ignored. The mean gain under option u_i $(i = 1, 2)$ is thus $\gamma_i = p_i e_i$. The values of e_1 and e_2 are fixed and known to the animal. The values of p_1 and p_2 change independently of one another over time. For each i $(i = 1, 2)$ the resulting mean rate γ_i switches between 1.5 and 3.5. The probability of a switch occurring in unit time is p_{change}. The animal gains information on γ_1 and γ_2 only by sampling the appropriate options. Its objective is to exploit the three options so as to maximise the long-term average rate at which it obtains energy.

McNamara (1996) takes $e_1 < e_2$, so that whenever $\gamma_1 = \gamma_2$, option u_2 has a higher variance in energetic gain than option u_1. McNamara showed that a rate-maximising animal spends a greater proportion of time on option u_1 than on option u_2 and that the difference in the proportions of time increases as the variability under option u_2 increases (Figure 5.7).

It is also possible to look at how the need to gain information affects risk-sensitive behaviour when there is a day–night cycle and survival is maximised. McNamara (1996) illustrates that uncertainty about the current value of γ increases risk-averse behaviour above the level found when γ is known. This is true even when there is no informational asymmetry between options. The informational asymmetry further increases risk-aversion.

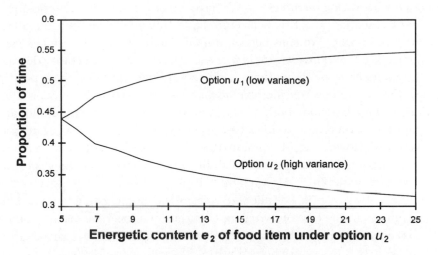

Figure 5.7. Risk-sensitivity when an animal is maximising its mean rate of energetic gain (Example 5.2). There are three options. The figure shows the proportions of time spent on option u_1 and on option u_2 (the options that can change). Results are shown for a range of values of the energetic content e_2 of items found under option u_2 ($e_1 = 5$ throughout). (From McNamara 1996.)

5.8 The selective advantage associated with risk-sensitive foraging

As well as finding the optimal response to variability, it is of interest to assess the advantage in terms of reproductive success that results from following this optimal strategy. The general techniques that we use in such an investigation were introduced in Section 3.6. Houston & McNamara (1986a) used these techniques to evaluate the selective advantage of risk-sensitive behaviour when there is a day-night cycle. When the choice is between two equal positive means, the canonical cost of choosing the option with the higher variance when this is not optimal could be of the same magnitude as the canonical cost of choosing the low-variance option when this is not optimal. In contrast, the selection pressure in favour of choosing the high-variance option was found to be much less than the selection pressure in favour of choosing the low-variance option. The reason for this difference is that when the common mean is positive, the animal is unlikely to be in the region where risk-prone behaviour is optimal, and the selection pressure takes this into account.

We can also evaluate selective advantage by looking at the mortalities associated with various strategies. (We have already given one example in the context of the models of Stephens (1981) and Houston & McNamara (1982); see Appendix 5.1). We consider an animal that must always use the same foraging option, and investigate how the long-term rate of mortality depends on the parameters of the option. If all the options have the same positive mean, then there is a considerable advantage associated with having a low variance. This holds for both continuous foraging and when the animal is attempting to survive a period of several days and foraging is interrupted by darkness (Houston & McNamara 1986a). When the options all have the same negative mean, the rate of mortality may be lowest at some intermediate level of variability (McNamara & Houston 1992a). Now assume that switching between options is allowed, so that the animal can adopt the optimal strategy. In the case of switching between equal positive means, there is at most only a small decrease in mortality compared to always choosing the option with the lower variance (Houston & McNamara 1986a). When the means are negative, switching may confer a considerable advantage when the variances are very different (McNamara & Houston 1992a).

We have emphasised that a realistic model of an animal's environment should allow feeding options to change over time (Section 5.6). Consider Example 5.1, which incorporates such a change. Suppose that the overall mean net gain $\bar{\gamma}$ is positive. Then McNamara (1996) found that an animal that could use only the low-variance option survived an extended period almost as well as an animal that made optimal use of both the low- and high-variance options (Figure 5.8). In contrast, an animal that could use only the high-variance option did very badly. Results were highly robust to the rates at which the environment changed (Figure 5.8).

McNamara *et al.* (1991) investigated the advantage that results from state-dependent decisions in the model that involves reproduction (Section 5.4). They compared the optimal strategy with strategies that choose the low-variance option with probability q and the high-variance option with probability $1-q$, regardless of reserves (see Section 3.6 for a discussion of strategies of this form). These strategies achieve a lifetime reproductive success that is much lower than that of the optimal strategy.

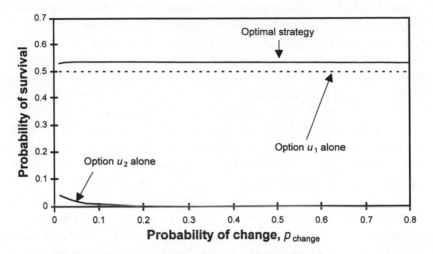

Figure 5.8. The strength of the selection pressure to make use of two rather than one option, in the model of Example 5.1. For each value of p_{change} a time T has been chosen such that, in equilibrium, the probability of surviving a period of length T is 0.5 if option u_1 alone is used (e.g. when $p_{change} = 0.071$, $T = 127$ days while if $p_{change} = 0.8$, $T = 468$ days). The probabilities of surviving for a time T under the two strategies of using option u_2 alone and of exploiting options u_1 and u_2 optimally are shown. (From McNamara 1996).

5.9 Variability in delay

The theoretical work on risk-sensitive foraging has concentrated on variability in the amount of food that an animal obtains. There is, however, a considerable body of evidence that animals are also sensitive to variability in the delay to food. A typical experiment involves giving an animal a choice between two alternatives, one of which provides a certain amount of food after a fixed delay d. The other alternative leads to the same amount of food, but the delay is either d_1 with probability p or d_2 with probability $1 - p$, where $pd_1 + (1 - p)d_2 = d$. In other words, one alternative has a fixed delay, and the other has a variable delay with a mean delay equal to the fixed delay. Studies of rats and pigeons in experiments involving this sort of choice have found a consistent preference for the variable delay (e.g. Herrnstein 1964, Davison 1969).

Although some discussions of risk-sensitive foraging have treated the response to variation in amount of food and the response to variation in the delay to food as if they were both part of a single risk-sensitive effect, they are logically distinct. McNamara & Houston (1987b) made this point by noting that the optimal risk-sensitive response to variation in amount of food depends on the sign of $\partial^2 V/\partial^2 x$, whereas the optimal risk-sensitive response to variation in the delay until food depends on the sign of $\partial^2 V/\partial^2 t$. They presented a simple example of a single choice made by an animal with reserves x at time t, after which the animal gains energy at a given mean rate until final time T. This rate has a fixed variance per unit time. The final time T is taken to be dusk, and the animal survives the night if its reserves are above some critical value, i.e. the terminal reward is a step function. McNamara and Houston showed that, depending on x and t, it is possible to have (i) $\partial^2 V/\partial^2 x > 0$, $\partial^2 V/\partial^2 t > 0$, so that it is optimal to maximise variability in amount and delay or (ii) $\partial^2 V/\partial^2 x > 0$, $\partial^2 V/\partial^2 t < 0$, so that it is optimal to maximise variability in amount and minimise variability in delay or (iii) $\partial^2 V/\partial^2 x < 0$, $\partial^2 V/\partial^2 t > 0$, so that it is optimal to minimise variability in amount and maximise variability in delay or (iv) $\partial^2 V/\partial^2 x < 0$, $\partial^2 V/\partial^2 t < 0$, so that it is optimal to minimise variability in both amount and delay.

Preference for immediacy

The model that we have just described assumes that a choice can be characterised in terms of the amount of food obtained and the associated delay. Such a characterisation is not necessarily adequate. For a given amount of food and delay, Logan (1965) found that rats preferred the food to come before rather than after the delay. This 'preference for immediacy' can result in an animal's choosing an alternative with a lower ratio of reward to delay if it results in food before the delay (see Logue 1988 for a review). McNamara & Houston (1987b) suggested that such a preference can be understood in terms of the uncertainty associated with obtaining rewards in the future (see also Logue 1988). As a simple illustration, assume that a foraging animal is 'interrupted' at a constant rate α. An animal that is interrupted does not obtain the item for which it is waiting. If there is a fixed delay d before the item becomes available, then the probability of obtaining the food item is

$$P_d = e^{-\alpha d}.$$

If the delay is a random variable D, then the associated probability of obtaining the item is

$$P_D = \mathbb{E}(e^{-\alpha D}). \tag{5.4}$$

If the random variable D has mean d and non-zero variance then, by Jensen's inequality,

$$P_D > e^{-\alpha \mathbb{E}(D)} = e^{-\alpha d} = P_d,$$

i.e. variability increases the probability of obtaining the item. To give some idea of the possible magnitude of this effect, assume that $\alpha = 0.05$ and $d = 30$. Then $P_d = 0.22$. If there are two equally likely delays, $d_1 = 5$ and $d_2 = 55$, then $(d_1 + d_2)/2 = 30 = d$, but $P_D = 0.42$. Some other examples are given in Table 5.1.

Table 5.1. Effect of variance in delay when foraging can be interrupted. The probability P_D of obtaining a delayed food item, as given by equation (5.4). There are two possible delays: delay d_1 occurs with probability p and delay d_2 occurs with probability $1-p$, where $pd_1 + (1 - p)d_2 = 30$. Results are shown for two values of p and two values of the discount factor α, resulting from interuption.

P_D, $p = 1/3$				P_D, $p = 2/3$			
d_1	d_2	$\alpha = 0.05$	$\alpha = 0.025$	d_1	d_2	$\alpha = 0.05$	$\alpha = 0.025$
1	44.5	0.389	0.544	1	88	0.638	0.687
2	44	0.375	0.539	2	86	0.608	0.673
5	42.5	0.339	0.525	5	80	0.525	0.633
10	40	0.292	0.505	10	70	0.414	0.577
20	35	0.238	0.480	20	50	0.273	0.500
30	30	0.223	0.472	30	30	0.223	0.472

5.10 Mechanisms underlying risk-sensitive foraging

So far in this chapter we have discussed models of risk-sensitive foraging based on optimisation. An alternative approach is to look for mechanistic models. We agree with Kacelnik & Bateson (1996, 1997) that these sorts of explanation are not mutually exclusive. We believe that a full treatment requires both an understanding of the mechanism and an account of why the mechanism has evolved.

Simple learning rules

A possibility that has been raised by several people is that risk-sensitive behaviour is a side effect of rules which enable an animal to learn about the environment. For example, March (1996) shows that a variety of simple learning rules result in risk-averse behaviour when mean gains are positive (see also Regelmann 1986). When mean gains are negative there is an initial period of risk-prone behaviour, but the long-term choice is risk-neutral.

Kacelnik & Bateson (1996, 1997) outline two possible mechanisms that can give rise to risk-sensitive foraging, which we now describe.

Associative learning.   Kacelnik & Bateson (1996, 1997) suggest that risk-proneness for variability in delay and risk-aversion for variability in amount can be understood as by-products of associative learning. The form of the argument is the same for delay and amount, being based on a non-linear relationship between the parameter and learning performance. We summarise the argument in the case of delay.

Kacelnik & Bateson (1996) argue that the experience of animals in foraging experiments can be viewed as a standard associative learning procedure in which a previously neutral stimulus, the conditioned stimulus or CS, is paired with a biologically relevant stimulus, the unconditioned stimulus or US. In a foraging experiment, the CS may be a light, and it is followed by food as the US. Kacelnik and Bateson claim that the relationships between the CS–US interval (i.e. the delay) and both the rate of learning and its asymptotic strength are non-linear and probably hyperbolic (i.e. drop off as the reciprocal of the delay). Kacelnik and Bateson then argue that it follows from Jensen's inequality that an animal should learn 'more effectively' about a variable delay than about a fixed delay that is equal to the mean of the variable delay. The final stage of the argument is that a greater efficiency of learning results in a greater preference, from which they conclude that preference for variable delays is a consequence of the general principles of associative learning. This argument is intriguing and the conclusions may well be true, but we do not find the logic of the argument totally convincing. The step that worries us is the use of Jensen's inequality to predict the effect of variable delays. Jensen's inequality (see Box 5.1) is a mathematical result about taking averages of the values of a function. It cannot be assumed to hold when we are considering the output of a mechanism that has a variable input. The consequences of such an input will depend on the details of the mechanism and need not take the form of a simple average of the response to the components of the input.

Scalar expectancy theory (SET).   The explanation of risk-sensitive behaviour on the basis of SET is based on how events are stored in an animal's memory. SET was developed by Gibbon (1977) in the context of errors in the ability of animals to reproduce time intervals. The data suggest that the error is proportional to the length of the interval, i.e. the relative error is constant (cf. Weber's law, which says that the just-noticeable difference in a stimulus divided by the stimulus value is constant). SET assumes that the memory associated with a given time interval consists of a normal distribution of times, with a mean and standard deviation proportional to the interval to be remembered. If there are two intervals to be remembered, then there will be a normal curve for each interval, resulting in an overall distribution that is skewed. A basic assumption in applying this framework to risk-sensitive foraging is that when a subject compares two options it draws a random sample from the memory store for each option. When the time until food has the same mean for each option but one is variable, then on more than half the choices the sample drawn from the memory for the variable option will be shorter than the sample drawn from the memory for the fixed option (Gibbon *et al.* 1988). In other words, animals will prefer variability in delay. Reboreda & Kacelnik (1991) showed that applying the scalar memory property results in risk-aversion for variability in the amount of food – see Kacelnik & Bateson (1996, 1997) for reviews.

Evolution of general rules

We believe that it is of considerable interest to establish that simple and plausible rules may be risk-sensitive (Regelmann 1986, March 1996). We do not feel, however, that the investigation should stop at this point. What is necessary if we are to understand the evolution of learning rules is to carry out a theoretical exploration of the performance of rules in relatively complex environments that are closer to those in which the rules evolved than are the simple laboratory environments often used to assess these rules. McNamara (1996) takes a step in this direction by considering environments in which the availability of food may change over time (see Sections 5.6 – 5.8). Although such environments are more complicated than those usually investigated in the context of risk-sensitive foraging, they are still much less complicated than the real world. In particular, McNamara's model is based just on risk-sensitive foraging, and even within this topic it ignores much of the biological detail such as the patchy distribution of food and the existence of cues in the environment that provide information about food availability.

Although an animal's environment is complex, there is a level at which we can take the animal to be using a single rule to cope with this environment. It will be hard to find the optimal behaviour in such environments, but this does not present a problem if we are interested in a restricted class of rules. In principle, we could evaluate the performance of all possible rules within the class and hence find the best one. In practice, there may be too many possible rules for this approach to be feasible. One way to procede in such circumstances is to simulate the evolution of rules. This involves a genetic representation of the rules, in other words, a coding. It is important to realise that both the class of possible rules and the way in which they are coded may influence the outcome of the evolutionary process.

Because the approach that we are advocating searches for good rules rather than optimal solutions, it may come up with rules that perform well in certain environments but less well, or even badly, in other environments. It is plausible that many aspects of animal behaviour are a consequence of animals' following rules with this property. For example, the matching law (Herrnstein 1970) says that the proportion of time or responses allocated to an option equals the proportion of food obtained from the option. Although this sort of rule performs well in some contexts, it performs badly in others. We believe that rather than investigating the optimality of matching in particular foraging environments based on laboratory procedures, we need to establish whether matching is likely to have evolved because it performs well in a variety of reasonably realistic environments.

A consequence of the assumption that evolution has acted on possible rules is that animals are expected to have specific abilities, e.g. the ability to forage efficiently in certain types of environment, rather than the general computational abilities that human beings use to find optimal solutions.

The approach based on the evolution of rules constitutes an ambitious research programme, but it is required if we are to understand the evolutionary significance not just of risk-sensitive foraging but of all foraging decisions (indeed, of all behavioural decisions). When viewed in this light the study of risk-sensitive foraging is still in its infancy, and the main interest in its study is the light that it may shed on rules.

Appendix 5.1 The diffusion approximation

Suppose that an animal can switch freely between two options. Under option u_i $(i = 1, 2)$ its mean rate of increase in reserves is γ_i and the variance in the increase per unit time is σ_i^2. We consider the strategy which will maximise the probability that the animal's reserves at time T exceed x_{night}.

McNamara (1984) modelled this decision problem by assuming that the change in the animal's energy reserves over time could be described by a diffusion process with the appropriate mean and variance. Any upper or lower boundaries were ignored so that reserves could lie in the full range $-\infty < x < \infty$. Under these simplifying assumptions it is possible to find the animal's optimal strategy analytically rather than by computation. Let $\sigma_1 < \sigma_2$ so that option u_2 is more variable. McNamara (1984) showed that when reserves are above a particular switching line, it is optimal to be risk-averse and choose option u_1, and when reserves are below the switching line it is optimal to be risk-prone and choose option u_2. The switching line is a straight line with slope

$$\beta = \frac{\gamma_1 \sigma_2 - \gamma_2 \sigma_1}{\sigma_2 - \sigma_1}, \tag{A5.1.1}$$

which passes through x_{night} at time T. When $\gamma_1 = \gamma_2$ we have $\beta = \gamma_1 = \gamma_2$. When $\gamma_1 > \gamma_2$ we have $\beta > \gamma_1 > \gamma_2$ and when $\gamma_1 < \gamma_2$ we have $\beta < \gamma_1 < \gamma_2$. When $\gamma_1 = \gamma_2 \, (= \gamma$ say) the switching line is given by

$$x + \gamma(T - t) = x_{night}.$$

This is also the critical line predicted by the model of Stephens (1981), where the animal is not allowed to switch between options; see Section 5.5. To quantify the advantage of flexible behaviour over the constrained behaviour required by Stephens' model, we note that an animal on the critical line survives the night with probability $\sigma_2/(\sigma_1 + \sigma_2)$ under the optimal sequential strategy, whereas it survives the night with probability 0.5 if it is not able to switch (Houston & McNamara 1982).

6

The energy–predation trade-off

6.1 Introduction

A simple way to compare the different foraging options that an animal can adopt is to use their resulting mean net rate of energetic gain γ. Rate maximisation (as discussed in Chapter 4) assumes that the best option is the one with the highest value of γ. A limitation of this approach is that it ignores the fact that the foraging options may also differ in terms of the probability that the forager will be killed by a predator. Foraging options differ both in terms of energetic gain and danger of predation, and animals are sensitive to these differences (e.g. Sih 1987, Lima & Dill 1990, Lima 1998). If high rates of energetic gain also incur high levels of predation, then the animal is said to be able to trade off energetic gain against the danger of predation.

In principle, it is straightforward to model the trade-off between the advantage of gaining energy and the disadvantage of being killed by a predator. If the animal gains energy, its reserves increase and this will typically increase its reproductive value. If the animal is killed, it loses the reproductive value associated with its current state.

The animal should behave so as to maximise its expected reproductive value. Thus reproductive value acts as a common currency for evaluating options that differ in terms of energetic gain and predation risk. Section 2.2 illustrated this in the context of a single decision. The model *Surviving a single day* (Section 3.2) shows how the energy–predation trade-off can be incorporated in a dynamic optimisation problem.

The models of Sections 2.2 and 3.2 consider decisions that are made at discrete time points $t = 0, 1, 2, \ldots$. Because of this, the parameters describing an option specify the probability of death between successive decision times and the amount of food obtained between successive decision times.

116

It is natural to formulate models that have to be solved by numerical techniques such as dynamic programming in terms of discrete decision times. In this chapter, however, we are concerned not with the details of computation but with providing a unified conceptual framework for understanding the energy–predation trade-off. To do this, it is clearer to assume that the animal can make a decision at any time rather than just at a discrete set of times. Within this framework, the instantaneous choice of an option at time t determines the predation rate and the net rate of energy intake at this time. Thus in this framework we are concerned with rates of predation and intake rather than with amounts.

The energy–predation trade-off is a central issue in behavioural ecology and it has implications for a variety of topics. Even though we are not attempting to discuss all the work that has been done on this trade-off, the chapter is quite long. To give the reader an overview, we now outline the topics covered and how they are related. In Section 6.2 we introduce the framework for the chapter, characterise the optimal action and describe a fundamental concept, the marginal rate of substitution of predation risk for energy. The analysis in this section depends on a knowledge of reproductive value V. In Section 6.3 we illustrate how an animal's biology influences the form of V, and hence the marginal rate of substitution and the optimal choice. In Section 6.4 we discuss how the behaviour of an animal depends on time if the animal is following the optimal strategy. The starting point for the analysis is the risk-spreading theorem, which provides sufficient conditions for optimal behaviour to be constant over time. We then look at what happens when the theorem does not hold, with particular emphasis on whether foraging intensity increases or decreases with time. Section 6.5 is concerned with how a change in the foraging options changes behaviour, and hence intake rate and predation rate. Options may change as an animal grows, and a consequence of growth may be a change in behaviour over time, so both Sections 6.4 and 6.5 are relevant to growth decisions when there is a trade-off between energetic gain and predation. In Section 6.6 we look at models with explicit starvation, so that the trade-off is between rate of starvation and rate of predation. We look at absolute levels of these factors, and also look at how the levels change as the options are changed (cf. Section 6.5, where we look at how intake rate and predation rate change). Both Sections 6.6 and 6.5 are relevant to models of the functional response, and we show that counter-intuitive effects are possible.

In Section 6.7 we explicitly incorporate density-dependent effects on predation rate and intake rate and analyze the distribution of a group of
animals between two patches. What we are doing is to determine ideal
free distributions when there is a risk of predation.

The effects of predation risk on the standard models of prey choice and
patch use are considered in Section 6.8.

6.2 A modelling framework

We consider an animal that has a choice of foraging options. A foraging
option u results in a mean net rate of energy intake $\gamma(u)$ and a predation
rate $M(u)$. Thus if an animal has reserves x at time t and adopts option
u for a small time δt, then the probability that it is alive at time $t + \delta t$ is
approximately equal to $1 - M(u)\delta t$. If it is alive, then its mean reserves at
$t + \delta t$ are $x + \gamma(u)\delta t$.

It is useful to represent options graphically by plotting their intake rate
and predation rate in the γM-plane. The options might be discrete, as in
Figure 6.1. An example would be a set of possible habitats in which to
forage. In other circumstances, the options might form a continuum as in
Figure 6.2. For example, many animals improve their chance of detecting
predators by interrupting foraging to scan the environment. This 'vigilance
behaviour' is reviewed by Elgar (1989) and McNamara & Houston (1992b).
In this case a foraging option is characterized by the proportion of time
u that the animal devotes to foraging as opposed to being vigilant. One
possibility is that

$$\gamma(u) = au - c,$$

where a is the gross intake when the animal devotes all its time to foraging and c is the animal's metabolic expenditure. $M(u)$ is likely to be an
increasing function of u with $M'' > 0$.

Regardless of whether the set of available foraging options is discrete or
continuous, some general features emerge. Suppose that the time taken for
an animal to switch between two options is very short and can be neglected,
and that the metabolic expenditure and predation risk while switching are
also negligible. Then, by allocating time to two options, the animal can
achieve various values of γ and M that are between those associated with
the points that represent these options (cf. Gilliam & Fraser 1987). This
means that values of γ and M on the straight line joining the points that
represent the two options can be regarded as representing further options
available to the animal. Extending this argument to the allocation of time

between more than two options, it can be seen that possible values of γ and M lie in a convex set in the γM-plane. This set is illustrated for the case of a discrete set of foraging options in Figure 6.1. The optimal choice will

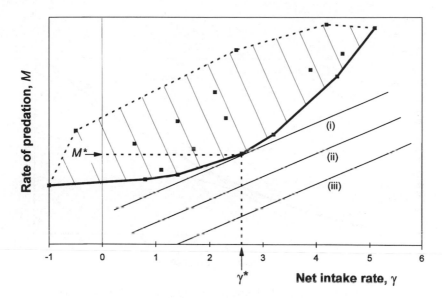

Figure 6.1. A graphical representation of a discrete set of foraging options. Each option is characterised by its net intake rate γ and predation rate M and is represented by a square. By allocating its time between options, an animal can achieve any of the combinations of γ and M shown in the shaded region. Only combinations of γ and M on the lower right-hand boundary (heavy curve) are optimal. The animal should behave so as to maximise $\dot{V}_{\text{forage}}$ (equation (6.3)). Three lines of constant $\dot{V}_{\text{forage}}$, labelled (i), (ii) and (iii), are illustrated. $\dot{V}_{\text{forage}}$ increases as γ increases and M decreases. The optimal option in this case has net intake rate γ^* and predation rate M^*.

be confined to the lower right-hand boundary of this area, since if an option is not on this boundary the animal can either increase γ without increasing M or decrease M without decreasing γ (cf. Section 4.7). On this boundary, the intake rate γ can only be increased by changing to an option with higher predation rate M. Thus, restricting attention to these options, we may think of the options as being parametrised by the intake rate γ and write the predation rate as a function $M(\gamma)$ of the intake rate. It can be seen that provided this is a sufficiently smooth function we must have

$$M'(\gamma) > 0 \tag{6.1}$$

and

$$M''(\gamma) \geq 0. \tag{6.2}$$

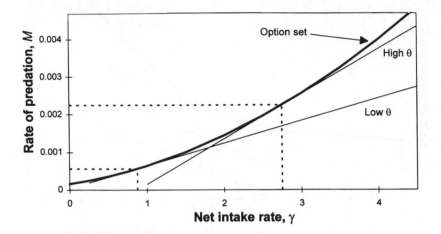

Figure 6.2. Graphical illustration of the effect of state on the optimal foraging option. The optimal option maximises $\dot{V}_{\text{forage}} = \gamma\, \partial V/\partial x - MV$. Lines of equal $\dot{V}_{\text{forage}}$ are straight lines in the γM-plane with slope $\theta = V^{-1}\partial V/\partial x$. As the reserves x increase, θ decreases and hence the new optimal option has a lower value of γ and M.

Optimal choice

The optimal trade-off between γ and M depends on the contribution that a gain in energy will make to the animal's reproductive value and on the loss in reproductive value that will result if the animal is killed. As is shown in Appendix 6.1, the optimal choice of option maximises the net rate of increase of reproductive value with time, which is equivalent to maximising

$$\dot{V}_{\text{forage}} = \gamma\frac{\partial V}{\partial x} - MV \tag{6.3}$$

(see also Appendix 4.1 and Houston & McNamara 1989, Ludwig & Rowe 1990). Expression (6.3) has a simple interpretation. γ is the rate at which x

increases and $\partial V/\partial x$ is the rate at which reproductive value increases with x, so $\gamma \partial V/\partial x$ is the rate of increase of reproductive value as a result of the intake of food. M is the rate of mortality and V is the loss in reproductive value as a result of being killed, so MV is the rate of decrease of reproductive value as a result of predation. Thus $\gamma \partial V/\partial x - MV$ is the net rate at which foraging increases reproductive value. The best option maximises the net rate of increase.

The marginal rate of substitution, θ

It can be seen from equation (6.3) that lines of equal $\dot{V}_{\text{forage}}$ in the γM-plane are straight lines with slope θ, where

$$\theta = \frac{1}{V} \frac{\partial V}{\partial x}. \tag{6.4}$$

Lines of constant $\dot{V}_{\text{forage}}$ are shown in Figure 6.1. Animals should be indifferent between two options with the same value of $\dot{V}_{\text{forage}}$, and hence should be indifferent between options that lie on one of these lines.

Given a set of options between which an animal should be indifferent, we can determine the small change in predation rate that is needed to offset a small change in gain rate. Caraco (1979) introduced this way of looking at the trade-off between predation and energy, pointing out that in economics such an equivalence is known as a marginal rate of substitution (MRS). The MRS of predation risk for energy is the slope $dM/d\gamma$ of an indifference curve, which is equal to θ.

The optimal choice of option maximises $\dot{V}_{\text{forage}}$. Let γ^* be the resulting net rate of gain and M^* be the resulting rate of predation. If we consider the set of achievable options in the γM-plane then the point (γ^*, M^*) lies on the bottom right-hand boundary of this set, and is the point at which lines of constant $\gamma \partial V/\partial x - MV$ are tangent to the set (see Figure 6.1). As we have said, M can be expressed as a function of γ along the bottom boundary (see equations (6.1) and (6.2)). It therefore follows that along this boundary

$$\frac{dM}{d\gamma} = \theta. \tag{6.5}$$

Equation (6.5) can be used to show how a change in the marginal rate of substitution θ changes the intake rate and predation rate under an optimal strategy. An increase in θ necessarily means an increase in $dM/d\gamma$ at the optimum. Because $d^2M/d\gamma^2 \geq 0$ along the bottom boundary of the set of achievable options, it follows that γ^* must have increased and, because $dM/d\gamma > 0$, M^* must have increased (see Figure 6.2).

6.3 Where does V and hence $\partial V/\partial x$ and θ come from?

In order to make use of equation (6.3), we need to know V and $\partial V/\partial x$. As we have emphasised (see Chapters 2 and 3), V depends on the animal's future expectations. In the current context, V depends on future levels of predation, the future availability of food and the use to which the animal puts this food. For example, we might expect V to depend on whether the animal is using food to avoid starvation or to reach a state at which reproduction is possible. In fact we do not need to know both V and $\partial V/\partial x$ since, from equations (6.3) and (6.4), maximisation of $\dot{V}_{\text{forage}}$ is equivalent (for a given V) to maximisation of $\gamma\theta - M$. Thus all we need to know is θ.

We now illustrate how θ depends on reserves x for three different models of the future. These models are based on different biological contexts and hence different uses of the energy obtained. For each context, we first present $\theta(x)$ for the case when the animal has only a single foraging option. We then comment on optimal behaviour when the animal has a range of foraging options.

(i) Deterministic foraging with reproduction at a critical level of reserves.
Consider first the case in which the animal gains energy deterministically at rate $\gamma > 0$ and reproduces when its reserves reach some critical level L. There is a constant rate of mortality M. Then an animal with reserves x reaches L after time $(L - x)/\gamma$, and so the probability of reaching L is $\exp[-M(L-x)/\gamma]$. V is proportional to this probability, and it follows that $\theta = M/\gamma$, so that θ is constant (see Figure 6.3).

Gilliam (1982) generalised this analysis to the case of a range of foraging options that differ in gain rate γ and predation rate M. Gilliam allowed M and γ to depend on state. He assumed that, under each option, foraging is deterministic (this applies to the option used to generate line (i) of Figure 6.3). He also assumed that the fitness of the animal does not depend on the time at which L is reached (again, this is assumed in Figure 6.3). Given these assumptions, Gilliam showed that the animal should always choose the option that minimises M/γ. We refer to this principle as the Gilliam criterion. θ is the resulting minimum value M^*/γ^*. Gilliam derived this result in the context of a growing animal, so that the state variable was size rather than reserves. The result does not hold if foraging is stochastic or the fitness of the animal depends on the time at which L is reached (for the effect of time constraints see Ludwig & Rowe 1990 and Houston *et al.* 1993; for the effect of stochasticity see Houston *et al.* 1993).

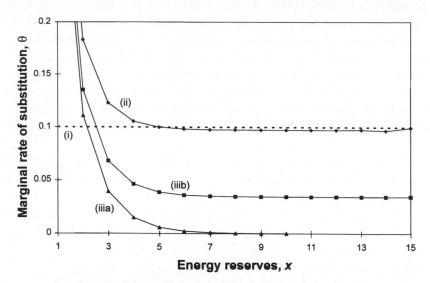

Figure 6.3. The marginal rate of substitution of predation risk for energy $\theta = V^{-1}\partial V/\partial x$ as a function of reserves. (i) Deterministic foraging with reproduction at the critical level L; $\theta = M/\gamma$. (ii) Stochastic foraging with reproduction at the critical level L. (iii) Stochastic foraging with no reproduction: (a) food that would take the reserves above L is lost; (b) food that would take the reserves above L is consumed in a safe refuge. $L = 15$, $\gamma = 0.5$, $M = 0.05$ in all cases.

(ii) Stochastic foraging with reproduction at a critical level of reserves.
We modify case (i) to make foraging stochastic. Now the animal can die either as a result of predation or as a result of starvation when x falls to zero. When the animal has a single option with gain rate γ and predation rate M, a diffusion approximation gives an analytic expression for θ – see Houston *et al.* (1993) for details.

It can be observed from Figure 6.3 that, unlike case (i), θ is not constant but increases as the lower boundary is approached because of the danger of starvation. θ asymptotes to a constant value at high reserves, but this value is less than M/γ. As Houston *et al.* (1993) showed, in this context a stochastic foraging option with gain rate γ is essentially equivalent to a deterministic option with a higher gain rate.

Merad & McNamara (1994) showed that when the animal has a range of foraging options, and behaves optimally, the form of θ remains the same. The animal should thus always be prepared to take risks to obtain food, but the acceptable risk is higher when reserves are low.

(iii) Stochastic foraging with no reproduction. We now consider a relatively long period during which there is no reproduction (e.g. winter). The animal's energy reserves are bounded above by a limit L. The animal dies either as a result of starvation (if reserves fall to zero) or predation. When food that would take its reserves above L cannot be exploited and is lost, θ decreases to zero as reserves increase. If, however, the animal can carry items that would take its reserves above L to a safe refuge and can remain in this refuge until they are consumed, then θ asymptotes to a positive value. An example is given in Figure 6.3.

McNamara (1990a) considered the maximisation of the probability of over-winter survival when the animal has a range of foraging options. He proved that θ decreases as reserves increase. As a consequence, both the resulting intake rate γ^* and the predation rate M^* decrease as reserves increase. The animal should not, however, seek a safe refuge unless it is 'within reach' of the upper boundary. For a fuller description of this result and how it is modified by state-dependent effects, see Houston *et al.* (1997).

6.4 Time-dependent behaviour

Once θ is known, the method shown in Figure 6.1 can be used to find the optimal action when the animal has a range of foraging options. Recall that only foraging options on the bottom right boundary in the γM-plane are ever used. Parametrise these options by a parameter u such that both $\gamma(u)$ and $M(u)$ increase with increasing u. In general, $\partial V(x,t)/\partial x$, $V(x,t)$ and their ratio are functions of state x and time t. Thus the value of u maximising

$$\gamma(u)\frac{\partial V}{\partial x}(x,t) - M(u)V(x,t)$$

will depend on x and t. We denote this optimal u by $\pi^*(x,t)$.

The state of an animal following the optimal strategy will typically change over time. We denote the state at time t by $x^*(t)$. We will refer to this function as the optimal trajectory of the animal.

To predict behaviour at a particular time t one must know both the optimal strategy π^* and the state of the animal $x^*(t)$. The optimal behaviour at t is then given by $u^*(t) = \pi^*(x^*(t),t)$. This function u^* predicts how the behaviour of an animal following the optimal strategy will change over time. Since u^* depends on x^* and π^*, behaviour can change over time even though the strategy π^* does not depend on time. Conversely, optimal behaviour $u^*(t)$ may be independent of time even when $\pi^*(x,t)$ depends on both x and t. This can arise if changes in state under the optimal trajectory

$x^*(t)$ exactly compensate for the change in $\pi^*(x, t)$ with time. These issues will now be analysed in more detail.

Behaviour just depends on state

In cases (i), (ii) and (iii) of Section 6.3, the marginal rate of substitution θ is a function of x alone. Thus the optimal choice of action $\pi^*(x)$ is independent of t. In case (i) an animal must raise its reserves to some critical level L before reproduction can occur. Under the restricted circumstances of this case, this is best done by minimising M/γ in every state. Case (i) considers M and γ to be independent of state. θ is then also independent of state and π^* is a constant. This implies that $u^*(t)$ is independent of time. If M and γ depend on state, e.g. on size, then M/γ is still minimised at every time; the result is that $\pi^*(x)$ depends on state. Then, as state increases from its initial value to L the optimal behaviour will change. Indeed the model was introduced in the context of changes in habitat choice by growing bluegill sunfish (Werner *et al.* 1983, Werner & Gilliam 1984, Werner & Hall 1988).

In case (iii), $\pi^*(x)$ decreases with increasing x. The animal's optimal trajectory $x^*(t)$ is, however, stochastic, so that the resulting behaviour $u^*(t)$ fluctuates as reserves fluctuate. In the long run, each of a set of possible foraging options is chosen for a certain proportion of the time. This process is described in detail in Section 3.5. It is important to remember that the optimal choice of foraging option is a deterministic function of state. The outcome of foraging, and hence the trajectory of reserves, is uncertain, so that if we look at an animal without knowledge of its reserves, our description of its behaviour must be probabilistic. A strategy that chooses between two habitats as a function of reserves results in a much higher survival probability than a strategy that, at each decision point, chooses one habitat with probability q and the other habitat with probability $1 - q$ independently of its state (McNamara 1990b, Houston *et al.* 1992). An example is given in Figure 3.3. It can be seen from the figure that any state-independent strategy results in a higher level of mortality than the optimal strategy. We can compare the optimal strategy with the state-independent strategy that spends the same proportion of time on each option as the optimal strategy. Mortality may be far greater under the state-independent strategy.

Behaviour independent of time: risk-spreading

In the three cases considered in Section 6.3 there were no time constraints on behaviour, nor did the payoffs received depend on time. For example, in case (ii) the animal had unlimited time to reach the critical state L, and its reproductive success on reaching L did not depend on the time taken to get there. There are many biological contexts in which it is important for an animal to raise its state to a certain value by a certain time. For example, a small bird in winter may have to raise its energy reserves to at least a certain level by dusk if it is to survive the night. Sometimes the final time is not fixed, but reproductive value depends on the time at which a certain state is reached (e.g. growth to a critical size for reproduction in a seasonal environment, Ludwig & Rowe 1990). In general, the optimal choice of foraging option will depend on both state and time, i.e. $\pi^*(x, t)$ is a function of both x and t. But $u^*(t) = \pi^*(x^*(t), t)$ depends on both the strategy π^* and the resulting trajectory x^*. We have emphasised that u^* may be constant over time even though π^* depends on state and time. We now describe a result, the risk-spreading theorem (Houston *et al.* 1993; see also Sibly *et al.* 1985), which gives sufficient conditions for u^* to be constant. This provides a baseline from which to understand systematic changes in u^* in other cases.

The risk-spreading theorem states that if the following conditions are met, then the optimal behaviour will not change.

(RS1) The foraging process is deterministic.

(RS2) The foraging process is not subject to interruptions.

(RS3) Neither the intake rate γ nor the predation rate M depends on state.

Let t_0 and t_1 be times with $t_0 < t_1$, and let x_0 and x_1 be any two states. We can compare ways of getting from state x_0 at time t_0 to state x_1 at time t_1. Given that each way ends at the same value of x and that the above assumptions hold, the ways can only differ in terms of their predation. The best way will be the one with the lowest level of predation. The theorem tells us that the best value of u over the interval satisfies $u^*(t) = $ constant $(= u^*$ say). Thus u^* must satisfy

$$x_1 - x_0 = \gamma(u^*)(t_1 - t_0).$$

This result holds regardless of whether the target state x_1 at time t_1 is the optimal state.

There is a simple argument that helps us to understand why the risk-spreading theorem holds. The requirement that the state is x_1 at time t_1 puts a constraint on possible values of u during the interval $[t_0, t_1]$. Compare the optimal behaviour with an alternative pattern in which the animal adopts two values of u during $[t_0, t_1]$. Because of the requirement that the state at final time t_1 must be x_1, one of these values of u must be less than u^* and the other must be greater than u^*. Recalling that $M(u)$ is an accelerating function of u, it follows from Jensen's inequality (see Box 5.1) that the average mortality when u takes two values is greater than when u takes one value. For a more rigorous and general argument, see Houston et al. (1993). Houston et al. consider all possible patterns of foraging intensity during the interval $[t_0, t_1]$ which have average intensity u^*. They show that the cost of deviating from u^* is approximately equal to

$$\tfrac{1}{2} M''(u^*) \operatorname{Var}(U),$$

where U is the foraging intensity at a randomly chosen time during the interval. This shows that the cost increases with both the acceleration of the predation function and the variance of the foraging intensity.

The risk-spreading theorem tells us that along a trajectory behaviour does not change, but it does not identify the constant behaviour. It is, nevertheless, a powerful tool in finding optimal trajectories. Suppose that the conditions of the risk-spreading theorem hold. Then, given any final state x_1 and final time t_1, we know the best trajectory that reaches x_1 at t_1; hence it is easy to find the payoff under this trajectory. We can then find the optimal trajectory by maximising the payoff over all x_1 and t_1. The results presented in Appendix 6.2 are based on this procedure. There we show how the optimal trajectory during a time interval $[0, T]$ depends on the initial state at time 0 and the time available, T.

Deviations from risk-spreading

From the baseline of the risk-spreading theorem, we can look at how the optimal foraging intensity depends on time when assumptions (RS1)–(RS3) are relaxed.

Foraging is stochastic. The optimal strategy $\pi^*(x, t)$ is a deterministic function of state and time. If, in following the optimal strategy, the foraging options chosen are stochastic, then neither the optimal trajectory $x^*(t)$ nor the foraging intensity $u^*(t)$ is a deterministic function of time. One way to describe behaviour over time is to use the mean value of $u^*(t)$, which we denote by $\mathbb{E}[u^*(t)]$.

To examine the effects of stochasticity we compare optimal behaviour in two environments. In one, all foraging options are deterministic. In the other, the mean gain rate $\gamma(u)$ and the predation rate $M(u)$ have the same dependence on u as in the first environment, but foraging is stochastic. We

can now compare $u^*(t)$ in the first environment with $\mathbb{E}[u^*(t)]$ in the second
environment. Typically the optimal foraging intensity at initial time t_0 is
higher in the stochastic case. The reason for this is that the foraging intensity that raises the state to its target value of x_1 by time t_1 in a deterministic
environment may not do so in a stochastic environment. Runs of bad luck in
which the achieved rate of gain is less than the mean may occur. To counter
this possibility, an initially higher foraging intensity is adopted. The foraging intensity that is chosen later in the interval will typically depend on both
state and time. Animals that have had good luck can reduce their foraging
intensity and may actually cease to forage (Houston *et al.* 1993, McNamara
et al. 1994a). Animals that have had bad luck must forage at a high intensity.

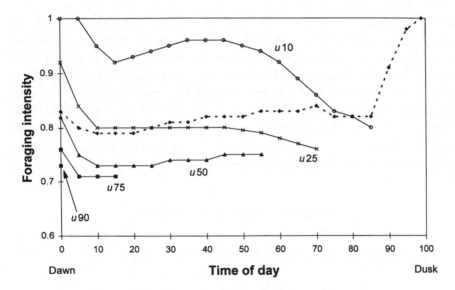

Figure 6.4. Foraging intensity as a function of time of day in the
model of foraging routines presented by McNamara *et al.* (1994a).
We consider a large population of birds, and at each time of day
show how the foraging intensity of a bird depends on how high
its reserves are compared with the rest of the population. For
example, $u75$ gives the foraging intensity of a bird whose reserves
are higher than those of 75% of the population. Birds with the
highest reserves cease foraging earliest in the day. The broken line
gives the mean foraging intensity $\mathbb{E}[u^*(t)]$ of birds that are still
foraging at time t. (The model *continuous foraging* with $G = 450$,
$\alpha = 1.4$; see McNamara *et al.* 1994a for further details.)

The mean level of foraging $\mathbb{E}[u^*(t)]$ reflects both these possibilities. If the contribution from animals that have had bad luck is substantial, $\mathbb{E}[u^*(t)]$ may increase with time (e.g. Houston *et al.* 1993). This increase will be accentuated if animals that have good luck cease to forage and the averaging process is confined to animals that are still foraging. An illustration is given in Figure 6.4. Towards the end of the period, only animals that have had bad luck will still be foraging. They will typically choose options with a high gain rate. As a result, over the day there may be a negative correlation between the number of animals foraging and their intake rate, even though there is no interference (e.g. McNamara *et al.* 1994a).

Foraging is subject to interruptions. An interruption of the foraging process can be modelled in various ways. One possibility is that once an interruption occurs, foraging cannot be resumed during the period under study. If it is necessary for the state at final time T to be above some critical level in order to obtain any reproductive value, then this sort of interruption may result in zero reproductive value and hence amounts to an extra source of mortality that affects all options equally. As such, it tends to increase foraging intensity (Lima 1987b, McNamara & Houston 1992b, Houston *et al.* 1993). Another possibility is that the interruption is permanent, but an interrupted animal has a reproductive value that depends on its state when interrupted (e.g. McNamara & Houston 1992b, Houston *et al.* 1993). The usual effect of these interruptions is to increase foraging intensity at the start of the interval. The mean foraging intensity may decrease over the interval.

Effects of state. The risk-spreading theorem breaks down if γ or M depends on state. We now look at two cases.

(i) *Gain rate depends on state.* If γ depends on state but M does not, then certain situations are clear cut. Let

$$\gamma(x, u) = a(x)u - c(x).$$

Houston *et al.* (1993) show the following:

if $c(x)/a(x)$ is an increasing function of x, then $u^*(t)$ will increase with t;

if $c(x)/a(x)$ is a decreasing function of x, then $u^*(t)$ will decrease with t.

(If $c = 0$, then u^* is constant.)

(ii) *Predation depends on state.* Suppose that M depends on state but that γ does not. The dependence of M on state gives rise to conflicting pressures on behaviour. Suppose that the state of an animal has to pass through a range of values for which predation is high. If on the one hand the animal forages at a high intensity it will pass through the range quickly but the predation rate will be very high. On the other hand, if the animal forages at a low intensity it will reduce the predation rate, but will incur this rate for longer. Which option is better depends on how state and behaviour interact to determine predation, as is illustrated by the following two examples.

(a) The effect of x and u is additive, i.e.

$$M(x, u) = \beta(x) + N(u). \qquad (6.6)$$

Costs of state and behaviour of this form have been assumed by Sibly & McFarland (1976) and Heller & Milinski (1979). Houston *et al.* (1993) show the following:

if $\beta(x)$ is increasing in x, then $u^*(t)$ is non-decreasing in t;

if $\beta(x)$ is decreasing in x, then $u^*(t)$ is non-increasing in t.

(b) The effect of x and u is multiplicative, i.e.

$$M(x, u) = \beta(x)N(u). \qquad (6.7)$$

In contrast to the additive case, the behaviour of $u^*(t)$ cannot be specified just in terms of $\beta(x)$; it also depends on the initial value of x (see Houston *et al.* 1993 for details).

6.5 The effects on intake rate and predation rate of changes in the foraging options

If an animal responds adaptively to a change in its environment, what are the consequences for its intake rate and predation rate? It is plausible to assume that if the availability of food is increased then the optimal intake rate γ^* will increase, but Abrams (1989) shows that in fact γ^* may decrease. It is also possible for an increase in predation rate to decrease M^* (e.g. Abrams 1993a). McNamara & Houston (1994) review these 'paradoxical' effects and show that the way in which the options are changed and whether the change is temporary or relatively permanent are both important in determining how γ^* and M^* change.

Temporary changes in food availability

The importance of how the foraging options are changed can be seen from analysing the consequences of a temporary change in food availability. Such a change will not alter how future reproductive value depends on energy reserves and hence will not alter θ. We can therefore find the best choice of foraging option by means of the graphical method shown in Figure 6.1 (see also McNamara & Houston 1994). In Figure 6.5(a) the increase in the availability of food is greater for the better foraging options. The figure shows that the effect of the increase in availability is to increase both the intake rate and the level of predation. In Figure 6.5(b) the effect of the increase in availability is greater for the poorer option. The effect of the change is a decrease in both the intake rate and the level of predation.

In the examples that we have just given, when the improvement in food availability has a bigger effect on the better options γ^* increases, whereas when the improvement has a bigger effect on the worse options, γ^* decreases. To see whether this effect is general, we can represent the quality of the environment by a parameter α. An increase in α increases the availability of food under all options, but the increase may be greater for some options than for others. An increase in α does not change the predation rate. Because predation rate is independent of α, α constitutes a label for the options, so that intake rate can be written as $\gamma(M, \alpha)$. Let us assume that the animal has a continuum of foraging options available. If

$$\frac{\partial^2 \gamma}{\partial M \partial \alpha} > 0,$$

the increase in food availability has a bigger effect on the options with high gain, whereas if

$$\frac{\partial^2 \gamma}{\partial M \partial \alpha} < 0,$$

then the effect of the increase is bigger on the options with low gain. The option that maximises

$$\gamma \frac{\partial V}{\partial x} - MV$$

is given by

$$\frac{\partial \gamma}{\partial M}(M^*(\alpha), \alpha) = V \bigg/ \frac{\partial V}{\partial x},$$

where $M^*(\alpha)$ is the predation rate if the optimal option is chosen when the food availability parameter is α. Differentiating with respect to α, it follows

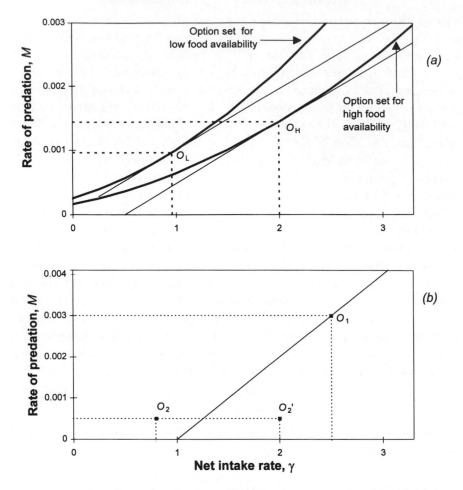

Figure 6.5. The effect of a change in the foraging options brought about by a change in the current availability of food. Future availability is not changed, so that θ does not change. (*a*) The increase in the availability of food is greater for the better foraging options. As a result, the set of foraging options is shifted more to the right at higher values of γ. The optimal option changes from O_L to O_H; as a result both γ and M increase. (*b*) There are two options O_1 and O_2. The solid line has slope θ, showing that the optimal option is O_1. The availability of food is increased in only the option O_2 with the lower net rate of gain, producing a new option O_2'. Because this option is to the right of the line with slope θ going through O_1, O_2' is the optimal option. It can be seen that the result of the increase in food availability is a decrease in γ from 2.5 to 2.

that

$$M^{*\prime}(\alpha) = -\left(\frac{\partial^2 \gamma}{\partial M \partial \alpha}\right) \Big/ \left(\frac{\partial^2 \gamma}{\partial M^2}\right). \tag{6.8}$$

It follows from equations (6.1) and (6.2) that

$$\frac{\partial^2 \gamma}{\partial M^2} \leq 0,$$

and so (ignoring the case when this second derivative is zero)

$$\frac{\partial^2 \gamma}{\partial M \partial \alpha} > 0 \quad \text{implies} \quad M^{*\prime}(\alpha) > 0.$$

This result tells us that if improving the availability of food has a greater effect on the options with high gain, then the optimal predation rate $M^*(\alpha)$ increases. Because $\gamma(M, \alpha)$ is an increasing function of both M and α, the optimal intake $\gamma(M^*(\alpha), \alpha)$ increases. If the improvement has a bigger effect on the options with low gain, i.e.

$$\frac{\partial^2 \gamma}{\partial M \partial \alpha} < 0,$$

then it follows from equation (6.8) that the optimal predation rate $M^*(\alpha)$ decreases. Because α is increasing and $M^*(\alpha)$ is decreasing, it is not possible to establish in general how $\gamma(M^*(\alpha), \alpha)$ will change. The decrease in $M^*(\alpha)$ means that the animal changes to a safer option with lower gain, but the increase in food availability (as represented by the increase in α) may compensate so that intake rate does not necessarily decrease. The example given in Figure 6.5(*b*) shows that the adaptive response to an increase in the availability of food may result in a decrease in intake rate. This sort of effect was obtained by McNamara & Houston (1987c) and Abrams (1989). Similarly, an increase in predation risk may decrease the predation rate under the optimal strategy because the animals adopt anti-predator behaviour.

Long-term changes in food availability

A change in current feeding opportunities may be taken by an animal as an indication of a change in its long-term prospects. For example, if current food availability increases, the animal may expect this increase to persist. This will mean that the animal's reproductive value V is increased, and will usually mean that the current value $\partial V / \partial x$ of food is decreased because food is expected to be more plentiful in the future. It follows that the ratio θ decreases and so the new best option, given expectation of a long-term improvement in food availability, has a gain rate and predation rate less than or equal to that under the same short-term improvement in food. A similar result holds if a change leaves the gain rates fixed, but decreases the predation rates associated with some options.

McNamara & Houston (1994) used a simple model for the allocation of time between feeding and vigilance to compare the effects of short- and long-term changes. They showed that a short-term increase in the amount of food found per unit time spent feeding should lead to an increase in the proportion of time spent feeding and hence a decrease in the proportion of time spent in vigilance. As a consequence both the animal's intake and predation rates increase. In contrast, a long-term increase in food decreases the proportion of time spent feeding. In this particular model the decrease exactly compensates for the increased food, so that intake remains constant. The decrease in time spent feeding means that more time is devoted to vigilance and so the predation rate is decreased.

Some examples

(*i*) *Changes in predation risk.*   There has been considerable interest in behavioural responses to an increase in the danger of being killed by a predator (e.g. Fraser & Gilliam 1992, Werner 1991, Sih 1987). As we have shown, the predicted effect of such an increase depends on both how the various foraging options are changed and whether the change is perceived as temporary or permanent. The importance of how the foraging options are changed can be seen from the following example. Assume that the animal has to get to a critical state in order to reproduce and that the predation rate $M(u)$ is given by

$$M(u) = m_0 N(u) + \mu,$$

where $m_0 N(u)$ is influenced by the animal's foraging intensity u and μ is independent of u. A change in environmental predation risk is represented by a change in m_0 or μ. To analyse the effect of a change, we assume that net rate of energetic gain $\gamma(u)$ is given by

$$\gamma(u) = au - c.$$

In the absence of stochasticity and time constraints (see Section 6.4) the optimal value, u^*, of u can be found by minimising M/γ. McNamara & Houston (1994) show that an increase in m_0 decreases u^*, whereas an increase in μ increases u^*. (Werner & Anholt 1993 obtain the same effect in a special case.) An increase in m_0 increases the predation rate associated with high gain rates more than the predation rate associated with low gain rates. The best response is to reduce predation (and gain rate) by reducing u. In contrast, an increase in μ increases the predation rate equally on all the options. The best response is to hurry through the phase of increased danger by increasing u and hence increasing the gain rate.

(*ii*) *Growth.* M and γ are likely to depend on an animal's size (Werner & Gilliam 1984), so growth can bring about changes in the foraging options. In discussing deviations from risk-spreading in Section 6.4 we showed that the manner in which u^* changes with time depends on how both state and behaviour influence predation rate. This has not always been appreciated. Sibly *et al.* (1985) analysed optimal growth using a graphical argument which involves an implicit assumption that the relationship between mortality (predation in our analysis) and behaviour does not change as the animal's size changes. This implicitly assumes that $M(x, u) = \beta(x) + N(u)$. There are, however, other possible forms of $M(x, u)$. When $M(x, u) = \beta(x)N(u)$, then an increase in size changes the shape of the relationship between M and u, and the argument used by Sibly *et al.* is incorrect.

(*iii*) *Open and closed economies.*   Experimental investigations of foraging behaviour vary in the extent to which the experimenter controls the animal's intake. Hursh (1980) introduced the terms 'open economy' and 'closed economy' to denote two extremes in the range of control by the experimenter. In an open economy experiment, daily intake is determined by the experimenter; in a closed economy experiment daily intake is determined by the animal. A typical open economy procedure involves a relatively short daily test period in which the animal can work for food. After the test, the animal is given an amount of food that keeps its daily intake at a constant level. In a typical closed economy procedure, the animal spends all its time in the experimental apparatus. Apart from establishing the relationship between the animal's behaviour and the delivery of food, the experimenter has no influence on consumption.

It has often been argued that the dependence of behaviour on the availability of food is different in the two economies (e.g. Hursh 1980, 1984, Collier 1982, 1983). In an open economy, animals often work harder when availability is increased, whereas in a closed economy they typically work less hard when availability is increased. Using an analogy with microeconomics, Hursh argues that in an open economy the demand for food is elastic, whereas in a closed economy it is inelastic. Loosely speaking, demand is inelastic if consumption declines slowly with proportionately large increases in price. Price can be manipulated in operant experiments by manipulating the number of responses that must be made in order to obtain an item of food. A high price can be thought of as a low availability. The food that is given to the animal after the test session is a substitute for the food that the animal can work for during the test session. The effect of a substitute is to increase the elasticity of demand.

The economic argument implies that the value of food is different in the two economies. Houston & McNamara (1989) explored this difference using a model of a relatively short test session in an open economy procedure and a model of an animal that spends all its time in a closed economy experiment. In the open economy model, the animal chooses its foraging intensity u at each level of reserves so as to maximise

$$\gamma(u)\frac{\partial V}{\partial x}(x,t) - M(u)V(x,t),$$

where

$$\gamma(u) = au.$$

Houston and McNamara show that $u^*(t)$ decreases with time in a session, i.e. satiation occurs. This effect occurs because $\theta(x)$ decreases as x increases, so gaining food becomes relatively less important. They also show that if u^* is less than its maximum possible value then, at a given level of reserves, u^* increases as the availability a increases.

In the closed economy model, the animal decides on the proportion of time to allocate to various activities, subject to the requirement that the energy gained is equal to the energy expended. This constraint means that reserves are on average constant; this is typical of closed economy experiments.

A major difference between the two economies is the effect of a change in the availability a of food. In the open economy model, an increase in a results in an increase in u^*, whereas in the closed economy model an increase in a results in a decrease in u^*. Houston & McNamara (1989) argued that this difference can be understood in terms of how a change in a changes future expectations. In both economies an increase in a lasts long enough for both current and future foraging options to be improved. As a consequence, θ is reduced. Because, however, the test session of an open economy experiment is typically short, the reduction in θ is less marked in the open economy model than in the closed economy model. In a closed economy, the effect of an increase in a is to produce a large enough reduction in θ to reduce u^*.

Houston and McNamara argued that their model of a closed economy is essentially an open economy with an infinite time horizon but with the possibility of starvation during the session. They used different models for the two economies, but the general argument suggests that a difference in the effect of a can be found by using just one model but varying the time horizon.

Optimal functional response. A predator's functional response is its intake rate as a function of the density or availability of its prey. It follows that what we have been discussing is relevant to finding the functional response of an animal that responds adaptively to changes in its environment. For other models of adaptive functional responses see e.g. Abrams (1982, 1987, 1989, 1990), Emlen (1984), Mitchell & Brown (1990), Getty & Pulliam (1991). The responses of predators to their prey and prey to levels of predation determine the interaction between populations of predators and prey and hence the stability of the populations' dynamics. For illustrations of the resulting effects and further references see Abrams (1992, 1997), Abrams & Matsuda (1997a,b), Matsuda & Abrams (1994).

6.6 Levels of starvation and predation

When an animal is not reproducing it may be reasonable to assume that the optimal strategy minimises the total rate of mortality. In this section we assume that starvation and predation are the only sources of mortality and investigate the relative levels of starvation and predation under the optimal strategy.

To model the trade-off between starvation and predation, assume that the starvation rate S and the predation rate M depend on some aspect u of the animal's behaviour. For example, u might be the proportion of time foraging as opposed to being vigilant or it might be an animal's target level of fat reserves. A necessary condition for the value u^* to minimise overall mortality $S + M$ is

$$-S'(u^*) = M'(u^*), \tag{6.9}$$

i.e. at the optimum the marginal rates of starvation and predation are equal (McNamara & Houston 1987c, McNamara 1990b, Abrams 1991). The relative magnitudes of $S(u^*)$ and $M(u^*)$ depend on the functions relating S and M to u. For example, computations based on the fat reserves of small birds in winter (McNamara & Houston 1990b, Houston & McNamara 1993) show that if overall mortality is small, then the optimum starvation rates are much less than predation rates. These results have some important implications.

(i) It is sometimes suggested that starvation and predation will be 'balanced' by natural selection, i.e. $S(u^*)$ will be approximately equal to $M(u^*)$. This is not necessarily the case.

(ii) The importance of a factor cannot be inferred from its magnitude (e.g. McNamara & Houston 1987c, McNamara 1990b, McNamara & Houston 1990a, Abrams 1993b). It is not valid to conclude that because the rate of starvation is low, food is not important. The benefit of gaining food may be so important that animals are prepared to incur considerable levels of predation in order to obtain it (Gibb 1954, Jansson *et al.* 1981). When extra food is provided, its contribution to survival may be a reduction in predation (e.g. Jansson *et al.* 1981). As is illustrated by the results of McNamara *et al.* (1987) the possibility of starvation may drive optimal routines, even though the probability of starvation under the optimal strategy is very low. Furthermore, as we show below, an increase in the availability of food may not decrease starvation.

(iii) Following on from point (ii), it is not meaningful to ask whether starvation or predation limits a population. As several workers have realised (e.g. Emlen 1973, Newton 1980, Lima 1986, McNamara & Houston 1987c, Walters & Juanes 1993), the interaction between starvation and predation means that the two factors cannot be decoupled.

Effects of a change in the environment

So far, we have looked at levels of starvation and predation under given circumstances. If the foraging environment changes, the optimal strategy will change. Since the levels of starvation and predation depend on both the environment and the animal's behaviour, it is not obvious how these levels will change. We have already shown in Section 6.5 that an increase in food availability may decrease intake rate. This suggests that the relationship between food availability and starvation rate may not be straightforward. We now illustrate the effects of environmental change on starvation rate and predation rate in state-dependent models.

Suppose that an animal can choose between two patches as a function of its energy reserves. When food availability in one patch is increased, starvation rate under the optimal strategy may increase (McNamara 1990b, McNamara & Houston 1990a). This effect arises because of changes in the proportion of time spent in the patches. For example, let patch 1 give a net gain of γ_1, and have a predation rate of zero, while patch 2 gives a net gain of γ_2 and has a predation rate greater than zero. McNamara (1990b) looks at the starvation rate under the optimal state-dependent strategy as a function of γ_1. When γ_1 is very much less than γ_2, patch 1 is hardly ever used. Most of the time is spent on patch 2, despite the predation risk. As

γ_1 is increased, more time is spent on patch 1, but as γ_1 is still less than γ_2 starvation increases. Because patch 1 is safe, predation is reduced. The total mortality rate also falls. If γ_1 increases sufficiently, the starvation rate decreases.

Now assume that patch 1 is safe, but γ_1 is much less than γ_2. As the predation rate M_2 while on patch 2 is increased, the predation rate under the optimal strategy may increase and then decrease (McNamara 1990b). When M_2 is first increased, the proportion of time spent on patch 2 is still high enough for the overall level of predation to increase. As M_2 becomes higher, the proportion of time spent on patch 2 drops so strongly that the overall level of predation falls. Because more time is now being spent on the option with the lower net intake rate, starvation increases.

The importance of a habitat or resource

It is tempting to assume that the frequency with which a habitat or resource is used provides a measure of the importance of this habitat in terms of the animal's fitness. This assumption is not necessarily valid. Consider an animal making a state-dependent choice between two habitats. In habitat 1 the animal is relatively safe from predators but has little food. Habitat 2 has more food but also a higher predation risk. It is optimal to choose habitat 1 when reserves are high and habitat 2 when reserves are low (see Sections 3.2, 6.3 and McNamara 1990a). It is easy to find parameter values under which habitat 2 is very rarely used and most of the predation occurs in this habitat. Because it is the habitat chosen when reserves are low, all the starvation occurs in habitat 2. If this habitat is no longer available to the animal, predation decreases (because all the time is now spent in the relatively safe habitat) but starvation is very much larger (because the food supply in habitat 1 is poor). As a result the total mortality may increase substantially (e.g. McNamara & Houston 1990a).

We have shown that removing a rarely used habitat may substantially increase an animal's mortality. This illustrates an important point about assessing the consequences of the destruction of part of an animal's habitat. An accurate assessment requires a knowledge not just of how often an animal uses a given area but also of the rôle that the area plays in the animal's life history and furthermore of how the animal's behaviour will change when the area can no longer be used. A similar point applies to arguments about welfare: the importance of an activity cannot be determined from how often it is performed (Dawkins 1990).

6.7 Distribution of animals across habitats

So far in this chapter we have ignored any explicit interactions between members of a species. When such interactions are present, a game-theoretic analysis may be required (e.g. Maynard Smith 1982; see also Chapter 7).

The basic paradigm for density-dependent habitat choice is the ideal free distribution (IFD). Since this term was introduced by Fretwell & Lucas (1970), it has been the focus of many empirical and theoretical investigations (see Milinski & Parker 1991, Kacelnik *et al.* 1992 and Tregenza 1995 for reviews). What Kacelnik *et al.* call the 'classical' IFD model involves the following assumptions, which underlie the ideal free model.

(i) A habitat contains a number of animals that are equal in terms of their ability to compete with each other for resources.

(ii) The habitat contains a number of resource patches that differ in terms of quality. Animals have perfect information about habitat quality.

(iii) The animals can move freely from patch to patch without incurring a cost.

(iv) The rate of increase $\dot{V}$ of reproductive value on a given patch decreases as the number of animals using it increases.

(v) Animals distribute themselves in such a way that no animal can increase $\dot{V}$ by moving.

Then, provided that the number of animals is large, all animals have the same $\dot{V}$, regardless of the patch or patches in which they are found. A distribution of animals that satisfies this condition is known as an ideal free distribution (IFD).

Much of the work on IFDs has equated the rate of gain of reproductive value with the rate of energetic intake. Given this identification, the classical IFD predicts that at equilibrium all animals will have the same rate of energetic gain. Assume that there are two patches, with food arriving at rate r_1 in patch 1 and r_2 in patch 2. Each item is consumed as it arrives and if there are n_i animals in patch i then each animal achieves a rate r_i/n_i. Then it follows that at equilibrium

$$\frac{r_1}{n_1} = \frac{r_2}{n_2} \tag{6.10a}$$

or

$$\frac{r_1}{r_2} = \frac{n_1}{n_2}. \tag{6.10b}$$

Equation (6.10b) says that the ratio of animals matches the ratio of inputs, and hence is known as input matching (Parker 1978).

Moody *et al.* (1996) investigate how this distribution is changed by the danger of predation. They consider a group of animals which have the same foraging ability and internal state and which can choose between two patches. If there are n_i animals in patch i, then each animal has a net rate of gain $\gamma_i(n_i)$ and suffers a predation rate $M_i(n_i)$. Given these assumptions, there are two possible types of stable equilibria, i.e. distributions in which no animal can do better by changing its location. In one type, all the animals are found in one of the patches. In the other type, the animals are found in the two patches. At this second type of equilibrium, the net rate of increase of reproductive value must be the same for each animal, i.e. from equation (6.3)

$$\gamma_1(n_1)\frac{\partial V}{\partial x} - M_1(n_1)V = \gamma_2(n_2)\frac{\partial V}{\partial x} - M_2(n_2)V. \tag{6.11}$$

To illustrate some implications of this equation, we can follow Moody *et al.* in investigating the distribution of animals when patch 1 is safe and patch 2 is dangerous and when

$$\gamma_i(n_i) = \frac{r_i}{n_i} - c, \tag{6.12}$$

where c is the metabolic rate. It is obvious that under these circumstances the animals on the dangerous patch must have a higher intake rate in order to compensate for the danger of predation to which they are exposed. We can, however, be more specific than this. It follows from equations (6.11) and (6.12) that the distribution of animals must satisfy

$$\frac{r_1}{n_1} = \frac{r_2}{n_2} - \frac{M_2(n_2)}{\theta}, \tag{6.13}$$

where θ is the marginal rate of substitution of predation risk for energy, given by equation (6.4). Two special cases of the predation rate $M_2(n_2)$ were considered by Moody *et al.* At one extreme, the *per capita* predation rate multiplied by the number of animals present is equal to some constant m_0, which means that

$$M_2(n_2) = m_0/n_2.$$

Moody *et al.* refer to this as full dilution of risk. At the other extreme, there is no dilution: the *per capita* predation rate is independent of the number of animals present, i.e.

$$M_2(n_2) = m_0.$$

These two assumptions about predation result in different general properties of the equilibrium distribution. When there is full dilution,

$$\frac{n_1}{n_2} = \frac{r_1\theta}{r_2\theta - m_0},$$

(6.14)

which shows that the relative distribution of the animals does not depend on the total number of animals. In contrast, when there is no dilution,

$$\frac{r_2}{n_2} - \frac{r_1}{n_1} = \frac{m_0}{\theta},$$

(6.15)

which implies that the difference in intake rate between the two patches does not depend on the total number of animals.

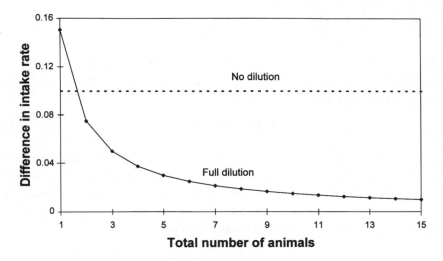

Figure 6.6. Ideal free distributions when patch 1 is safe and patch 2 is dangerous. The figure shows the difference in intake rate $r_2/n_2 - -r_1/n_1$ as a function of the total number of animals $n_1 + n_2$ for two cases: full dilution, in which the predation rate on patch 2 is m_0/n_2, and no dilution, in which the predation rate on patch 2 is m_0. $\theta = 0.01$, $m_0 = 0.001$, $r_1 = 6$, $r_2 = 12$.

In practice it is hard to estimate θ and m_0. We can, however, use the qualitative trends described above to discriminate empirically between full dilution and no dilution. Figure 6.6 shows the predicted difference in intake rate between patches, given no dilution or full dilution, when the total number of animals in the environment is changed. It can be seen from the figure that the dependence of the difference in intake rate on the number of animals is very different in the two cases. Similarly, the dependence of the proportion of animals in a patch on the total number of animals is also different in the two cases. It must be remembered, however, that these two cases do not exhaust the possible ways in which the predation rate depends on the number of animals.

Titration procedures

Abrahams & Dill (1989) introduced a powerful way of analysing the distribution of animals between two patches in the presence of predation risk. We feel, however, that their theoretical analysis is not as clear as it might be. We now present an analysis, based on the above formalism, that clarifies the logic of their procedure, and hence establishes the rôle of various assumptions.

The procedure of Abrahams & Dill (1989) titrates the food delivered at one patch to establish a certain distribution of animals between the patches. We view the procedure as consisting of three experiments, which we refer to as Exp 0, Exp 1 and Exp 2.

Exp 0. There is no predation risk in either patch. It follows from equation (6.10a) that the numbers of animals $\tilde{n}_1$ on patch 1 and n_2 on patch 2 satisfy

$$\frac{r_1}{\tilde{n}_1} = \frac{r_2}{\tilde{n}_2}. \tag{6.16}$$

Exp 1. Patch 1 is safe, but there is a danger of being killed by a predator in patch 2. If n_2 animals use this patch, then the rate of predation is $M_2(n_2)$. Thus from equation (6.13) the numbers of animals n_1^* on patch 1 and n_2^* on patch 2 satisfy

$$\frac{r_1}{n_1^*} = \frac{r_2}{n_2^*} - \frac{M_2(n_2^*)}{\theta}. \tag{6.17}$$

Following Abrahams & Dill (1989), we define

$$E = \frac{r_2}{n_2^*} - \frac{r_1}{n_1^*}. \tag{6.18}$$

Abrahams and Dill refer to E as the energetic equivalent of predation risk per individual. It follows from equations (6.17) and (6.18) that

$$E = \frac{M_2(n_2^*)}{\theta}. \tag{6.19}$$

Exp 2. The input rate at patch 2 is increased by an amount Δ such that the distribution of animals returns to that found in Exp 0, i.e. $n_1 = \tilde{n}_1$ and $n_2 = \tilde{n}_2$. From equation (6.13),

$$\frac{r_1}{\tilde{n}_1} = \frac{(r_2 + \Delta)}{\tilde{n}_2} - \frac{M_2(\tilde{n}_2)}{\theta}. \tag{6.20}$$

It follows from equations (6.16), (6.19) and (6.20) that

$$\Delta = \tilde{n}_2 E \frac{M_2(\tilde{n}_2)}{M_2(n_2^*)}. \tag{6.21}$$

Equation (6.21) simplifies under two particular assumptions about how predation risk depends on the number of animals in a patch.

No Dilution: Predation risk is constant regardless of group size, i.e. $M_2(\tilde{n}_2) = M_2(n_2^*)$ and so

$$\Delta = \tilde{n}_2 E. \tag{6.22}$$

Full Dilution: $M_2(n_2) = m_0/n_2$, where m_0 is a positive constant. It follows that

$$\Delta = \tilde{n}_2 E \frac{n_2^*}{\tilde{n}_2} = n_2^* E. \tag{6.23}$$

Equation (6.22) is the result given by Abrahams & Dill (1989), but they say that it depends on a linear relation between reproduction value (fitness in their terms) and energy reserves, and they do not make clear that it depends also on no dilution of predation risk. What we have done is not to offer an alternative model to that of Abrahams and Dill, but to formalise their verbal argument. The formalisation shows that the relationship between intake and fitness is irrelevant. This is a consequence of our assumption that experiments are brief, so that state does not change during the experiment. If the experiment is long enough for reserves to change, then neither the verbal argument of Abrahams and Dill nor our formal argument suffices to determine the critical amount of food Δ. This follows from the fact that Δ will change as reserves increase during the experiment. This is true even if reproductive value is a linear function of reserves. When the relationship is linear, $\partial V/\partial x$ does not depend on reserves but V increases with reserves and hence the marginal rate of substitution θ decreases as reserves increase.

State-dependent IFD models

By considering instantaneous decisions by a group of foragers that are all in the same state, Moody *et al.* (1996) were able to avoid an explicit consideration of state-dependent behaviour. In contrast, McNamara & Houston (1990c) investigated the evolutionarily stable distribution of a group of foragers that are making state-dependent decisions about which of two patches to exploit. The patches may differ both in terms of food and predation risk. In each patch food is continually renewing and is depleted by the animals in the patch, so that the equilibrium level of food availability, and hence the equilibrium rate of energetic gain, depends on the number of animals in the patch. (For a slightly different approach, see Mangel 1990.)

Given that the patches differ in the parameters that determine food availability and predation rate, animals will distribute themselves across patches in such a way that the mean rate of gain is higher on a patch with a higher rate of predation. An animal's decision about which patch to visit depends on its energy reserves. Animals with high reserves will choose a safer patch, and animals with low reserves will choose a patch with better food but higher predation risk. If a large group of animals is following a particular state-dependent strategy, this will determine their distribution between the patches and hence the availability of food in the patches. Given this resident population strategy, dynamic programming can be used to find the state-dependent strategy for an individual animal that results in the lowest long-term rate of mortality. This strategy is the best response to the strategy adopted by the population. At the evolutionarily stable (ES) equilibrium, the best response to the population strategy is itself the population strategy, i.e. no strategy can invade by the action of natural selection (for further information about the concept of evolutionary stability, see Chapter 7).

There are several general and interesting features of the ES distribution.

(i) Animals switch not because they need to sample other patches but because the choice of patch depends on state, and state changes stochastically as a result of foraging.

(ii) In contrast to the usual approach to IFDs, nothing is equalised across the patches. In particular, intake rate is not equalised, i.e. equation (6.10a) does not hold, and net rate of increase of reproductive value is not equalised, i.e. equation (6.11) does not hold. Equation (6.11) does not hold because animals on the two patches have different levels of reserves and hence different values of $\partial V/\partial x$ and V.

Houston & McNamara (1997) build on the state-dependent IFD to investigate the implications of state-dependent patch choice for population dynamics. They assume a density-independent fecundity in summer and a density-dependent mortality in winter. They compare the population dynamics that results if animals follow the state-dependent IFD (the 'optimal strategy') with the dynamics that results if animals distribute themselves in such a way as to equalise the intake rate in each patch (the 'equal-rates' strategy). The optimal strategy results in a larger equilibrium population size than the equal-rates strategy, but the difference in population size may not be large. The difference in population size tends to increase as the variability of one of the patches increases, the duration of winter increases or the fecundity decreases. The strategy used may have a significant effect on the population dynamics.

For an application of state-dependent IFDs to parasitoids, see Sirot & Bernstein (1996).

6.8 The effect of predation on standard foraging models

In Chapter 4 we discussed the standard foraging models for patch use and prey choice. These models are based on the maximisation of the net rate of energetic gain, and hence do not include the fact that behavioural options may differ in terms of predation risk. There have been several attempts to incorporate predation into patch use (e.g. Houston & McNamara 1986b, Brown 1988, 1992, Gilliam 1990, Newman 1991) and prey choice (e.g. Lima 1988, Gilliam 1990, Godin 1990). We now provide a unified account of this work and establish general principles.

Our approach is based on the fact that when the predation risk differs between the various phases of a foraging cycle, an animal can control both its gain rate γ and its overall predation rate M by means of the time it spends in a patch or the prey items that it accepts. We can use established results (e.g. Houston et al. 1993) to see how predation risks should influence these foraging decisions.

Patch use with predation when there are no effects of state

We assume a simple model of a patchy environment (cf. Chapter 4). All patches are the same, and the food intake rate after a time s has been spent on a patch is $r(s)$. The intake rate is a continuous and monotone decreasing function of s.

The animal's cumulative gain if it stays on the patch for a time s is

$$G(s) = \int_0^s r(v)dv.$$

The mean time to travel from one patch to another is τ, so the mean rate of gain if the animal stays on each patch for a time s is

$$\gamma(s) = \frac{G(s)}{\tau + s}.$$

As the following examples show, when γ is of the above form and predation risk does not depend on state, it is possible to obtain analytic results regarding optimal behaviour. In all these examples, we ignore the fact that the animal is repeating a cycle in which it only obtains food when on a patch. We make the approximation that if the animal chooses to spend a time s on a patch, then its energy reserves increase deterministically with a constant rate $\gamma(s)$.

Finite time horizon. We investigate the animal's behaviour over a period $[0, T]$. The animal has a level of reserves x_0 at time 0. If its patch residence time is s, then its reserves at final time T are $x_0 + \gamma(s)T$. This model provides us with a convenient starting point from which to investigate the effects of predation. Assume that the predation rate is μ while the animal is on the patch and ν while it is travelling between patches. We now consider the foraging behaviour during $[0, T]$ that maximises the animal's energy reserves at time T multiplied by the probability that the animal has not been killed by a predator during $[0, T]$. We assume that the animal does not have the option of spending part of the period in a refuge where it is safe from predators. (We describe below how conclusions are altered by the presence of a refuge.) The animal's choice of s determines its gain rate and predation rate. Because the conditions of the risk-spreading theorem are met, it is optimal to use the same behaviour throughout the interval, i.e. to use the same patch residence time on all patches. Denoting the optimal patch residence time by s^*, it can be shown (see Appendix 6.3) that

$$r(s^*) = \gamma(s^*) - [x_0 + \gamma(s^*)T](\nu - \mu)\left(\frac{\tau}{s^* + \tau}\right). \qquad (6.24)$$

We note that the right-hand side of this equation is overall gain rate minus the product of three terms. These terms are

$x_0 + \gamma(s^*)T$,	level of reserves at final time T
$\nu - \mu$,	predation rate while travelling minus predation rate while on patch
$\tau/(s^* + \tau)$,	proportion of time spent travelling.

When $\nu - \mu = 0$ (i.e. the predation rates are equal) the product is zero and it is optimal to leave when

$$r(s^*) = \gamma(s^*),$$

which is the marginal value theorem (MVT, Section 4.3). Thus when the two rates of predation are equal, it is optimal to use the patch residence time s_r that maximises the rate of energetic gain. When the difference $\nu - \mu$ is not zero then there will be departures from s_r. The direction of these departures depends on the sign of $\nu - \mu$:

$$\text{when} \quad \nu > \mu, \qquad s^* > s_r$$
$$\text{when} \quad \nu < \mu, \qquad s^* < s_r.$$

The results given in Appendix 6.2 can be applied to patch use by taking a foraging option to be specified by a patch residence time. We are concerned with a simple case in which there is no metabolic expenditure, so that the gain rate is non-negative for every patch residence time. Thus, provided that it is always worth foraging, the optimal gain rate γ^* and associated predation risk M^* both decrease as either x_0 increases or T increases. When the predation rate ν while travelling is greater than the predation rate μ while on a patch, the decrease in γ^* is achieved by spending a greater time on a patch. When $\nu < \mu$, the decrease in γ^* is achieved by spending a shorter time on a patch.

In Appendix 6.3 we investigate how the optimal patch residence time s^* depends on $\alpha = \nu - \mu$. It is shown that when $\alpha = 0$

$$\frac{ds^*}{d\alpha} = \frac{\text{final reserves} \times \text{proportion of time travelling}}{\text{deceleration in gain}}, \qquad (6.25)$$

where

$$\text{deceleration in gain} = -G''(s_r) = -r'(s_r).$$

Effect of a refuge.   In many circumstances it is reasonable to assume that the forager has the option of spending a proportion of its time in a refuge where it is safe from predators. Adding this option changes the above conclusions. When there is no refuge we have shown that if either x_0 or T is increased, γ^* decreases. The results given in Appendix 6.2 show that when a refuge is present there is a minimum rate $\hat{\gamma}$ below which the animal should not simply forage but should instead spend part of its time foraging at $\hat{\gamma}$ and its remaining time in the refuge.

Newman (1991) investigated patch use when foraging is stochastic and when the animal dies of starvation if reserves fall to 0 during the interval $[0, T]$. He assumed the same rate of predation while foraging and travelling (i.e. $\mu = \nu$) but included a refuge where the animal is safe from predators. Newman found the optimal state-dependent behaviour using dynamic programming, and then used Monte Carlo simulations to find the resulting expected behaviour. He concluded that the predation rate has little effect on the mean patch residence time. It is clear from our analysis that, in the absence of stochastic effects, the optimal patch residence time is independent of predation in this case. Newman found that increasing predation decreases the animal's level of reserves at T. This is what we would expect from the analysis presented in Appendix 6.2, since an increase in the overall level of predation decreases the target state at T. This is achieved by spending more time in the refuge.

Reaching a critical state. Gilliam (1990) applied his criterion of minimising M/γ to the sort of patch problem that we have been discussing. The mortality rate is $(\mu s + \nu\tau)/(s+\tau)$ and the gain rate is $G(s)/(s+\tau)$, so the patch residence time should minimise

$$\frac{\mu s + \nu\tau}{G(s)}.$$

Gilliam (1990) points out that this is equivalent to maximising

$$\frac{G(s)}{s + (\nu/\mu)\tau}, \tag{6.26}$$

which shows that the optimal solution is given by a modified version of the rate-maximising solution in which the travel time is multiplied by the ratio ν/μ. Another way of characterising the solution is to note that the animal should leave the patch when

$$r(s^*)\left(\frac{s^* + (\nu/\mu)\tau}{s^* + \tau}\right) = \gamma(s^*). \tag{6.27}$$

Comparing equation (6.24) with equation (6.27), it can be seen that both are modified versions of the MVT that reduce to the standard MVT when $\mu = \nu$. It can also be seen that in the finite-time-horizon case the effect of the predation rates depends on the difference between ν and μ, whereas in the case where a critical level is reached the effect depends on their ratio.

Houston & McNamara (1986b) investigated the effect of predation on patch use when $G(s)$ is the contribution to reproductive success from a patch visit of duration s. They assumed that the animal continues to exploit patches and travel between them until it dies, and found the patch residence time that maximises the total expected lifetime reproductive success. They showed that when μ and ν were very small, the best patch residence time is given by maximising expression (6.26). This means that the patch residence time that maximises total lifetime reproductive success is also the patch residence time that minimises M/γ. The reason for this can be seen from the following argument. We focus on the case considered in Houston & McNamara (1986b) and define M to be the predation rate and γ to be the rate of reproduction. Ignoring the details of patch residence time and travel time, we take M and γ to be instantaneous rates controlled by behaviour.

The probability that the animal survives until time t is $\exp(-Mt)$, and its total expected lifetime success is

$$\int_0^\infty \gamma e^{-Mt} dt = \frac{\gamma}{M}.$$

Thus total expected lifetime reproductive success is maximised by maximising γ/M, i.e. by minimising M/γ.

Patch use with predation: the general case

There is a general approach to the problem of when to leave a patch that applies both to the cases that we have just considered and to cases that involve state-dependent effects. Let $V(x,t)$ be the reproductive value of an animal with reserves x, given that it has just left a patch at time t. In this context t measures the time spent in the environment, not the time spent on the patch. Now consider an animal that is on the patch at time t_0 and whose reserves are x_0. In deciding whether to leave the patch now or remain for a short additional time the animal must take into account

$$\frac{\partial V}{\partial x}(x_0, t_0)$$

which is the value of gaining additional food before leaving the patch, and

$$C = -\frac{\partial V}{\partial t}(x_0, t_0)$$

which is the opportunity cost of wasting time on the patch before leaving. It is shown in Appendix 6.1 that if the animal's current net gain on the patch is γ and its predation rate is M, it is optimal to leave if

$$\gamma \frac{\partial V}{\partial x} < MV + C. \tag{6.28}$$

This equation has the following interpretation: $\gamma(\partial V/\partial x)$ is the rate of increase of reproductive value as a result of food intake on the patch; MV is the rate of decrease of reproductive value as a result of predation on the patch; C measures the cost of wasting time on the patch. It is optimal to leave if benefits are less than costs. Condition (6.28) is analogous to equation (7) of Brown (1988).

Prey choice with predation

As in the case of patch use, an animal may be able to control both its gain rate and its predation rate by its choice of prey items. In this sub-section we illustrate some of the effects that can occur, using a simple example based on one prey type. The type has an energetic content e, a handling time h and an encounter rate λ. The predation rate is μ while the animal is handling an item and ν while it is searching for items. Neither the predation rates nor the intake rate depend on state. Let p be the probability that an item is accepted. Then

$$M(p) = \frac{\nu/\lambda + ph\mu}{\lambda + ph}$$

and

$$\gamma(p) = \frac{pe}{1/\lambda + ph}.$$

Finite time interval. We consider optimal prey choice over an interval $[0, T]$. As in the case of patch use, we maximise the level of energy reserves at T multiplied by the probability that the animal is alive at this time. The risk-spreading theorem applies to this case, so p must be constant over $[0, T]$. We therefore seek the value of p that maximises

$$F(p) = [x_0 + \gamma(p)T]\exp[-M(p)T],$$

where x_0 is the level of reserves at the start of the interval. Assume for the moment that $x_0 = 0$. Then if $p = 0$, fitness is zero, and hence some value of $p > 0$ must be optimal. If μ is very much larger than ν, then $p = 1$ may result in a very small probability of being alive at T. This argument suggests that $F(p)$ may be maximised by some value of p between 0 and 1. This is indeed the case, i.e. partial preferences can be optimal.

The intuitive reason for the optimality of partial preferences in this model is that when $\mu > \nu$ the only way in which the animal can reduce the predation rate M is by refusing to accept (and hence handle) some of the items. This is no longer true if the animal can stop foraging and spend some of the interval in a refuge where the predation rate is zero. It is then optimal to spend some of the time in the refuge and the rest of the time foraging with $p = 1$.

When there is just one prey type and the animal has a refuge, the above argument means that the animal should accept all items (i.e. $p = 1$ is optimal). This results in a higher gain rate than might have been optimal in the absence of the refuge, but the animal does not forage at this rate for the whole interval, and is thus able to reduce its predation rate.

Reaching a critical state. Gilliam (1990) looked at optimal prey choice when there are several prey types and an animal can use a refuge. Under the assumption that the rate of energy expenditure is the same inside and outside the refuge, the optimal behaviour minimises M/a, where a is the gross rate of gain (cf. Gilliam & Fraser 1987). Gilliam presented a graphical argument which shows that when the refuge is used there are no partial preferences. If the refuge is not used, then it may be optimal to have a partial preference for one of the prey types.

In our analysis and that of Gilliam (1990), the predation rates for handling and searching are fixed. Lima (1988) presented a model in which predation rates are determined by how frequently the animal scans for predators. The optimal behaviour now specifies both the probability of accepting a prey type and the frequency of scanning during searching and handling. Lima obtains partial preferences. He does not, however, allow the animals a refuge.

6.9 Static versus dynamic models

Abrams (1991) and Leonardsson (1991) raised the question whether a dynamic approach to the energy–predation trade-off is necessary.

When we are concerned with a single choice of foraging option, after which behaviour is specified, then it is relatively easy to find V and $\partial V/\partial x$. Equation (6.3) then gives us a static optimisation principle for determining optimal choice. We may, however, be concerned with an extended period of time during which an animal makes a sequence of decisions. The best action at any time is evaluated on the assumption that behaviour in the future is optimal. Given this dependence on the future, working backwards from some final time using dynamic programming is a natural procedure that exposes the underlying logic of sequential choice. This does not, of course, mean that this procedure is always required. Furthermore, when it is not required simpler analytic methods are to be preferred. For example, Ydenberg (1989) investigated the age at which young seabirds should leave their nest site and go to sea. On the assumption that growth rate and predation rate were higher at sea than at the nest site, he found the optimal departure age numerically using dynamic programming. Growth at sea is given and the growth rate while still in the nest is given and deterministic. It follows that an optimal strategy can be specified in terms of a single number, the age (or size) at which to leave the nest (Byrd *et al.* 1991). The best value of this departure age can be found without recourse to dynamic programming. Once this is realised it is possible, for some forms of growth curve, to obtain an analytic solution to the problem. Such solutions give more insight into the problem than numerical computations.

There are two obvious general circumstances in which problems involving a sequence of decisions do not require dynamic optimisation but can be reduced to a single decision.

(i) *Reaching a critical level.* Gilliam (1982) considered the problem of an animal growing deterministically and reproducing at a critical size. The optimal solution is given by the static rule: minimise the mortality per unit increase in size. Gilliam's analysis assumes that reproductive value does not depend on the time at which the critical size is reached. When reproductive value decreases exponentially with this time, a static analysis is still possible (Leonardsson 1991, Houston *et al.* 1993). If the time penalty is not exponential, or there are time constraints, then a static analysis is in general not possible (Ludwig & Rowe 1990, Houston *et al.* 1993).

(ii) *Risk-spreading.* When foraging is deterministic and intake rate and predation rate are independent of state and time, then from the risk-spreading theorem an animal should always choose the same option during the time period under consideration (Sibly *et al.* 1985, Houston *et al.* 1993). The problem is thus reduced to a single choice, i.e. it is effectively a static optimisation. This is the sort of case that is investigated by Abrams (1991) (see also Schmitz & Ritchie 1991).

In some cases all that might be important is the time that an animal devotes to various activities during a given period. Such an allocation problem can often be formulated as a constrained optimisation problem and solved using Lagrange multipliers (e.g. Brown 1988). Reducing the problem to one of allocation requires that the order in which the various activities are performed is irrelevant. For obvious reasons, this will not be the case if rate of gain, metabolic expenditure or predation depend on state. It will also not be the case when foraging is stochastic. For example, if an animal has to accumulate a certain critical amount of energy by the end of the time period, then the proportion of time that it should spend foraging cannot be specified at the start of the time period. If the choice is between foraging and resting, the uncertainty associated with foraging makes it advantageous to forage at the start of the period and then rest (e.g. Houston *et al.* 1993, see also Sections 3.2 and 9.3).

Appendix 6.1 The optimal energy versus predation trade-off

Let an animal have a given 'background' behavioural strategy, which may or may not be optimal. Under this strategy the animal has reproductive value $V(x, t)$ at time t if its reserves at this time are x. We consider three scenarios.

Under each scenario the animal has reserves x_0 at time t_0 and we characterise the animal's optimal choice of foraging behaviour at this time. In scenario (i) the animal must choose a foraging option from some set of available options for a short time δ before reverting to the background behavioural strategy. In this case we are interested in the best choice of option. In scenario (ii) we look at whether it is better for the animal to adopt the background strategy immediately at time t_0, or to choose an option with net rate of gain γ and predation rate M for some time before reverting to the background strategy. In (iii) we suppose that the background strategy is the optimal strategy and find the dynamic programming equations determining V.

If the animal chooses a foraging option that delivers energy as a smooth flow with mean net rate γ between times t_0 and $t_0 + \delta$, its reserves at time $t_0 + \delta$ are $x_0 + \gamma\delta$, provided it is not killed during the interval. Its reproductive value if it survives until time $t_0 + \delta$ is thus $V(x_0 + \gamma\delta, \, t_0 + \delta)$. Assuming δ is small and ignoring terms of order δ^2 we can write

$$V(x_0 + \gamma\delta, t + \delta) \simeq V(x_0, t_0) + \gamma\delta\frac{\partial V}{\partial x}(x_0, t_0) + \delta\frac{\partial V}{\partial t}(x_0, t_0). \qquad \text{(A6.1.1)}$$

Let M be the predation rate under the option; then, again ignoring terms of order δ^2, the probability that the animal survives the time interval of length δ is $1 - M\delta$. Thus, under this option its reproductive value at time t_0 is $(1 - M\delta)V(x_0 + \gamma\delta, t_0 + \delta)$, which, by equation (A6.1.1), is given by

$$V(x_0, t_0) + \delta\frac{\partial V}{\partial t}(x_0, t_0) + \delta\left[\gamma\frac{\partial V}{\partial x}(x_0, t_0) - MV(x_0, t_0)\right], \qquad \text{(A6.1.2)}$$

where terms of order δ^2 are again ignored.

Scenario (i). We now suppose that the animal has a range of foraging options available at time t_0, where each option is characterised by its own γ and M. The best option maximises expression (A6.1.2). Looking at this expression it can be seen that the foraging option only affects the terms within the square bracket on the right hand side. Thus the optimal choice maximises

$$\dot{V}_{\text{forage}} = \gamma\frac{\partial V}{\partial x}(x_0, t_0) - MV(x_0, t_0) \qquad \text{(A6.1.3)}$$

(cf. equation (6.3)).

Scenario (ii). Now suppose that the animal can either choose an option with gain rate γ and predation rate M between times t_0 and $t_0 + \delta$ or can immediately revert to the background strategy at time t_0. The background strategy has reproductive value $V(x_0, t_0)$. Thus the background strategy is the best choice if $V(x_0, t_0)$ exceeds expression (A6.1.2), or equivalently

$$0 > \frac{\partial V}{\partial t}(x_0, t_0) + \gamma \frac{\partial V}{\partial x}(x_0, t_0) - MV(x_0, t_0). \qquad \text{(A6.1.4)}$$

Set

$$C = -\frac{\partial V}{\partial t}(x_0, t_0)$$

as the cost of lost time under the background strategy. Then inequality (A6.1.4) shows that it is better immediately to adopt the background strategy at time t_0 if the gain rate under the alternative satisfies

$$\gamma \frac{\partial V}{\partial x} < MV + C. \qquad \text{(A6.1.5)}$$

Scenario (iii). As a final application of expression (A6.1.2) we suppose that the animal has a range of foraging options and that the background strategy is the optimal strategy. Then expression (A6.1.2) is maximised when γ and M are the gain rate under the background strategy and the maximum value of expression (A6.1.2) is $V(x_0, t_0)$, i.e.

$$V(x_0, t_0) = \max \left[V(x_0, t_0) + \delta \frac{\partial V}{\partial t}(x_0, t_0) + \delta\gamma \frac{\partial V}{\partial x}(x_0, t_0) - \delta MV(x_0, t_0) \right]$$

where the maximum is taken over all available options. Cancelling $V(x_0, t_0)$ from both sides of this equation and dividing through by δ gives

$$-\frac{\partial V}{\partial t}(x_0, t_0) - \max \left[\gamma \frac{\partial V}{\partial x}(x_0, t_0) - MV(x_0, t_0) \right], \qquad \text{(A6.1.6)}$$

which is the dynamic programming equation for this continuous-time problem (see also e.g. Ludwig & Rowe 1990).

Appendix 6.2 Effect of the time horizon and initial state

Assume that conditions (RS1) – (RS3) of the risk-spreading theorem hold. At time 0 the animal has state x_0. If the animal has state x_T at final time T, then its reproductive success is $R(x_T)$. It is assumed that $R'(x_T)/R(x_T)$ is a strictly decreasing function of x_T. Here we describe qualitatively how optimal behaviour during the time interval $[0, T]$ depends on x_0 and T.

By the risk-spreading theorem optimal behaviour during $[0, T]$ is constant. If the animal uses a given option with gain rate γ and predation rate M throughout the interval, its reproductive value at time zero is

$$e^{-MT}R(x_0 + \gamma T).$$

The optimal option maximises this quantity.

Gain rate and predation rate. Let γ^* be the gain rate and M^* be the predation rate under the optimal option. It can be shown that γ^* and M^* both decrease as either x_0 or T increase.

Approach to the target state. We now introduce a 'target' state x^{**}. To do so, recall from Section 6.2 that we can write M as a function $M(\gamma)$ of the intake rate γ provided that we restrict attention to options on the lower right hand boundary of the option set. Since $R'(x)/R(x)$ decreases with increasing x there is a unique level of reserves x^{**} such that

$$\frac{R'(x^{**})}{R(x^{**})} = M'(0).$$

Let $x_T^* = x_0 + \gamma^* T$ be the final state given the initial state x_0 and the time horizon T. Then it can be shown that

(i) the final state x_T^* is always nearer to the target state x^{**} than is the initial state x_0. In particular

$$x_0 < x^{**} \quad \Rightarrow \quad x_0 < x_T^* \leq x^{**}$$
$$x_0 = x^{**} \quad \Rightarrow \quad x_T^* = x^{**}$$
$$x_0 > x^{**} \quad \Rightarrow \quad x^{**} \leq x_T^* < x_0$$

(ii) as x_0 gets closer to x^{**} or as T increases x_T^* gets closer to x^{**}.

In addition, as the time available tends to infinity, x_T^* tends to x^{**} regardless of the initial state.

The effect of a refuge. The effect of a refuge on the optimal choice of foraging options has been considered by Gilliam & Fraser (1987) and Houston *et al.* (1993). We now extend these results.

Suppose that while in the refuge the predation rate is zero and the metabolic rate is c_{r}. Consider the choice of active foraging option that minimises

$$\frac{M}{\gamma + c_{\mathrm{r}}}.$$

Let the net rate of energetic gain under this option be $\hat{\gamma}$. It is clear from Gilliam & Fraser (1987) that it is never optimal to choose an active foraging option with rate less than $\hat{\gamma}$. (An analogous result is illustrated in Figure 4.5). Gilliam and Fraser consider a problem with a finite time horizon in which there is a step-function terminal reward. They show that it is optimal either to switch between the refuge and foraging at the rate $\hat{\gamma}$ or to forage at a rate $\gamma^* > \hat{\gamma}$ and not use the refuge.

Results extend to the case of a general reward function as follows. Let x^{**} be the target level of reserves defined above. Then

(i) if $x_0 + \hat{\gamma}T < x^{**}$ then the animal forages at a rate $\gamma^* > \hat{\gamma}$ and the refuge is not used.

(ii) if $x_0 + \hat{\gamma}T \geq x^{**}$ the animal spends a proportion of the time in the refuge and the rest of the time foraging at rate $\hat{\gamma}$. The proportion of time in the refuge is chosen so that final state $x_T^* = x^{**}$.

Appendix 6.3 Patch use with predation over a fixed time interval

Suppose that all patches are identical. Each gives a continuous delivery of food, and the net rate of food delivery after time s on a patch is $r(s)$. This rate decreases with time on a patch. The predation rate while on a patch is μ. The mean travel time between patches is τ and the predation rate while travelling is ν. Metabolics are ignored. The animal has energy reserves of x_0 at time 0. It forages for a time T during which it visits many patches. By the risk-spreading theorem we can assume that the residence time on each patch is the same. We seek the patch residence time that maximises its reserves at time T (conditional on survival) multiplied by the probability that it survives until this time.

If the animal stays on each patch for time s, its long-term rate of energy gain is

$$\gamma(s) = \frac{G(s)}{s + \tau}, \tag{A6.3.1}$$

where $G(s) = \int_0^s r(y)dy$. Thus its reserves at T are $x_0 + \gamma(s)T$. The probability that the animal survives until T is $\exp[-\omega(s)T]$, where $\omega(s)$ is the mean predation risk per unit time and is given by

$$\omega(s) = \frac{\mu s + \nu\tau}{s + \tau}. \tag{A6.3.2}$$

If reproductive value at time T is proportional to energy reserves then the best choice of patch residence time, s^*, maximises $[x_0 + \gamma(s)T]\exp[-\omega(s)T]$. Substituting for $\gamma(s)$ and $\omega(s)$ from equations (A6.3.1) and (A6.3.2) into this expression, differentiating with respect to s and setting the derivative equal to 0 at $s = s^*$ gives equation (6.24).

We now treat s^* as a function $s^*(\alpha)$ of the parameter $\alpha = \nu - \mu$ in equation (6.24). Differentiating with respect to α gives

$$\frac{ds^*}{d\alpha}[r'(s^*)(s^* + \tau) + \alpha\gamma'(s^*)T\tau] = -[x_0 + \gamma(s^*)T]\tau. \tag{A6.3.3}$$

When $\alpha = 0$ there is no difference in predation, so that $s^* = \hat{s}$. Thus, setting $\alpha = 0$ and $s^* = \hat{s}$ in equation (A6.3.3) we obtain equation (6.25).

7

Dynamic games

7.1 Introduction

In this chapter we are concerned with cases in which the fitness of an organism depends on both the action that it performs and the behaviour of other members of the species. The appropriate modelling approach in such cases is evolutionary game theory. The central concept in evolutionary game theory is that of an evolutionarily stable strategy (ESS). Maynard Smith (1982) defines an ESS as a strategy which, if adopted by all members of a population, cannot be invaded under natural selection by another strategy. The idea behind this definition is that an ESS is an endpoint of the evolutionary process in the same way as an optimal strategy is the endpoint when there is no dependence of fitness on the behaviour of conspecifics (Eshel 1996, Hammerstein 1996, Weissing 1996, Eshel *et al.* 1998). We will not attempt to review this area. Instead, we will outline the basic ideas and then concentrate on games that involve a sequence of state-dependent interactions between organisms. These dynamic games are the game-theoretic counterpart of dynamic optimisation problems.

The Hawk–Dove game (Maynard Smith & Price 1973, Maynard Smith 1982) was the first evolutionary problem to be analysed using ESS theory, but the basic idea was used by Fisher (1958) and by Hamilton (1967) in discussions of the ratio of male to female offspring (the sex ratio) that a mother should produce. There is now a large literature on the theory of ESSs and their application – see for example Maynard Smith (1982), Riechert & Hammerstein (1983), Parker (1984), Hines (1987), Vincent & Brown (1988). Some idea of the range of possibilities can be obtained from the following examples.

(*i*) *Sequential assessment game.* Two animals are engaged in a contest. The state of an animal is its estimate of its fighting ability relative to that of its opponent and the reliability of this estimate. By interacting, the animals revise their estimates and the estimates become more reliable. After each interaction, an animal can continue the contest or can give up. This sequential assessment game has been analysed by Enquist & Leimar (1983, 1987, 1990) and will be discussed in more detail later in this chapter (Section 7.6).

(*ii*) *Parental effort games.*   A male and a female have mated and produced offspring. Each parent can decide on the amount of parental effort that it contributes to the young. The success of the current breeding attempt increases with the sum of the efforts of both parents. Each parent's future reproductive success decreases with its level of effort, and each attempts to maximise its own total lifetime reproductive success. Problems of this type are discussed by Houston & Davies (1985) and Motro (1994). This parental effort game is further analysed in Section 7.9. For a review of games between the sexes see Hammerstein & Parker (1987).

(*iii*) *Sperm competition games.*   A female may be mated by more than one male. A given male can increase the probability that he fertilises the female by increasing the number of sperm that he transfers, but this reduces his expected reproductive success from future matings. Parker (1993), Parker & Begon (1993) and Ball & Parker (1996, 1997) analyse such sperm competition games.

(*iv*) *Size-dependent competition.*   Males compete with each other for access to females. If the outcome of a contest depends on the relative size of the contestants, then a male's reproductive success depends not on his absolute size but on his size compared to males in the population. Maynard Smith & Brown (1986) analyse this game under the assumption that a male's probability of surviving to a given size is a decreasing function of this size. They show that a single body size cannot be evolutionarily stable, but under certain conditions a distribution of body sizes can be stable. For related models see Mäkelä (1985) and King (1990).

7.2 Strategies and payoffs

As before, we use the term 'strategy' to mean a rule that specifies how behaviour depends on state and time. Here the state of an organism could include information on other population members. We will classify strategies in the following three ways.

Pure strategies.   A strategy is *pure* if at any given time the relationship between an individual's state and the action that it takes is deterministic. When considering state-dependent decision-making, such a strategy is also referred to as a pure conditional strategy.

Mixed strategies. Suppose that $\pi_1, \pi_2, \ldots, \pi_n$ are pure strategies. Consider the strategy π, which consists of choosing strategy π_i with probability p_i. The organism then follows this strategy during the time interval of interest. We refer to π as a mixed strategy and write π schematically as

$$\pi = p_1\pi_1 + \cdots + p_n\pi_n. \tag{7.1}$$

Randomised strategies. Suppose that the relationship between the state of an organism at any given time and the action it takes is probabilistic. In other words, each time the organism has to choose an action it does so probabilistically. Then we refer to the rule specifying how the probabilities of taking each possible action depend on state and time as a randomised strategy.

When there is just one state and one action there is no logical distinction between a mixed strategy and a randomised strategy. The distinction arises when an organism takes a sequence of actions and is important in the analysis of dynamic games (see Section 7.5).

These distinctions can be illustrated in the context of variants on the Hawk–Dove game. In the standard Hawk–Dove game (Maynard Smith & Price 1973, Maynard Smith 1982), two individuals from a population engage in a contest for a resource. The Hawk strategy requires an animal to persist with aggressive behaviour until it wins the contest or is injured. The Dove strategy is to display and then retreat if the opponent behaves aggressively. An animal that always plays Hawk is adopting a pure strategy. An animal that plays Hawk with probability p and Dove with probability $1 - p$ is adopting a mixed strategy (or randomised strategy; there is no distinction here).

In the standard Hawk–Dove game all animals are assumed to be in the same state and each animal makes a single decision. Houston & McNamara (1988) put the Hawk–Dove game into an ecological setting by considering a population of conspecifics that are trying to survive an extended period of time such as winter. Animals search for food items, some of which are uncontested while access to the remainder involves a contest with a randomly selected member of the population. In a contest, each animal can either play Hawk or Dove. In this model, each animal is characterised by its level of energy reserves, x. A strategy specifies the choice of action in a contest as a function of reserves and time of day. Houston and McNamara show that the ESS strategy can be formulated in terms of a critical reserve level, $x_c(t)$, which can depend on the time of day, t. Under this strategy, on each day in winter, an animal plays Hawk if reserves are below $x_c(t)$ at time of day t and plays Dove if reserves are greater than $x_c(t)$. The strategy is thus a pure conditional strategy.

For this game, let π_0 be the strategy whereby a constant critical reserve level $x_c(t) = 10$ is used at all times t during each day. Thus an animal following π_0 plays Hawk if reserves are below 10 and Dove if reserves are above 10. Let π_1 be the strategy whereby a critical level $x_c(t) = 15$ is used throughout each day. Let $0 < \lambda < 1$, and define π_λ as the strategy under which π_0 is followed with probability $1 - \lambda$ and π_1 is followed with probability λ. Thus, in the notation of equation (7.1), $\pi_\lambda = (1-\lambda)\pi_0 + \lambda\pi_1$. Then π_λ is a mixed strategy. For example, under strategy $\pi_{1/2}$ an animal tosses a fair coin at the beginning of winter. If a head is obtained, a critical level $x_c(t) = 10$ is used each day, while if a tail is obtained, a critical level $x_c(t) = 15$ is used. In this example, the probability that an individual with reserves $x = 12$ plays Hawk is 0.5, but it either always plays Hawk in this state during the winter or it always plays Dove.

Finally, suppose that an animal adopts the following strategy. At any given time, if reserves are below 10 the animal plays Hawk, if reserves are between 10 and 15 it plays Hawk with probability 0.5 and Dove with probability 0.5 independently of behaviour before and if reserves are above 15 it plays Dove. Then this strategy is a randomised strategy. Under it the animal will decide its choice of action by the toss of a coin every time that it contests an item and its reserves lie between 10 and 15. This means that, for example, the behaviour of an animal with reserves $x = 12$ is uncorrelated with its previous behaviour in this state. This is in contrast to the mixed strategy, where there is perfect correlation.

In the above example, under the randomised strategy the probability that an action is chosen depends only on energy reserves. If probabilities are also allowed to depend on what has happened in the past then correlations between behaviours chosen at different times can be obtained.

Payoffs

For simplicity we consider a large population. We refer to a strategy π as the resident population strategy if almost all population members are following π. We can think of the resident strategy together with the physical environment as creating a background that is experienced by all individuals within the population. Suppose that the resident strategy is π and that a single 'mutant' within this population plays strategy π'. We denote the fitness of this mutant strategy by $W(\pi', \pi)$.

We illustrate the payoff function W with an example based on feeding and vigilance behaviour in a group of foraging animals. This example will later be used to illustrate various stability concepts.

Example. A vigilance game. Members of a group of n animals are foraging. Each animal in the group chooses the proportion of time that it spends feeding as opposed to being vigilant for predators. Each animal attempts to maximise

$$\theta \times \text{net intake rate} - \text{predation rate},$$

where θ is the marginal rate of substitution of predation risk for energy gain (Section 6.2). The parameter θ is the same for all group members. If an animal spends a proportion of time u feeding (where $0 \le u \le 1$), its net intake rate is au where a is constant. If a predator attacks the group, then the animal detects the attack before the predator can get within range with probability $1 - u^2$. Detection is independent of whether other group members detect the predator. If at least one group member detects the predator all group members survive the attack. If no group member detects the predator, one animal, chosen at random from the group, is killed.

Suppose the resident population strategy is to spend a proportion of time v feeding when in a group of size n. Consider a mutant (an animal following a mutant strategy) which spends a proportion of time u feeding when in such a group. Then the probability that the mutant is killed in an attack is $u^2 v^{2(n-1)}/n$. Thus, assuming that attacks occur at rate α, the payoff to the mutant is

$$W(u, v) = a\theta u - \alpha u^2 v^{2(n-1)}/n. \tag{7.2}$$

Typically, when a strategy specifies one or more numerical traits, the payoff to an individual is a non-linear function of these traits and also those of resident population members. This is certainly true of the above example, where $W(u, v)$ is a non-linear function of both u and v. The exception to this non-linear dependence occurs when traits give probabilities of choosing strategies, as we now discuss.

Mixed strategies and payoffs. Let $\pi_1, \pi_2, \ldots, \pi_n$ be strategies and let π' be the mixed strategy which specifies that strategy π_i is chosen with probability p'_i. Thus, schematically, $\pi' = p'_1\pi_1 + p'_2\pi_2 + \cdots + p'_n\pi_n$. Let π be the resident strategy and suppose that a mutant adopts strategy π'. Then since the mutant's expected payoff is a weighted sum of conditional expectations we have

$$W(\pi', \pi) = p_1' W(\pi_1, \pi) + p_2' W(\pi_2, \pi) + \cdots + p_n' W(\pi_n, \pi), \qquad (7.3)$$

so that the mutant's payoff depends linearly on the p_i'.

Now suppose that the resident strategy π is the mixed strategy given by equation (7.1). Then since the population is large we can assume that a proportion p_i of this population chooses strategy π_i. Thus the background created when the resident strategy is given by equation (7.1) is the same as when the resident strategy is polymorphic, with a proportion p_i of the population deterministically adopting strategy π_i. Consider a single organism that follows a mutant strategy π' within this population. Then in general the payoff to the mutant is not linear in the p_i. That is, in general the equation

$$W(\pi', \pi) = p_1 W(\pi', \pi_1) + p_2 W(\pi', \pi_2) + \cdots + p_n W(\pi', \pi_n) \qquad (7.4)$$

does not hold. To illustrate this suppose that males compete for access to females and that all females are monopolised by only the largest males. Let π_1 be the strategy of being a small male and π_2 the strategy of being a large male. Let π be the strategy $\frac{1}{2}\pi_1 + \frac{1}{2}\pi_2$. Let π' be the strategy of being a male of intermediate size. Then $W(\pi', \pi_1)$ is high since the male of intermediate size is the largest male when the resident strategy is π_1. Similarly $W(\pi', \pi_2) = 0$ since the intermediate male is now smallest. However, when the resident strategy is π, half the males are larger than the intermediate male and again $W(\pi', \pi) = 0$. Thus $W(\pi', \pi) \neq \frac{1}{2}W(\pi', \pi_1) + \frac{1}{2}W(\pi', \pi_2)$.

Although equation (7.4) does not hold in general, it does hold in games between two players where an opponent is chosen from the population at random, again by conditional expectation arguments. Simple two-player games are discussed in Section 7.4.

7.3 Evolutionary stability

In defining an ESS, Maynard Smith's idea was to concentrate on the evolutionary stability of a population if it adopted a particular strategy. A strategy is evolutionarily stable if, given that almost all population members adopt the strategy, then there is a selection against mutations that code for other strategies. Thus if the population arrives at an ESS, it will remain there. The motivation behind this approach is a desire to avoid dealing with the details of the evolutionary process (including genetics), instead merely giving a simple phenotypic characterisation of the endpoints of evolution by natural selection. The problem with this approach is that it establishes whether a strategy is stable, but it does not establish whether evolution by natural selection will lead to such a strategy. To establish this, we need to consider evolutionary dynamics, and this leads to stability concepts other than the ESS conditions. In this section we give a brief description of various stability concepts.

We start with the idea of a Nash-equilibrium strategy.

Nash equilibrium

Suppose that the resident population strategy is π, i.e. almost all population members are adopting π. Again focus on a single 'mutant' organism within the population. The strategy adopted by the mutant is said to be a best response to the resident population strategy if it maximises the mutant's fitness within the population. Formally, a strategy $\hat{b}(\pi)$ is a best response to π if

$$W(\hat{b}(\pi), \pi) = \max_{\pi'} W(\pi', \pi). \tag{7.5}$$

There may be more than one strategy that achieves the maximum on the right-hand side of equation (7.5). When this occurs, each maximising strategy is a best response.

If a strategy π' has higher fitness than the resident population strategy, then this population is invadable by mutants adopting π'. Thus a necessary condition for a resident strategy π^* to be evolutionarily stable is that

$$W(\pi^*, \pi^*) \geq W(\pi', \pi^*) \qquad \text{for all } \pi'. \tag{7.6}$$

A strategy π^* satisfying this condition is said to be a Nash-equilibrium strategy. The condition is equivalent to demanding that

$$\pi^* \text{ is a best response to itself.} \tag{7.7}$$

When best responses are unique, so that $\hat{b}(\pi)$ is well defined for each resident strategy π, we may also write the condition as

$$\hat{b}(\pi^*) = \pi^*. \tag{7.8}$$

We illustrate the above concepts by returning to the vigilance game of Section 7.2. Suppose that members of the resident population spend a proportion v of their time feeding as opposed to being vigilant. The fitness of a mutant that spends a proportion u of time feeding is given by equation (7.2), and the best response $\hat{b}(v)$ to the resident population is the value of

u that maximises this fitness. This can be found by differentiating W with respect to u for fixed v, and using the constraint that $u \leq 1$. It follows that

$$\hat{b}(v) = \min\left(1, \frac{a\theta n}{2\alpha v^{2(n-1)}}\right). \tag{7.9}$$

This unique best response is illustrated in Figure 7.1. It can be seen from the figure that as the resident proportion of time spent feeding increases (and hence the vigilance of residents decreases), the best strategy for a single mutant is to increase its own vigilance by decreasing the proportion of time that it spends feeding.

The Nash-equilibrium level of the proportion of time spent feeding v^* is given by $\hat{b}(v^*) = v^*$. The graphical solution of this equation is given in Figure 7.1.

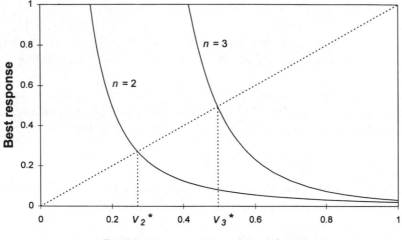

Resident proportion of time feeding, v

Figure 7.1. Best-response function and Nash equilibrium for the vigilance game of Section 7.2. The best-response function $\hat{b}(v)$ is given by equation (7.9). The Nash-equilibrium proportion of time feeding, v^*, is the solution of the equation $\hat{b}(v^*) = v^*$. This proportion is the value of v at which the curve $\hat{b}(v)$ intersects the 45° line. Two cases are shown: group size $n = 2$ and group size $n = 3$. The parameters a, θ and α are such that $a\theta = 0.02\alpha$.

The Bishop–Cannings theorem. Suppose that a Nash-equilibrium strategy π^* is a mixed strategy $\pi^* = p_1\pi_1 + p_2\pi_2 + \cdots + p_n\pi_n$, where $p_1 > 0, p_2 >$

$0, \ldots, p_n > 0$. Then it is easy to show (e.g. Bishop & Cannings 1978) that

$$W(\pi_1, \pi^*) = W(\pi_2, \pi^*) = \ldots = W(\pi_n, \pi^*) = W(\pi^*, \pi^*). \qquad (7.10)$$

Thus, given that π^* is the resident strategy, not only is π^* a best response to the resident strategy but so are the strategies $\pi_1, \pi_2, \ldots, \pi_n$. Since an ESS is a Nash equilibrium (see below), this result also holds for mixed ESSs. In the context of ESSs, it is known as the Bishop–Cannings theorem.

Conditions for evolutionary stability

As we have said, (7.6) gives a necessary condition for evolutionary stability. It may, however, not be sufficient. There are two reasons for this. One is that the condition does not exclude mutants that are equally fit as members of the resident population. Such mutants could increase in numbers as a result of drift. The other is that condition (7.6) does not tell us what happens if the whole population is perturbed away from π^*. We now discuss both of these issues in turn.

Evolutionarily stable strategies. Maynard Smith's idea was to consider a population at a Nash equilibrium π^* and to ask whether this resident strategy was stable against invasion by small numbers of a mutant strategy. His condition for evolutionary stability (Maynard Smith 1982) is as follows. The strategy π^* is an ESS if and only if for every strategy $\pi \neq \pi^*$ one of the following holds.

(ES1) $W(\pi^*, \pi^*) > W(\pi, \pi^*)$

(ES2) $W(\pi^*, \pi^*) = W(\pi, \pi^*)$ and there exists $\delta > 0$ depending on π, such that for all η in the range $0 < \eta < \delta$
$W(\pi^*, (1 - \eta)\pi^* + \eta\pi) > W(\pi, (1 - \eta)\pi^* + \eta\pi)$.

The motivation behind this definition is that no strategy $\pi \neq \pi^*$ can spread in the population when the resident strategy is π^*. If a mutant adopting strategy π arises then either (ES1) it does less well than the resident strategy π^* or (ES2) it does equally well unless mutant numbers increase to some small proportion η of the population (so that the resident strategy is the polymorphic strategy $(1 - \eta)\pi^* + \eta\pi$), when it does less well. These conditions for evolutionary stability are described by Maynard Smith as 'playing the field'.

Local convergence stability. Assume that a strategy is given by the numerical value of a single trait that can take a continuous range of values. Suppose that π^* is a Nash equilibrium, and consider a resident population strategy π that is close to, but not equal to, π^*. We can then ask whether the action of natural selection will change the value of the resident population strategy so that it moves towards π^*. This question was analysed for two-player games by Eshel (1983); see also Taylor (1989). In the general context of games against the field, Christiansen (1991) defines a Nash-equilibrium strategy π^* to be (local) convergence stable if, given that the resident strategy π is close to π^*, then for any mutant strategy π' sufficiently close to π the following conditions hold.

(C1) The mutant has greater fitness than the resident if π' lies between π^* and π, i.e.

$$W(\pi', \pi) > W(\pi, \pi) \qquad \text{if either} \quad \pi^* < \pi' < \pi \quad \text{or} \quad \pi < \pi' < \pi^*,$$

and

(C2) the mutant has lower fitness than the resident if π lies between π^* and π', i.e.

$$W(\pi', \pi) < W(\pi, \pi) \qquad \text{if either} \quad \pi^* < \pi < \pi' \quad \text{or} \quad \pi' < \pi < \pi^*.$$

Taylor (1989) refers to convergence stability as m-stability. For a review of terminology and stability concepts see Eshel (1996).

The conditions for evolutionary stability and convergence stability can be illustrated using the vigilance game of Section 7.2. In this game, the unique best response to the Nash-equilibrium strategy v^* is v^*. Thus if the resident population adopts strategy v^*, then every mutant adopting a different strategy has lower fitness than members of the resident population. It follows that condition (ES1) holds for every $v \neq v^*$ and so the Nash-equilibrium strategy v^* is an ESS. To show convergence stability, suppose that the resident population strategy is v, where $v > v^*$. Then since the best-response function $\hat{b}(v)$ is a decreasing function of v, we have $\hat{b}(v) \leq \hat{b}(v^*) = v^*$. For given v, it can be seen that the payoff function $W(u, v)$ is a unimodal function of u with a maximum when $u = \hat{b}(v)$. In particular $W(u, v)$ is a strictly decreasing function of u for $u > \hat{b}(v)$. Thus, since $\hat{b}(v) \leq v^*$, $W(u, v)$ is a strictly decreasing function of u for $u \geq v^*$. It follows that if $v^* < v' < v$, then $W(v', v) > W(v, v)$. Similarly if $v^* < v < v'$, then $W(v', v) < W(v, v)$. A similar argument can be applied when $v < v^*$. Taken together, these arguments show that the Nash equilibrium v^* is convergent stable. The crucial feature of this argument is that the best-response function is decreasing in the neighbourhood of the Nash equilibrium (see Eshel 1983).

What types of stability are possible?

It is reasonable to require that an endpoint of evolution by natural selection should be both evolutionarily stable and convergence stable. A Nash equilibrium that satisfies both these conditions is called a continuously stable strategy or CSS (Eshel 1983).

Given a Nash equilibrium, a variety of different degrees of stability can hold. For a discussion, see Geritz *et al.* (1998) and references therein. In particular we note the following. (i) A Nash equilibrium may be neither an ESS nor convergence stable. (ii) A Nash equilibrium may be an ESS but not convergence stable, although this combination is not possible for two-player games where each player has only a finite number of possible actions. Evolution by natural selection will not lead a population to such an equilibrium. (iii) A Nash equilibrium may be convergence stable but not an ESS. Geritz *et al.* (1998) argue that convergence stability without evolutionary stability can lead to evolutionary bifurcations (see also Eshel 1996, Christiansen 1991). (iv) A Nash equilibrium may be both evolutionarily stable and convergence stable, i.e. be continuously stable. The vigilance game falls into this category.

The analysis of convergence stability that we have given is based on a one-dimensional trait. The idea that if there is a convergence-stable equilibrium then it is an attractor for the evolutionary dynamics is based on the 'phenotypic gambit'. But, even in one dimension the details of the genetics might be important and a convergence-stable equilibrium might not be reached (e.g. Weissing 1991, but see also Eshel *et al.* 1997). The situation is much more complicated when there is more than one dimension. (See Matessi & Di Pasquale 1996 and Leimar (unpub.) and the references therein.)

For many games, particularly dynamic games, there is no ESS. This absence is often the result of the unrealistic assumption that animals do not make errors. Introducing errors leads to the concept of a limit ESS. We discuss this concept in the context of dynamic games in Section 7.5.

7.4 Simple two player games

Suppose that in a game each individual in a population interacts with just one other individual chosen at random from the class of potential opponents. Suppose, additionally, that the payoff to an individual depends solely on its strategy and that of the opponent it meets. In this game each individual makes a single choice of action from some finite set of available actions. The two players make simultaneous choices, so that when an individual decides on its action it does not know the decision of its opponent. We begin by considering symmetric games. In other words, the individuals do not have different rôles; they have the same actions available to them and there are no payoff asymmetries.

If the available actions are $u_1, u_2, \ldots, u_n$ then a strategy specifies the probability that each of these actions is chosen. A pure strategy makes a deterministic choice. Let π_i denote the pure strategy of choosing action u_i with probability 1. Let π be the strategy of choosing actions $u_1, u_2, \ldots, u_n$ with probabilities $p_1, p_2, \ldots, p_n$ respectively. Then π is the mixed strategy given by equation (7.1).

For a two-player game the payoff function W satisfies both conditions (7.3) and (7.4). As a consequence, the conditions for evolutionary stability can be simplified to the following.

The strategy π^* is an ESS if and only if for every strategy $\pi \neq \pi^*$ one of the following holds

(ES1) $W(\pi^*, \pi^*) > W(\pi, \pi^*)$

(ES2') $W(\pi^*, \pi^*) = W(\pi, \pi^*)$ and $W(\pi^*, \pi) > W(\pi, \pi)$.

In other words, if there is a rare mutant strategy π in a population with a resident strategy π^*, then either mutants do worse than residents in contests against residents, or if mutants do equally well in such contests, they do worse than residents against other mutants playing π.

For a two-player game with a finite set of available actions, an ESS is automatically convergence stable and hence a CSS.

The Hawk–Dove game

We now describe the Hawk–Dove game (Maynard Smith & Price 1973) in detail. This game will be used to illustrate both the above stability conditions and the theoretical developments that arise later in the chapter.

The game involves two individuals from a population engaged in a contest for a resource of value V. Each contestant can either choose to play Hawk or Dove. Choices are made before individuals know the choice of their opponent – effectively there is a simultaneous choice. If one contestant plays Hawk and the other plays Dove, the Hawk behaves aggressively and the Dove runs away. The Hawk thus obtains the resource and the Dove gets nothing. If both contestants play Dove, there is costless ritualised display and each obtains the resource with probability 0.5. The expected payoff to each contestant in this interaction is thus $0.5V$. When both contestants play Hawk, they fight. Each wins the fight with probability 0.5. The winner of the fight gets the resource. The loser pays a cost C of injury. The expected payoff to each contestant in this interaction is thus $0.5(V - C)$. We assume that $V < C$.

A strategy specifies the probability π that a contestant plays Hawk. Suppose the resident strategy is π. Consider a mutant that adopts strategy π'. The payoff to this mutant is

$$W(\pi', \pi) = (1 - \pi')(1 - \pi)0.5V + \pi'(1 - \pi)V + \pi\pi'0.5(V - C),$$

which can be expressed as

$$W(\pi', \pi) = K + 0.5(V - C\pi)\pi', \tag{7.11}$$

where $K = 0.5V(1 - \pi)$ does not depend on π'. From expression (7.11) we see that if $V - C\pi > 0$ then $W(\pi', \pi)$ is maximised over π' by setting $\pi' = 1$, while if $V - C\pi < 0$, $W(\pi', \pi)$ is maximised by setting $\pi' = 0$. If $V - C\pi = 0$, then $W(\pi', \pi)$ does not depend on π'. Thus the best response, $\hat{b}(\pi)$, to the resident strategy π is given by

$$\hat{b}(\pi) = \begin{cases} 1 & \text{if } \pi < V/C \\ \text{any strategy } \pi' & \text{if } \pi = V/C \\ 0 & \text{if } \pi > V/C. \end{cases} \tag{7.12}$$

It follows that a strategy π^* is a best response to itself if and only if $\pi^* = V/C$. Thus, $\pi^* = V/C$ is the unique Nash equilibrium in this game. This can also be seen graphically from Figure 7.2.

To establish that the Nash-equilibrium strategy $\pi^* = V/C$ is an ESS, we need to verify that, for every strategy $\pi' \neq \pi^*$, either condition (ES1) or condition (ES2') holds. By equation (7.11), $W(\pi', \pi^*) = W(\pi^*, \pi^*)$ for all π'. Thus when the resident strategy is π^*, all mutant strategies have fitness equal to the resident strategy π^*. Condition (ES1) thus fails to hold and we must verify that $W(\pi^*, \pi') > W(\pi', \pi')$. But this is easily established since, setting $V/C = \pi^*$ in equation (7.11), we obtain

$$W(\pi^*, \pi') - W(\pi', \pi') = 0.5C(\pi^* - \pi')^2,$$

which is positive if $\pi' \neq \pi^*$.

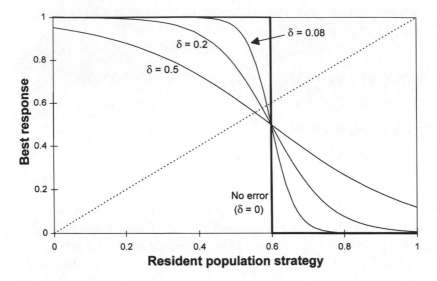

Figure 7.2. Best-response functions in the standard Hawk–Dove game. The best response without error is given by equation (7.12). Best-response functions with error are given by equation (7.23). The probability of error decreases with decreasing δ and is zero when $\delta = 0$. In each case the Nash equilibrium is the value of the resident strategy at which the best-response function intersects the 45° line ($V = 3$, $C = 5$).

Two-player games where individuals have different rôles

In the Hawk-Dove game, individuals are *a priori* the same, but in other games each contestant may have a clearly defined rôle that is known to all contestants. For example, when there is conflict between two individuals over territory defence, both individuals may be well aware which of them is the owner and which is the intruder. In games between members of a breeding pair, there are clear rôles: one individual is male, one is female. When such rôles exist they can influence solutions of the game even when the options and payoffs to all individuals do not depend on their rôles. This is so because when there are clearly defined rôles there can be asymmetric solutions (e.g. Hammerstein, 1981 Maynard Smith 1982). We now illustrate this phenomenon.

When there are clearly defined rôles, a strategy is a rule specifying the behaviour adopted in every possible rôle. In this sense, a rôle is essentially a state, and a strategy specifies what to do in every state. The component of a strategy that specifies behaviour in a particular rôle is often referred to as the local strategy in that rôle (e.g. Selten 1983).

It will be convenient for notational purposes to restrict attention to games between the male and female members of a breeding pair. Within this context a strategy specifies behaviour if male and behaviour if female, i.e. a strategy is given by a pair (π_m, π_f), where π_m is the local strategy of the male and π_f is the local strategy of the female. Suppose that within a breeding pair the male adopts local strategy π_m and the female adopts local strategy π_f. The payoffs to the male and female will then be denoted by $W_m(\pi_m, \pi_f)$ and $W_f(\pi_m, \pi_f)$ respectively. A strategy (π_m^*, π_f^*) is a Nash equilibrium if, given that all males use local strategy π_m^*, a best local strategy for the female is to use π_f^*, and given that all females use π_f^*, a best local strategy for the male is to use π_m^*. Formally,

$$W_f(\pi_m^*, \pi_f^*) \geq W_f(\pi_m^*, \pi_f) \qquad \text{for all} \;\; \pi_f \qquad (7.13)$$

$$W_m(\pi_m^*, \pi_f^*) \geq W_m(\pi_m, \pi_f^*) \qquad \text{for all} \;\; \pi_m. \qquad (7.14)$$

To illustrate formulae (7.13) and (7.14), suppose that each parent has two options: to care for the young or to desert. The parents choose simultaneously, so that each decides before it knows the decision of its partner. For each parent, if the parent cares it receives a payoff of 2, regardless of the behaviour of its partner. If the parent deserts, it receives a payoff of 3 if its partner cares and a payoff of 1 if its partner deserts. If there were no clearly defined rôles in this game, a Nash equilibrium (π_m^*, π_f^*) for this game would have to satisfy the constraint that $\pi_m^* = \pi_f^*$. That is, one would be restricted to symmetric solutions. Criteria (7.13) and (7.14) would then reduce to the criterion for a Nash equilibrium when there is no rôle asymmetry (inequality (7.6)). It can easily be verified that the only such symmetric Nash-equilibrium strategy is for each parent to care with probability 0.5 and to desert with probability 0.5. Because the game involves one male and one female, gender acts as a clearly defined label and there are now two further Nash-equilibrium strategies that are asymmetric. At one equilibrium the male always deserts and the female always cares. At the other the male always cares and the female always deserts.

Consider a Nash equilibrium in which at least one of the local strategies is a mixed strategy. Selten (1980) has shown that in games where there are clearly defined rôle asymmetries such an equilibrium pair cannot be an ESS. We can illustrate his argument in the context of the desertion game.

Suppose that (π_m^*, π_f^*) is a Nash-equilibrium for this game and that the local strategy π_m^* is a mixed strategy. By an extension of the Bishop–Cannings theorem (Section 7.3), both care by the male and desertion by the male are alternative best local responses to the female local strategy π_f^*. Suppose that the resident strategy is to play π_m^* if male and π_f^* if female. Introduce a mutant into this population, where the mutant strategy is to desert if male and play π_f^* if female. Then a single mutant does as well as members of the resident population. To see whether mutants are selected against as their numbers increase, we need to check whether mutants do worse against each other than residents do against mutants (ES2' of this section). But when mutants play each other, one is in the male rôle and one is in the female rôle, and the female plays the local female strategy π_f^*. It follows that mutants do as well against each other as residents to against mutants. Mutants are therefore not selected against and their numbers can increase by random drift.

Selten's result implies that when there are rôle asymmetries, a Nash equilibrium strategy π^* is an ESS if and only if in every rôle the local strategy specified by π^* is a unique best response to π^*. In the context of the desertion game this means that the symmetric Nash equilibrium in which each pair member deserts with probability 0.5 is not an ESS. In contrast, the two Nash equilibria based on the rôle asymmetry (one sex cares, the other deserts) are both ESSs.

When there are rôle asymmetries, Selten's result shows that a strategy that involves randomisation cannot be an ESS. This does not, however, mean that such a strategy cannot be the stable endpoint of evolution by natural selection. Selten's argument is based on the existence of certain mutants which alter behaviour in one rôle. Such mutants are not selected against. But if these mutants increase in frequency, they may introduce selection pressure on behaviour in other rôles. If behaviour in other rôles changes, this may lead to selection pressure against the original mutant (see e.g. the 'Battle of the Sexes', Hofbauer & Sigmund 1988).

7.5 Dynamic games

We consider a period $[0, T]$ in which decisions are made at times $t = 0, 1, \ldots,$ $T - 1$. At each of these times each organism in the population makes a decision. For a given organism the consequences of its decision depend on the behavioural strategies adopted by other population members. In this sense the organisms are interacting with each other. An organism may be involved in a series of interactions with one other organism (as in sequential assessment games) or may interact with several other organisms. The interaction may be direct, e.g. an animal may fight with another animal, or a plant may shade another plant. The interaction may also be indirect, e.g. an animal may compete with other animals to attract a mate by singing or calling. The only requirement is that the consequences of the organism's action depend on the behaviour of other members of the population. When we used dynamic programming to model the behaviour of an organism (Section 3.2), the organism's state included physiological variables and knowledge about the environment. Given its state, the action chosen at time t determined: (i) an immediate contribution to reproductive success, and (ii) its state, or probability distribution of states, at time $t + 1$. In dealing with dynamic games, state can again include physiological variables and information about the environment, but can now also include information gained about organisms with which the focal organism has interacted before t. Given its state, the quantities (i) and (ii) now depend on the action chosen by the organism and the strategy of other organisms. As before, the organism's reproductive value after time T is a function $R(x)$ of its state x at time T. Within this framework a strategy π is a rule that specifies the action $\pi(x, t)$ taken at each state x and time t. Then $W(\pi', \pi)$ is the reproductive value at time 0 of an organism that uses strategy π' while all other population members use π during $[0, T]$.

As we have seen from the Hawk–Dove game (Section 7.4), it may be important to allow some sort of probabilistic behaviour if an ESS is to exist. In Section 7.2, we distinguished between mixed strategies and randomised strategies. In the context of dynamic games, an individual following a mixed strategy uses a probabilistic rule to choose a pure strategy at time 0 and follows this strategy during $[0, T]$. In contrast, an individual following a randomised strategy uses a probabilistic rule to choose its action at every possible state and every time during $[0, T]$. These two classes of strategies (mixed and randomised) are equivalent in the following sense. Given any mixed strategy, there is a randomised strategy (perhaps involving recall of earlier behaviour) that produces the same behaviour, and vice versa (Kuhn 1953). Despite this equivalence, there are technical reasons for preferring randomised strategies. These arise because, on the one hand, for any state-dependent or dynamic game there are very many more mixed strategies than randomised strategies, and any given randomised strategy has a large number of mixed strategies that produce equivalent behaviour (e.g. Kuhn 1953). On the other hand, given a mixed strategy there is exactly one randomised strategy that produces this behaviour. One consequence of this has been emphasised by Selten (1983). If mixed strategies are allowed, there are

always other strategies that are behaviourally equivalent. As a result, the
formal definition of an ESS has to fail, because it cannot identify a unique
strategy. Another consequence concerns the practicalities of computing solutions to dynamic games. A high amount of computer memory may be
required to record even a single mixed strategy. For both these reasons, we
prefer to work with randomised strategies and will do so for the remainder
of this chapter.

Evolutionary stability in dynamic games

To illustrate both the concept of an ESS for a dynamic game and various
problems that can arise in the search for an ESS, we consider the following
desertion game. Suppose that young can potentially be cared for by both
parents. If at least one of the parents cares, the young survive; if neither
parent cares, the young die. By caring, a parent wastes time which could
be spent on other reproductive attempts. Each parent must decide whether
to care for the young or desert. The male chooses first. The female then
makes her decision knowing the decision that the male has taken.

In this game it is important to the parents that one of them cares for the
young, but each parent does better if it deserts and its partner cares. A pure
strategy for the male specifies whether to care or desert. A pure strategy
for the female specifies whether to care or desert if the male has decided to
care and whether to care or desert if the male has deserted.

We can use this game to illustrate some important general points.

(*i*) *Behaviour should be optimal given the current state.* It seems reasonable
to require that an animal should make the best decision given the current
state and future expectations. (This is known as modular rationality in the
context of human decision-making; see e.g. Skyrms 1996.) In the desertion
game a Nash equilibrium has the property that the male's local strategy
is a best response to the local strategy of the female, and vice versa. The
desertion game has two such equilibria. In one, the male's local strategy is
always to desert, and the female's local strategy is to desert if the male cares
and care if the male deserts. At the other equilibrium the male cares and the
female deserts regardless of the male's behaviour. These local strategies are
best-responses to each other, since, given that the female always deserts, it
is best for the male to care and given that the male always cares, the female
can always afford to desert.

The two Nash equilibria are fundamentally different. In both, the male
makes the best decision given future behaviour of the female. When the
female makes her decision, she can be in one of two states, corresponding

to whether the male has deserted. In the first Nash equilibrium, the behaviour of the female in each of the two states is optimal: if the male has deserted she should care, if the male cares she should desert. At the second Nash equilibrium, the female deserts if the male has deserted, so that her behaviour in this state is not optimal. The existence of this equilibrium relies on the fact that the female is making a threat to desert if the male does. This forces the male always to care, and hence the female is not called upon to carry out her threat. Thus although the female is not behaving optimally in all possible states, she is behaving optimally in the only state in which she can be at the equilibrium.

The second Nash equilibrium relies on the threat that the female makes. We do not regard this threat as biologically realistic. The threat violates the principle that an organism should always behave optimally given its current state. Reasons why this principle should hold are given below.

(*ii*) *Inaccessible states and neutrality.* Suppose that all population members are playing a given strategy, and that in this population they never reach a certain state. The population is then invadable under drift by a mutant that differs from the resident strategy only in the action taken in this state. The resident population is thus not evolutionarily stable under Maynard Smith's definition. This is one reason why some games have no ESS. The desertion game is an example. At the Nash equilibrium in which the male always deserts, the female never finds herself in the state in which the male cares. At the other equilibrium the male always cares and the female never finds herself in the state in which the male deserts. Neither of these equilibria is thus an ESS in the sense of Maynard Smith.

It can be argued that drift *per se* is not crucial, in that we may still be able to characterise a class of stable strategies (Thomas 1985a, b). For example, in the desertion game, if the male always deserts then variation in the female's behaviour given that the male cares may not be biologically relevant. In contrast to this, drift may have the effect of destabilising the game in a fundamental way. Consider the equilibrium in the desertion game in which the male cares and the female always deserts. In a population following this equilibrium strategy, a mutant female that cares if the male deserts, and deserts if the male cares, has the same fitness as the resident females. Since this mutant is not selected against, its numbers can increase by drift; if they reach a sufficient proportion, it is optimal for males to desert rather than care. The general point is that the threat is based on what would happen in a state that is not reached at the equilibrium. The neutrality of choice in this state allows drift, which in turn destroys the threat.

(*iii*) *Errors in decision-making.*   We believe that it is unrealistic to assume that animals can make error-free decisions (see Houston 1987, McNamara & Houston 1987d). Because of this, we can expect that there is a positive probability that all states will be reached during a game. It then matters what is decided in every possible state, and this is a reason for evolution to favour animals that always adopt the best option given their current state.

In the desertion game, errors by the male destabilise the equilibrium in which the female threatens to desert if the male does. In a population at this equilibrium, if no errors are made then the male always cares and the female loses nothing by making the threat since the threat is never carried out. In this population, a mutant female strategy, under which the female deserts if the male cares and cares if the male deserts, has the same fitness as the resident female strategy. However, if we allow males to make errors and desert occasionally then this mutant does better than the resident strategy since the young of resident females die when these females carry out their threat and desert. Thus the introduction of errors destabilises this Nash equilibrium.

Selten (1983) emphasises the importance of errors, and introduces them both for biological realism and to regain ESSs in games (such as the desertion game) that have no ESS. Selten analyses two-player games as follows. For a given game, we can perturb the game by assuming that for each state and time each possible action has a minimum probability of being chosen. Within the class of strategies that satisfy these constraints, he defines an ESS as one satisfying the Maynard Smith conditions (ES1) and (ES2') of Section 7.4. Such an ESS is referred to as an ESS for the perturbed game. Selten then looks at a sequence of perturbed games in which the minimum probabilities that actions are chosen is reduced to zero. A strategy is said to be a limit ESS for the original unperturbed game if it is the limit of a sequence of ESSs for such a sequence of perturbed games.

In the desertion game, we have pointed out that the Nash equilibrium in which the male deserts and the female cares if and only if the male deserts is not an ESS. This is so because at equilibrium there is no selection acting on what the female does if the male cares. Suppose now that the male always has a small probability of caring. Then the best thing for the female to do is to try (subject to errors in decision-making) to care if the male deserts and to desert if the male cares. Given that the female follows this strategy, it is best for the male to try and desert. It is easy to show that this pair of strategies is an ESS for the perturbed game. As the probability of error tends to zero, the ESS tends to the limiting strategy under which the male always deserts and the female cares if and only if the male deserts (cf. Selten 1983).

(*iv*) *Best responses and dynamic programming.* Given a resident population strategy π, we can seek the best response for a single mutant within this population. This best response is a strategy for behaviour over $[0, T]$. From remarks (i) – (iii) above, we can demand that the actions chosen under the best response strategy are always optimal given the organism's current state and time (and future behaviour). Dynamic programming yields just such a strategy. To be more specific, we can regard the resident population strategy as creating a fixed background. Given this background, finding the best response is an optimisation (as opposed to a game-theoretic) problem, and can be solved by working back from final time T using dynamic programming. Following van Damme (1991), we argue that attention should be restricted to only those best responses that are yielded by dynamic programming. In formal terms, a strategy $\hat{b}(\pi)$ is a best response to population strategy if equation (7.5) holds. A Nash equilibrium is then a strategy π^* satisfying equation (7.6), i.e. it is a best response to itself. Whether a Nash equilibrium is an ESS depends on our criteria for evolutionary stability. For a particular class of two-person games, Leimar (1997a) proved the following result. Let π^* be the resident population strategy. If, given the background created by π^*, dynamic programming yields the unique strategy π^* then π^* is a limit ESS.

In the context of the desertion game consider the resident population strategy π^* in which the male always desert and the female cares if and only if the male deserts. Dynamic programming yields this strategy as the unique best response to itself. (There are other best responses to π^*, but they are not given by dynamic programming.) It follows from Leimar's result that π^* is a limit ESS, as we concluded above.

7.6 A classification of dynamic games

We now offer a classification of dynamic games and give examples of each of our categories. Our classification is based on the nature of the interaction between individuals and the information that an individual could potentially have about the state of other individuals. This classification scheme gives us four categories. Examples from each category are given in Table 7.1.

Table 7.1. Classification of dynamic games.

Interaction	Solvable by working backwards only?[a]	
	Yes	No
Repeated interactions with one individual	Lima (1989) Kaitala *et al.* (1989) Clark & Ydenberg (1990) Leimar (1997a)	Enquist & Leimar (1983, 1987, 1990) Leimar & Enquist (1984) Leimar (1997b)
Interactions with several individuals	Iwasa & Odendaal (1984) Iwasa & Obara (1989) Crowley *et al.* (1991) McNamara & Collins (1990)	Houston & McNamara (1987, 1988) Mangel (1990) McNamara & Houston (1990c) Collins & McNamara (1993) Crowley & Hopper (1994) Lucas & Howard (1995) Johnstone *et al.* (1996) Johnstone (1997) Lucas et al. (1996) Bednekoff (1997) McNamara *et al.* (1999)

[a] Not all the games that are solvable by working backwards were in fact solved in this way; dynamic programming is not the only technique for solving dynamic games.

Nature of the interaction

During the period of interest an individual may either

(i) have repeated interactions with just one other individual, or
(ii) interact with several individuals.

This distinction corresponds to a difference in the relevant conditions for evolutionary stability. If an organism is involved in repeated interactions with the same individual, we can regard this set of interactions as one superinteraction, and apply the standard ESS condition for a pairwise interaction (Section 7.4) or, if appropriate, the condition for a limit ESS. This approach cannot be adopted when an organism interacts with several individuals. The non-linearity of the payoff means that the averaging used to obtain the ESS condition for a pairwise interaction is not valid. Instead, the organism may be regarded as playing against the field and an appropriate condition for evolutionary stability must be used.

When modelling repeated interactions between two organisms, care is necessary in identifying the state variables. We can illustrate this point in the context of a population in which each individual plays the Hawk–Dove game twice. Suppose first that each of the two rounds is played against different opponents. Then, with the usual cost structure (see Section 7.4), the ESS is to play Hawk in each round with probability V/C. In contrast, suppose that both rounds are played against the same opponent. The first round introduces an asymmetry between the contestants, because the two individuals differ in terms of the outcome of this round (and may also differ in terms of the action that they chose). Since an asymmetry has been established, no mixed strategy can be evolutionarily stable for the second round (see Section 7.4). Instead, at an ESS the action chosen in the second round must be a deterministic function of what has happened in the first round (Leimar pers. comm.). In this game, what has happened to an individual in the first round constitutes the state of that individual when it plays the second round. Given a resident population strategy, a best response can be found by working backward from the second round using dynamic programming. In doing so, we first find the best action in every possible state at the beginning of the second round; from this the best action in the first round can be found.

Information about state

Consider a focal organism in a given state at a given time. The best action for this individual depends on the states of other members of the population. In general, the organism may not know these states precisely, but will only have probabilistic information about them. We can make a logical distinction between games in which the strategy adopted by population members in the past provides information about the current states of population members and games in which it does not.

Games solvable by working backwards.   Suppose that an individual gains no additional information about the states of other population members from a knowledge of their past strategy. If this holds for all individuals in a population, then we can look for an ESS by working backwards over the time interval $[0, T]$. For each time t, we define a rule π_t^* that specifies the action taken in every state at this time. These rules are found by the following procedure. Suppose that we know $\pi_{t+1}^*, \pi_{t+2}^*, \ldots, \pi_{T-1}^*$. We can then let $V^*(x, t + 1)$ be the reproductive value of an individual in state

x at time $t+1$, given that the individual and the rest of the population adopt rule π_s^* at time s for each s such that $t+1 \leq s \leq T-1$. (Trivially, $V^*(x, T) = R(x)$ where R is the terminal reward.) Consider the actions chosen by population members at time t. Each individual chooses a single action at this time. This action and the actions chosen by other population members determine any immediate contribution to the reproductive success of the focal organism at t. They also determine the probability distribution of states of the focal organism at $t+1$, and hence via the function $V^*(x, t+1)$ determine the expected reproductive value of the organism at time $t+1$. Since each organism makes only one decision at time t we can treat behaviour at t as a simple game and find the ESS for this game. Call the rule specified at this ESS π_t^*. Having found $\pi_0^*, \pi_1^*, \ldots, \pi_{T-1}^*$ in this way we can then define the strategy π^*, specifying behaviour over the whole interval $[0, T]$, by taking behaviour at time t under π^* to be given by π_t^*.

Even though the strategies $\pi_0^*, \pi_1^*, \ldots, \pi_{T-1}^*$ are locally ESSs this does not guarantee that the strategy π^* is an ESS (see e.g. van Damme 1991). In two-person games, π^* is a limit ESS if it specifies a unique best action in each state at each time (Leimar 1997a, b).

When the strategy of other population members before time t provides useful information to an individual at time t, the decision problem at time t cannot be solved unless one specifies this strategy both before and after t. Thus one cannot find the ESS solution over $[0, T]$ by simply working backwards. Techniques based on alternating between working backwards and working forwards are described in Section 7.7.

Before describing these technical details we discuss a variety of examples. We begin by looking at two games that cannot be solved solely by working backwards. We then look at a variety of game-theoretic problems that arise in the context of mate choice, emphasising which problems can be solved solely by working backwards and which cannot.

Dependence on the future and the past

We now give examples of games where the strategy of population members in the past gives information on the current state of an opponent.

State-dependent Hawk–Dove game.   Houston & McNamara (1988) considered a population of animals that are trying to survive an extended period of time such as winter. Animals search for food items, some of which are uncontested, while access to the remainder involves a contest with a randomly selected member of the population. All food items have the same energetic content. In a contest each animal can either play Hawk or Dove. Each animal is characterised by its level of energy reserves x. A strategy specifies an animal's choice of action as a function of its reserves and time of day. An animal dies of starvation if its reserves reach zero and has a probability of being killed if it loses a Hawk–Hawk fight. Each animal tries to maximise its probability of surviving the winter.

We focus on an animal involved in a given contest and ask whether it should play Hawk or Dove. This decision should depend on the value of a food item, in terms of its contribution to future reproductive success, and on the probability that the opponent will play Hawk. The value of a food item depends on the reserves of the focal animal. It also depends on the expected energetic gains in the future. Future gains depend in part on the future behaviour of both the focal animal and other members of the population. The probability that the opponent will play Hawk depends on the opponent's current state and strategy. The probability that this animal is in a particular state depends on how much food it has obtained in the past, which in turn depends on its previous strategy and the previous strategy of other members of the population. Because of this dependence on the past, it is not possible to find the ESS in this case just by working backwards from final time.

Sequential-assessment game. Enquist & Leimar (1983) modelled the repeated interactions between two opponents labelled A and B during a contest for access to a resource. Contestants differ in their relative fighting ability θ. They have limited information about this parameter, but improve their knowledge of it as a result of observations during each of the interactions. After n interactions, contestant A bases its estimate of θ on the average of its observations so far, x_n^A, and contestant B bases its estimate on its average, x_n^B. The variables x_n^A and x_n^B act as the state variables for A and B respectively. After each interaction each animal must decide whether to continue or give up. A strategy for A specifies, for each n, the range of values of x_n^A for which A will give up. Similarly for B. At each n an animal's best choice of action depends on its own state, which determines its estimate of relative fighting ability, on the likelihood that the opponent will give up at this time and on the likelihood it will give up at each future time.

If both contestants make the same observations during each interaction, then $x_n^A = x_n^B$ for all n and both contestants always agree on their current estimate for θ. Contestant A would then know the state of its opponent and vice versa, and the additional knowledge of the strategy employed by opponents in the past would be irrelevant to future behaviour. This game could thus be solved by working backwards.

In the game analysed by Enquist and Leimar, however, there is observational error. Consequently, although the contestants' state variables x_n^A and x_n^B (and hence their estimates of θ) are correlated, they do not necessarily agree. Contestant A's state x_n^A provides some information on contestant B's state x_n^B. But if contestant B has not previously given up, B's previous strategy provides additional information on x_n^B. The reason is that B's previous decisions were made on the basis of previous values of B's state and B's strategy in the past; hence knowledge of B's previous strategy affects A's estimate of x_n^B. A similar remark holds for B's information on A's state. Consequently, in this case the ESS cannot be found just by working backwards from final time.

Mate choice

The topic of mate choice generates many game-theoretic problems. We now review some of the models that have been produced, emphasising whether the ESS can be found solely by working backwards. In all these models, an animal of one sex sequentially encounters members of the opposite sex and must decide whether to attempt to mate with these prospective partners. If a partner is rejected the animal cannot later decide to mate with this individual unless it happens to encounter it again. Prospective mates differ in quality, and this quality is immediately known when an encounter occurs. A searching individual's optimal strategy is based on a critical acceptance threshold; a mating should be attempted if and only if the quality of an encountered individual is above the threshold.

Mate-choice problems need not be game-theoretic. For example, if only one sex chooses and there is no depletion of potential mates, then we have a standard sequential decision problem on which there is an extensive mathematical literature. For a discussion in the context of mate choice, see Real (1990).

The problem becomes game-theoretic if the distribution of qualities of potential mates and/or the frequency with which potential mates are encountered depends on the behaviour of other population members. For example, if mated animals are temporarily or permanently removed from the pool of available mates, and if the time for an individual to find a mate depends on the quality of the individual, then individuals that take a long time to pair become over-represented in the pool of available mates as the breeding season progresses. This phenomenon occurs in the model of Collins &

McNamara (1993). In this model there is quality variation in one sex (say males) and the other sex is searching and choosing. Once a female finds a suitable mate, both the female and her mate no longer take any part in the process. It is clear from this that the strategy adopted by the population of females determines the distribution of available males. Consequently, the ESS cannot be found by just working backwards using dynamic programming. Since females will employ an acceptance threshold, high-quality males tend to be removed first, and the mean quality of available males tends to decrease over time. As a result, the longer that a female has been searching, the less choosy she should become, i.e. the critical acceptance threshold of females should decrease with time. Collins and McNamara give an analytic formula that determines the decrease.

A variety of models involve searching by both sexes. In these models, a pair is formed if and only if both members of the potential pair wish to mate with each other. This means that the best acceptance threshold for a member of one sex depends on how likely members of the other are to accept it as a mate. Consequently, the problem is game-theoretic. The best threshold for a male depends on his quality and the thresholds of all females, and the best threshold for a female depends on her quality and the thresholds of all males. Models of this type can be divided into those in which the distribution of qualities experienced by a searching animal is given and is hence independent of the behaviour of population members, and those in which it is generated by the behaviour of population members. The model of McNamara & Collins (1990) is of the first type. In this model, an animal is not constrained in the time available to find a mate, but the value of a mate is discounted by the time taken to find it, and there may be a cost of inspecting potential mates. McNamara and Collins prove that the ESS has the qualitative form shown in Figure 7.3. The result when population members follow this strategy is assortative pairing, in which males from the highest-quality class mate with females from the highest-quality class, males from the second-highest-quality class mate with females from the second-highest-quality class and so on. Because the distribution of qualities is given, the ESS can be found by working backwards using dynamic programming, although this is not the method employed by McNamara & Collins (1990).

When the distribution of qualities is generated by the behaviour of the population, the qualitative result proved by McNamara & Collins (1990) still holds, and the ESS when there are no time constraints has the form illustrated in Figure 7.3. Numerical procedures for finding the ESS cannot, however, be based just on dynamic programming. One method of solution is to alternately work backwards and forwards iteratively, as described in Section 7.7. Johnstone *et al.* (1996) present numerical results for a problem of this type. For a related problem based on divorce, see McNamara *et al.* (1999).

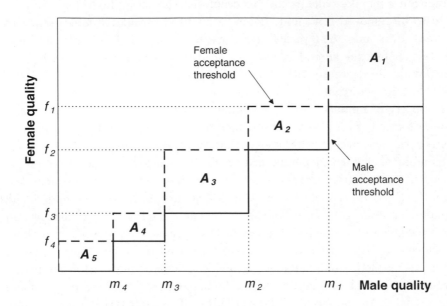

Figure 7.3. Male and female strategies for the mutual mate-choice game of McNamara & Collins (1990). A male will choose a female if and only if her quality exceeds his acceptance threshold. Similarly a female will choose a male if and only if his quality exceeds her threshold. (Thus, for example, if female quality y lies in the range $f_3 < y < f_2$ then the female will accept a male of quality x if and only if $x \geq m_3$.) Consequently, a male of quality x and a female of quality y will mate if and only if the point (x, y) lies in one of the sets A_1, A_2, A_3, A_4 or A_5.

When there is a finite time during which mate choice occurs, acceptance thresholds will typically depend on time, individuals becoming less choosy as they run out of time. Crowley *et al.* (1991) presented numerical results for the case in which there are two quality classes and the distribution of qualities is given. Johnstone (1997) presented numerical results for the case in which the pool of potential mates is depleted.

7.7 Computational procedures

We now concentrate on games where an individual's best current decision depends on both the past and future behaviours of other population members. As we have emphasised, such games cannot be solved by simply working backwards from final time. Here we describe iterative methods, based on alternatively working backwards and forwards, which can be used to compute solutions to these games.

The computational methods that we present seek to find a Nash-equilibrium strategy π^*, i.e. π^* is a best response to itself. Once such strategy has been found we can then investigate whether it also satisfies the appropriate conditions necessary for evolutionary stability (Section 7.3).

Iteration of the best-response map

A simple iterative scheme is to start with any strategy π_0 and then by iteration form a sequence of strategies $\pi_0, \pi_1, \pi_2, \ldots$, where each strategy in the sequence is the best response to the previous strategy. Thus $\pi_{n+1} = \hat{b}(\pi_n)$, where $\hat{b}$ is the best-response function. In calculating the sequence of strategies $\pi_0, \pi_1, \pi_2, \ldots$ the hope is that it will converge to a Nash-equilibrium strategy, i.e. $\pi_n \to \pi^*$ as $n \to \infty$, where $\hat{b}(\pi^*) = \pi^*$.

Given a strategy π_n the best response is usually found by first taking a large population whose members adopt strategy π_n and then following the population forwards in time. This generates, for each time t, the proportion of population members in each possible state at this time. These probabilities then form the background against which the fitnesses of mutant strategies are compared. Taking this background, the fittest mutant, i.e. the best response strategy $\hat{b}(\pi_n)$, is found by working backwards from final time using dynamic programming. To illustrate this process consider the state-dependent version of the Hawk–Dove game (Houston & McNamara 1988), described in Section 7.6. A strategy π_n for this game is a rule specifying the probability $\pi_n(x, t)$ of playing Hawk for every time of day t and level of energy reserves x. Suppose that almost all members of a large population follow strategy π_n. Then after several days the distribution of energy reserves of those members left alive will settle down to an equilibrium value that depends only on the time of day. This distribution and the state and time-of-day-dependent strategy π_n determine the probability $p_n(t)$ that a population member, selected at random at time of day t, will play Hawk. We now consider a single mutant individual within this population. The function p_n gives, for each time of day, the probability that any opponent of the mutant will play H. Using this, we can determine by dynamic programming back from the end of winter, the strategy that maximises the mutant's overwinter survival probability. This strategy is $\hat{b}(\pi_n)$.

The simple scheme based on iteration of the best-response function has been successfully used to calculate ESSs in several dynamic games. In particular it succeeds for a range of parameter settings in the games of both Enquist & Leimar (1983) and Houston & McNamara (1988). But, as Houston and McNamara found, the scheme fails to work for other parameter settings. Leimar & Enquist (1984), Lucas & Howard (1995) and Lucas *et al.* (1996) also experienced problems with certain parameter values. Typically in these problematic cases the sequence $\pi_0, \pi_1, \pi_2, \ldots$ does not converge at all and tends to oscillate.

The sequence of strategies $\pi_0, \pi_1, \pi_2, \ldots$. might loosely be regarded as specifying the time evolution of the resident population strategy over successive years. In this process the population present in year n is entirely replaced by a population of best mutants by year $n+1$. It is not surprising that this process with complete replacement does not converge. To give the process more chance of convergence and, perhaps, to be more realistic biologically the population in year $n+1$ can instead be formed by replacing a proportion $\lambda > 0$ of the resident strategy in year n, π_n, by the best mutant strategy $\hat{b}(\pi_n)$. The strategy in year $n+1$ is thus

$$\pi_{n+1} = (1 - \lambda)\pi_n + \lambda\hat{b}(\pi_n). \tag{7.15}$$

Although we are here interpreting this resident strategy as a polymorphism, from a mathematical point of view we can also regard the resident strategy as mixed, each individual adopting π_n with probability $1 - \lambda$ and $\hat{b}(\pi_n)$ with probability λ. Neither interpretation leads to computational tractability, and it is often best to regard equation (7.15) as schematic, saying that π_{n+1} is 'somewhere between π_n and $\hat{b}(\pi_n)$'. For example, in the state-dependent Hawk–Dove game (Houston & McNamara 1988) discussed in Section 7.6, suppose strategy π_n is to play Hawk if and only if reserves are below 50 and $\hat{b}(\pi_n)$ is to play Hawk if and only if reserves are below 100. Then instead of taking π_{n+1} to.be a polymorphic or mixed strategy, we might take it to be the pure contingent strategy under which Hawk is played if and only if reserves are below $(1 - \lambda)50 + \lambda100$. The above two ways of forming π_{n+1} from π_n are not equivalent either evolutionarily or mathematically, but the second way is computationally simpler than the first. If the second way works and leads to an ESS then it is irrelevant that the process cannot be given a direct evolutionary interpretation. We are here trying to find the endpoints of the evolutionary process (ESSs) rather than track evolutionary pathways.

Whatever the interpretation of equation (7.15), we will refer to the iterative scheme given by this equation as iteration of the best-response function with damping, and refer to λ as the replacement factor. Damping the best-response function increases the range of problems and parameters for which the sequence $\pi_0, \pi_1, \ldots$ converges to a Nash-equilibrium strategy. For example, Leimar & Enquist (1984) obtained convergence when there was sufficient damping, whereas their sequence of strategies failed to converge when there was no damping. There are problems, however, for which no fixed level of replacement λ will work. This usually occurs for one of the following two reasons.

Discontinuity of the best-response function. In many games the best response function is discontinuous at a Nash-equilibrium strategy. For example, in the Hawk–Dove game a strategy π specifies the probability of playing Hawk. As equation (7.12) and Figure 7.2 show, $\hat{b}(\pi)$ jumps from 1 to 0 as π increases through its ESS value of V/C. To illustrate how the discontinuity causes oscillations in this game, suppose first that the sequence of best responses is undamped. Let $\pi_0 = 0$ (i.e. the initial resident strategy is to play Dove at all times), then $\pi_1 = \hat{b}(\pi_0) = 1$ (i.e. play Hawk at all times), $\pi_2 = \hat{b}(\pi_1) = 0$ and so on. Thus the sequence of iterates is $0, 1, 0, 1, 0, \ldots$, which does not converge. Damping reduces the amplitude of the oscillations but does not produce convergence. For example, suppose that $V/C = 0.5$ and the replacement factor $\lambda = 0.2$. Then starting with $\pi_0 = 0$ the sequence of iterates is $0, 0.2, 0.36, 0.488, 0.5904, 0.4723, 0.5779, \ldots$, which settles down into an oscillation between 0.4444 and 0.5555.

Grid effects. When a model has to be solved by computation, state variables must take a discrete set of values. To illustrate why the resulting grid effects can lead to oscillatory behaviour, consider the state-dependent Hawk–Dove game (Houston & McNamara 1988) described in Section 7.6. Suppose for simplicity that there is no day–night cycle. Then, given a time-independent resident strategy π, the best response to π is to play Hawk if reserves are below a critical threshold x_π and play Dove if reserves are above x_π. In looking for a Nash-equilibrium strategy, we can therefore restrict attention to strategies of this threshold form and equate a strategy π with the threshold used under it. In this notation $x_\pi = \hat{b}(\pi)$. When reserves are modelled as taking a continuum of values, $\hat{b}(\pi)$ depends continuously on π, there is a solution to $\pi^* = \hat{b}(\pi^*)$ and furthermore we can find this π^* by damped iteration of the best response function. Computations, however, need to model energy reserves on a discrete grid. Suppose, for definiteness, reserves are modelled as taking integer values in the range $0, 1, 2, \ldots$. Suppose that the Nash-equilibrium threshold for the continuous-reserve model is, say, 10.4. Then it could happen in the discrete-state computer model that $\hat{b}(10) = 11$ and $\hat{b}(11) = 10$. Thus iterates of the best-response map can oscillate between 10 and 11. Furthermore, damping may fail to stop the oscillation.

When any fixed level of the replacement factor λ leads to oscillations, we can instead use a replacement factor λ_n that decreases with the number of iterations n. For example, we might take $\lambda_n = 1/n^2$. This might make the sequence $\pi_0, \pi_1, \pi_2, \ldots$ converge, but could force convergence to some strategy that is not a Nash equilibrium (see McNamara *et al.* 1997). This highlights another problem: if iteration of the best-response function with damping produces a sequence $\pi_0, \pi_1, \pi_2, \ldots$ that appears to be converging, how do we know that convergence is to a Nash equilibrium? Since a Nash-equilibrium strategy π^* satisfies $\hat{b}(\pi^*) = \pi^*$, we may have confidence that the sequence $\pi_0, \pi_1, \pi_2, \ldots$ is converging to a Nash-equilibrium strategy if the difference between $\hat{b}(\pi_n)$ and π_n tends to zero as n increases; schematically,

$$|\hat{b}(\pi_n) - \pi_n| \to 0 \quad \text{as} \quad n \to \infty. \tag{7.16}$$

However, because the best-response function $\hat{b}$ is discontinuous, relation (7.16) might not hold even when convergence is to a Nash-equilibrium strategy. For example, consider the Hawk–Dove game with $\pi^* = V/C = 0.5$. Suppose that a sequence of iterates of the best-response function with decreasing replacement factor λ_n produces the strategies $0.34, 0.58, 0.46, 0.52$, $0.49, 0.504, 0.498, 0.501$, which appear to be converging on 0.5. The sequence of best responses of these strategies is $1, 0, 1, 0, 1, 0, 1, 0$, which is not converging. The difference $|\hat{b}(\pi_n) - \pi_n|$ between $\hat{b}(\pi_n)$ and π_n does not converge to 0, thus condition (7.16) does not hold. If we did not already know that $\pi^* = 0.5$ we would not be able to tell from the numerical values of π_n and $\hat{b}(\pi_n)$ whether the sequence of strategies was converging to a Nash equilibrium.

Errors in decision-making as a computational tool

In Section 7.5 we argued that organisms make errors and that this needs to be taken into account when deciding whether a Nash-equilibrium strategy is biologically realistic. Here we describe a way of introducing errors into games that was proposed by McNamara *et al.* (1997). There is evidence that, in nature, costly errors are rarer than errors that result in small losses in reproductive success (see e.g. Houston 1987). McNamara *et al.* incorporate this feature into games and show that as a result the best-response function is better behaved than when there are no errors. As a consequence, many of the computational problems experienced in attempting to find solutions to complex games disappear.

Consider first a simple game in which each contestant chooses between one of K actions $u_1, u_2, \ldots, u_K$. A strategy for this game π specifies the probabilities $p_\pi(u_1), p_\pi(u_2), \ldots$ that an individual following π will choose action $u_1, u_2, \ldots$ respectively. We denote the payoff to an individual that chooses action u_i when the resident strategy is π by $H_\pi(u_i)$.

Now suppose that π is the resident population strategy. Define the best response with error to this resident strategy as follows. Let

$$H_\pi^* = \max_{1 \leq i \leq K} H_\pi(u_i) \tag{7.17}$$

be the maximum possible payoff to a mutant in this population. Let

$$C_\pi(u_i) = H_\pi^* - H_\pi(u_i) \tag{7.18}$$

be the loss in payoff that results from choosing action u_i rather than the action achieving the maximum payoff (cf. Section 3.6). Then $C_\pi(u_i) \geq 0$ for every u_i, and $C_\pi(u_i) = 0$ if and only if action u_i achieves the maximum payoff. We now introduce an error function L. This can be any function defined for $x \geq 0$ that satisfies $L(x) > 0$, $L'(x) < 0$ and $L(x) \to 0$ as $x \to \infty$. For example, the function $L(x) = e^{-x}$ satisfies these criteria. The weight assigned to action u_i is then

$$\beta_\pi(u_i) = L(C_\pi(u_i)). \tag{7.19}$$

Finally, let

$$\hat{p}_\pi(u_i) = \frac{\beta_\pi(u_i)}{\Sigma_{j=1}^K \beta_\pi(u_j)}. \tag{7.20}$$

Then the best response to strategy π with error function L is defined to be the strategy $\hat{b}_L(\pi)$ under which u_i is chosen with probability $\hat{p}_\pi(u_i)$.

Under strategy $\hat{b}_L(\pi)$ there is a non-zero probability of choosing each action. The optimal action (given resident strategy π) is the most likely to be chosen. The greater the fitness cost of a suboptimal action, the less likely is that action to be chosen under $\hat{b}_L(\pi)$.

The best response with error has two especially useful properties. First, $\hat{b}_L(\pi)$ is always uniquely defined and second, it depends continuously on π provided that the payoffs $H_\pi(u_i)$ are continuous in π. Neither of these properties needs to hold when there is no error, and in fact both fail to hold for the Hawk–Dove game.

A Nash equilibrium with error function L is defined to be a strategy π_L^* such that

$$\hat{b}_L(\pi_L^*) = \pi_L^*; \qquad (7.21)$$

i.e. π_L^* is the best response to itself with error function L. By the continuity of $\hat{b}_L$ there is always at least one such strategy. Continuity also allows us to recognise when a sequence of strategies is converging to a Nash equilibrium with error. Suppose that the sequence of strategies $\pi_0, \pi_1, \pi_2, \ldots$ converges to a strategy π_∞. Then π_∞ is a Nash equilibrium with error function L if and only if

$$|\hat{b}_L(\pi_n) - \pi_n| \to 0 \text{ as } n \to \infty. \qquad (7.22)$$

This is in contrast to the case without error, where condition (7.16) is not in general equivalent to the convergence of the sequence of strategies to a Nash-equilibrium (without error).

An obvious question is the limiting behaviour of the Nash equilibrium with error as the probability of error tends to zero. In particular, does the Nash equilibrium with error tend to a Nash equilibrium without error as errors disappear? To look at this limiting behaviour, we can introduce a parameter δ ($\delta > 0$) that controls the degree of error. Given an error function L, define a one-parameter family of error functions $\{L_\delta\}$ by setting $L_\delta(x) = L(x/\delta)$. Then $L_1 \equiv L$ and, as δ tends to zero, the probability of making an error tends to zero.

To illustrate the above concepts consider the Hawk–Dove game (Section 7.4). In this game there are two pure actions 'Dove' and 'Hawk', and a strategy specifies the probability π of playing Hawk. When the resident strategy π is less than its ESS value, V/C, the best response without error is to play Hawk with probability 1 (equation 7.12). The cost of playing Hawk is thus $C_\pi(\text{Hawk}) = 0$. By equation (7.11) the cost of playing Dove is $C_\pi(\text{Dove}) = 0.5(V - C\pi)$. Similarly, if $\pi > V/C$ then $C_\pi(\text{Dove}) = 0$ and $C_\pi(\text{Hawk}) = 0.5(C\pi - V)$. When $\pi = V/C$ both actions incur zero cost. Thus, if we take the error function $L_\delta(x) = e^{-x/\delta}$, it can be seen from equations (7.19) and (7.20) that the probability of playing Hawk under the best response with error to the strategy π is given by

$$\hat{b}_{L_\delta}(\pi) = \frac{1}{1 + \exp[-0.5(V - C\pi)/\delta]}. \qquad (7.23)$$

This function is illustrated in Figure 7.2. It can be seen that, as δ decreases, the best response with error function L_δ tends to the best response without error. Furthermore, the Nash equilibrium with error, $\pi_{L_\delta}^*$, tends to the ESS without error, $\pi^* = V/C$. This holds whatever error function L is initially chosen (McNamara *et al.* 1997).

McNamara *et al.* (1997) incorporate error-making into dynamic games by assuming that the formalism given above for a single decision is applied separately at every decision made by an organism, thus generating a randomised strategy. Incorporating errors into state-dependent dynamic games in this way makes the best response with error, $\hat{b}_L(\pi)$, a continuous function of π (McNamara *et al.* 1997). Consequently, condition (7.22) still holds, making it possible to determine whether an iterative scheme is converging to a Nash equilibrium with error. However, more theoretical work needs to be done on the topic of games with errors. In particular, we need general results on what happens when the probability of error tends to zero.

The technique has, however, proved useful in practice. In particular a number of workers have used the technique to solve complex games in cases that were previously problematic (Johnstone 1997, Henson Alonso unpub, Lucas, pers. comm., Székely *et al.*, unpub.).

Nash equilibria and evolutionary stability

This section has been concerned with methods by which we can find Nash equilibria for state-dependent dynamic games. The Nash condition is necessary for evolutionary stability but not sufficient. So having found a Nash-equilibrium strategy π^* how do we determine whether π^* is an ESS?

In many games π^* is the unique best response to itself and hence is an ESS. When π^* is not the only best response to itself the situation is less clear, but π^* cannot be an ESS if there are alternative local best responses in some situation involving clearly defined rôle asymmetries (Selten 1980, Leimar 1997a).

Evolutionary stability (in the sense of Maynard Smith 1982, or of a limit ESS) does not guarantee that the strategy is an endpoint of the evolutionary dynamics. When the set of strategies is one dimensional, as in the Hawk–Dove game, it is easy to determine whether the evolutionary dynamics returns a perturbed population back to a Nash equilibrium. The set of strategies for a state-dependent dynamic game is, however, typically a subset of a space of high dimension. The evolutionary dynamics may then be complex and depend on the genetics assumed (see e.g. Hofbauer & Sigmund 1988, Weissing 1991 and Weibull 1995).

Given the difficulties in determining the precise stability properties of a complex game, we might, in the first instance, adopt an heuristic approach. The numerical schemes to calculate a Nash equilibrium are usually iterative.

In such schemes, a resident population is modified to form a new population and so on. This process is loosely analogous to the evolutionary dynamics of the population. Thus if a sequence of strategies converges to a Nash equilibrium, we might regard this as giving support to the fact that the proper evolutionary dynamics would home in on this Nash equilibrium. Under this heuristic approach the exact form of the iterative scheme used suggests the stability property of any Nash equilibrium found by the scheme.

7.8 A dynamic game of brood care and desertion

To illustrate the detailed formulation of a dynamic game, we describe the game of parental brood care and desertion of Székely *et al.* (unpub). It looks at the breeding behaviour of birds over a breeding season of finite length. When a brood is produced each parent can either care for the brood or desert the brood. Care increases the number of young that survive to independence, but is time consuming. By deserting, a bird can attempt to remate and hence produce another brood. Each bird in the population thus faces a trade-off: it can either choose to produce a few broods in which each young has a high survival probability, or to produce many broods in which each young has a lower survival probability. The decision problem is game-theoretic for two reasons. First, the number of young that survive from a brood is assumed to depend on the care or desertion decisions of both parents. Thus the value of care by a parent (in terms of increasing the survival chances of its young) depends on the behaviour of its partner. Second, if a bird deserts, its chances of remating depend on the number and sex of other birds seeking mates, and hence depends on the past desertion-or-care decisions of all population members.

For simplicity, the detailed model that we analyse assumes a large population with equal numbers of breeding males and females. Energetic considerations are ignored, as are any quality differences between individuals. The breeding season lasts T days. At the start of day t, where $t = 0, 1, \ldots, T-1$, each bird must choose its behavioural action for that day. If a bird is single it either rests or searches for a mate. A bird that rests on day t is still single on day $t + 1$. A searching male is still single on day $t + 1$ with probability $1 - P_m(t)$ or is paired with a mate with probability $P_m(t)$. Similarly, a female that searches on day t is paired with a mate on day $t + 1$ with probability $P_f(t)$. The functions $P_m(t)$ and $P_f(t)$ depend in some appropriate way on the numbers of searching males and females at time t. For instance, we might expect that the probability that a bird will pair decreases as the number of searching birds of its own sex increases and increases as the number of searching birds of the opposite sex increases.

Once a pair is formed, both members of the pair spend τ_{brood} days building a nest and producing a clutch of eggs. Each parent must then decide whether to care for the brood until brood members become independent or to desert and become single again. Care of the young takes τ_{care} days, after which the surviving young are independent. If both parents desert, all the young die. If the male cares but the female deserts, S_{m} young survive to independence. Similarly, if there is care by the female alone, S_{f} young survive to independence. Finally, if both parents care then S_{mf} young survive. The reproductive value of an offspring that becomes independent on day t is $R_{\text{off}}(t)$. A decrease in reproductive value with time in the breeding season is a widely observed phenomenon in birds (Daan *et al.* 1989). We thus assume that $R_{\text{off}}(t)$ decreases as t increases and that $R_{\text{off}}(t) = 0$ for $t \geq T$.

Parents that care for their young become single again once the young become independent.

The process of desertion

So far we have not specified what information is available to a parent when it chooses whether to desert a brood. Once a brood has been produced, the two parents are effectively involved in a two-player game against each other that is embedded within the larger dynamic game described above. The details of the desertion process, and in particular the available information, are crucial to the outcome of the game between the parents. To highlight the importance of the desertion process, we compare the following two scenarios.

(i) *Simultaneous decisions.* Once the brood is produced, both parents make their desertion decisions independently of one another. In particular, in making a decision, neither parent knows the decision of its mate.

(ii) *Sequential decisions – male chooses first.* The male chooses whether to desert. His decision is known to the female when she makes her decision.

To illustrate why these two processes can lead to different care outcomes suppose that the payoffs to the male of each possible outcome are as given in Table 7.2(a) and the payoffs to the female of each outcome are as given in Table 7.2(b). Consider first the case of simultaneous choice. If the female deserts, the male obtains a payoff of 2 if he deserts and a payoff of 3 if he cares (Table 7.2(a)). Thus, if the female deserts the male's best action is to care. Similarly, if the female cares, the male's best action is again to care ($5 > 4$). Of course, the male does not know the female's decision, but since his best action is to care regardless of her choice, he will care. Given the male will always care, the female does best by deserting (Table 7.2(b)). Thus the resultant pattern of care when there is simultaneous choice is uniparental care by the male.

Table 7.2. Payoffs in a simple desertion game between members of a breeding pair. This example is used to illustrate that the outcome of the game may depend on whether choice is simultaneous or sequential.

(a) Payoff to the male

Action of male	Action of female	
	care	desert
care	5	3
desert	4	2

(b) Payoff to the female

Action of the female	Action of male	
	care	desert
care	5	4
desert	6	3

Now suppose that there is sequential choice with the male choosing first. If the male cares, the female will desert (Table 7.2(b)) and the male will obtain a payoff of 3 (Table 7.2(a)). If the male deserts, the female will care (Table 7.2(b)) and the male will obtain a payoff of 4 (Table 7.2(a)). Thus the male does best by deserting and the resultant pattern of care is uniparental care by the female. For the payoffs given in Table 7.2, for any fixed choice of the female, the male would choose to care, but since his decision affects the decisions of the female he must desert to prevent her from deserting.

The desertion decisions of birds may not be simultaneous or sequential. Instead, the desertion process may be much more complicated. If both parents are making foraging trips to provision the brood, each may be unable immediately to distinguish between the late return of its partner and the desertion of its partner. The desertion decision of a bird is liable to depend on its energy reserves and body condition, and on the past care-behaviour of its partner. It may also depend on whether a potential new mate has been found during brood provisioning trips. Since, as we have illustrated, the details of the desertion process may radically affect its outcome, we may need to know much more about these details to make good predictions or give satisfactory explanations about the behaviour of natural populations.

Outcomes when the male chooses first

The results presented in this section assume that there is sequential choice with the male choosing first. To find a Nash equilibrium in this case, in line with the arguments given in Section 7.5 we assume that the female always makes the best choice given the decision of the male and cannot make empty threats. We introduce the following notation. The outcome of the game between two parents is denoted by a pair in which the male's behaviour is given first, then the female's behaviour. Thus the outcome is (C, C) if both care, (C, D) if the male cares and the female deserts, (D, C) if the male deserts and the female cares and (D, D) if both parents desert. Let $H_m(C, C)$ be the payoff to the male if both parents care, $H_m(C, D)$ the payoff to the male if he cares and the female deserts, $H_m(D, C)$ and $H_m(D, D)$ being defined analogously. Similarly, let $H_f(C, C)$, $H_f(C, D)$, $H_f(D, C)$ and $H_f(D, D)$ be the payoffs to the female for the four outcomes of care (C, C), (C, D), (D, C) and (D, D) respectively.

Since the male decides first, the female knows his decision when she makes her own. Thus the female's decision is based on comparing, for a given behaviour of the male, the payoff to her if she deserts with the payoff if she cares. Let

$$D_f(D) = H_f(D, D) - H_f(D, C). \tag{7.24}$$

Then $D_f(D)$ is the net benefit of desertion to the female given that the male has deserted. Thus, if the male has deserted, the female should care if $D_f(D) < 0$ and desert if $D_f(D) > 0$. Similarly, let

$$D_f(C) = H_f(C, D) - H_f(C, C). \tag{7.25}$$

Then, given that the male has decided to care, the female should care if $D_f(C) < 0$ and desert if $D_f(C) > 0$.

The male bases his decision on the behaviour of the female given his decision. To illustrate this, suppose, for example, that $D_f(C) > 0$ and $D_f(D) < 0$. This is case 2 of Table 7.3. Then, if the male cares, the female will desert and the male will obtain a payoff $H_m(C, D)$. If the male deserts, the female will care and the male will obtain a payoff $H_m(D, C)$. Suppose that $H_m(D,C) < H_m(C,D)$ (case 2B of Table 7.3). Then the male should care. The Nash-equilibrium strategy is thus for the male always to care and the female to care if and only if the male deserts. The outcome when the pair uses this strategy is for the male to care and the female to desert, i.e. (C,D). This case and all other possible cases are summarised in Table 7.3.

The value U_m of the game to the male is the payoff under the form of care at the Nash equilibrium. The value U_f to the female is similarly defined. Thus, for example, in case 2B of Table 7.3 we have $U_m = H_m(C,D)$ and $U_f = H_f(C,D)$.

Table 7.3. *Relationship between payoffs and the resulting form of care in the desertion game in which the male chooses first*

Case	Female payoff		Male payoff	Form of care
1	$D_f(D) > 0,\ D_f(C) > 0$	A	$H_m(D,D) > H_m(C,D)$	(D,D)
		B	$H_m(D,D) < H_m(C,D)$	(C,D)
2	$D_f(D) < 0,\ D_f(C) > 0$	A	$H_m(D,C) > H_m(C,D)$	(D,C)
		B	$H_m(D,C) < H_m(C,D)$	(C,D)
3	$D_f(D) > 0,\ D_f(C) < 0$	A	$H_m(D,D) > H_m(C,C)$	(D,D)
		B	$H_m(D,D) < H_m(C,C)$	(C,C)
4	$D_f(D) < 0,\ D_f(C) < 0$	A	$H_m(D,C) > H_m(C,C)$	(D,C)
		B	$H_m(D,C) < H_m(C,C)$	(C,C)

Where do payoffs come from?

In the game between two parents discussed above, the four payoffs to a parent are the reproductive values of the parent under the possible patterns of care. Each payoff is the sum of the expected reproductive success from the current brood, the success from future broods during that breeding season and the success in future breeding seasons.

Consider first the reproductive success from the current brood. Suppose that a brood is produced at time t. If both parents care for the brood then S_{mf} young survive until maturity at time $t + \tau_{care}$. Since each of the surviving young has reproductive value $R_{off}(t + \tau_{care})$ at this time, the value of the brood to a parent is

$$R_{brood}(C,C;t) = \tfrac{1}{2} S_{mf}\, R_{off}(t + \tau_{care}). \tag{7.26}$$

Here the factor $\tfrac{1}{2}$ is the coefficient of relatedness of a parent to its young. Similarly, the expressions for the reproductive success, to a parent, of the current brood under the three other care combinations are

$$R_{\text{brood}}(C, D; t) = \tfrac{1}{2} S_m \, R_{\text{off}}(t + \tau_{\text{care}}) \tag{7.27}$$

$$R_{\text{brood}}(D, C; t) = \tfrac{1}{2} S_f \, R_{\text{off}}(t + \tau_{\text{care}}) \tag{7.28}$$

and

$$R_{\text{brood}}(D, D; t) = 0. \tag{7.29}$$

Let $W_m(t)$ be the expected future reproductive success of a male that is single on day t. Then a male that deserts a brood on day t has payoff

$$H_m(D, C; t) = R_{\text{brood}}(D, C; t) + W_m(t) \tag{7.30}$$

if the female cares, and payoff

$$H_m(D, D; t) = R_{\text{brood}}(D, D; t) + W_m(t) \tag{7.31}$$

if the female also deserts. If the male cares, he does not become single until time $t + \tau_{\text{care}}$. Thus payoffs are

$$H_m(C, C; t) = R_{\text{brood}}(C, C; t) + W_m(t + \tau_{\text{care}}) \tag{7.32}$$

if the female also cares, and

$$H_m(C, D; t) = R_{\text{brood}}(C, D; t) + W_m(t + \tau_{\text{care}}) \tag{7.33}$$

if she deserts. Analogous formulae hold for the payoffs to the female. Given these payoffs, the outcome of the game between two parents at time t can be found from Table 7.3. The values $U_m(t)$ and $U_f(t)$ of the game to the male and female respectively can also be found.

A single male has the option of resting or searching for a mate. If a male searches on day t he obtains a mate by time $t + 1$ with probability $P_m(t)$. Once a mate is obtained, a further time τ_{brood} is taken to produce the brood. He is then involved in a desertion game with his mate. Thus

$$W_m(t) = \max\{W_m(t+1),\ [1 - P_m(t)]W_m(t+1) + P_m(t)U_m(t + 1 + \tau_{\text{brood}})\}, \tag{7.34}$$

where $U_m(t + 1 + \tau_{\text{brood}})$ is the value of the game to the male. An analogous equation holds for females.

We assume, for simplicity, that behaviour in the focal breeding season does not affect breeding prospects in future seasons. We can then take $W_m(t) = R_m$ and $W_f(t) = R_f$ for $t \geq T$, where R_m and R_f are constants representing future reproductive success. Since they are constant and there is no mortality during the breeding season, the solution of the game over the breeding season does not depend on the values of R_m and R_f.

For given functions $P_m(t)$ and $P_f(t)$ we can solve the system of equations (7.26) – (7.34), together with the analogous equations for the female, by working backwards from final time T. As a result, we obtain a strategy that specifies how the behaviour of a bird depends on its sex, status (unpaired or paired), behaviour of its partner (in the case of the female that decides second) and on the time t in the breeding season.

Evolutionarily stable strategies

Suppose that at the beginning of the breeding season all population members are single. Then the numbers of males and females searching for mates on this day and subsequent days will depend on the strategies adopted by all population members. Thus, the remating probabilities $P_m(t)$ and $P_f(t)$ on day t, which depend on the numbers of searching individuals of each sex on that day, depend also on the past behaviour of the whole population. For the results given below we assume that

$$P_m(t) = 0.2\sqrt{\frac{F(t)}{M(t) + F(t)}\left(\frac{F(t)}{N}\right)}$$

and

$$P_f(t) = 0.2\sqrt{\frac{M(t)}{M(t) + F(t)}\left(\frac{M(t)}{N}\right)},$$

where $M(t)$ and $F(t)$ are the numbers of searching males and females respectively at time t and N is the total number of each sex in the population.

As in Section 7.2, we refer to a strategy as the resident population strategy if almost all population members adopt this strategy. Let π be the resident strategy and let the functions P_m and P_f be generated when all population members follow this strategy. If we take these remating probabilities as given, we can work backwards from the end of the season to solve equations (7.26) – (7.34) and the analogous equations for females. In doing so, we find a strategy that is the solution to the game in which P_m and P_f are given. We refer to this strategy as the best response to the resident strategy π, and denote it by $\hat{b}(\pi)$.

A necessary condition that a strategy π^* is an ESS is that π^* is a best response to itself. That is, π^* satisfies $\hat{b}(\pi^*) = \pi^*$. This condition for a Nash equilibrium can be spelled out as follows. Consider functions P_m^* and P_f^* which determine how remating probabilities vary over the breeding season. Then π^* is a best response to itself if

    (i) the resident strategy π^* generates remating probabilities P_m^* and P_f^*
      and

   (ii) given P_m^* and P_f^* the solution of the game with these remating probabilities is π^*.

Figure 7.4(a) illustrates the Nash-equilibrium strategy π^* and the corresponding remating probabilities P_m^* and P_f^*. To understand this ESS strategy, focus on decisions under this strategy on day t. Suppose that two parents have to decide whether to desert their brood at this time. For given payoffs to the male and female, we can understand the decision of the parents by using the analysis summarised in Table 7.3. There is, however, a fundamental difference between this game and most simple games analysed in the literature. The standard approach in much game theory is to assume payoffs at the start and then derive the solution of the game. This approach is not possible here, since payoffs depend on future expectations. These in turn depend both on the remating probabilities, which are a function of the past behaviour of the entire population, and on the future behaviour of the focal birds and their future partners. Payoffs at time t are thus not given in advance. Instead, we solve for a Nash-equilibrium strategy, specifying behaviour over the whole breeding season. In solving for a Nash-equilibrium strategy π^* the payoffs for a parent at time t emerge as the expected future reproductive success of the parent under the assumption that the behaviour of the rest of the population is given by π^* and the behaviour of the focal parent after time t is also given by π^*. In other words, payoffs emerge as solutions to equations such as (7.20) – (7.34) rather than being assumed at the start. Of course, once a Nash-equilibrium strategy π^* and the corresponding payoffs have been found, it is possible to understand the form of π^* by understanding the desertion-or-care decisions at each particular time in terms of the payoffs at that time, and to understand payoffs in terms of decisions taken under π^* over the whole season.

    Figures 7.4(b),(c) give the payoffs under the Nash-equilibrium strategy π^* shown in Figure 7.4(a). We can understand strategy π^* as follows. Near the end of the breeding season $D_f(D) < 0$ and $D_f(C) < 0$ (Figure 7.4(b)). Thus, regardless of her mate's decision, a female will get a higher payoff by caring for the brood than by deserting. This holds because a female that deserts does not have time to remate and raise another brood before the end of the breeding season. For similar reasons $H_m(C, C) > H_m(D, C)$ (Figure 7.4(c)). The male's best action is thus to care and the resultant care pattern is biparental care, (C, C) (case 4B of Table 7.3).

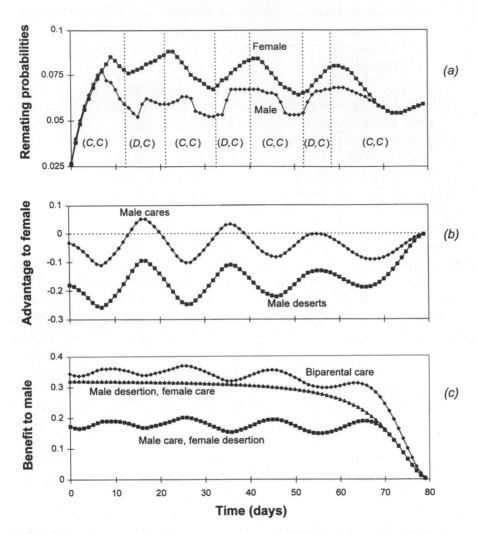

Figure 7.4. The pattern of parental care at the ESS in the model of Székely *et al.* (unpub.) when $S_m = S_f = 0.65 S_{mf}$, $\tau_{brood} = 4$ days and $\tau_{care} = 10$ days (Section 7.8). (*a*) The remating probabilities for males, $P_m^*(t)$, and females $P_f^*(t)$, at the ESS. The ESS is also shown; this strategy alternates between biparental care (C, C) and care by the female alone (D, C) over the breeding season (indicated by the vertical broken line). (*b*) The advantage of desertion over care to the female at the ESS. When the male cares this advantage is $D_f(C)$ and when the male deserts it is $D_f(D)$. (*c*) The payoff to the male for each pattern of care (at the ESS), shown as the advantage over biparental desertion. The three functions shown are thus $H_m(C, D) - H_m(D, D)$, $H_m(C, C) - H_m(D, D)$ and $H_m(D, C) - H_m(D, D)$.

Over the breeding season, the pattern of care under π^* oscillates regularly between care by the female alone, (D, C), and biparental care, (C, C). To understand this phenomenon we first note that there is desertion by males at certain times, but no desertion by females. Thus, there will be more males than females searching for mates and remating probabilities are lower for males than for females (Figure 7.4(a)). Now consider a period in the breeding season when there is biparental care and focus on the decision of a female at a time just before this period. Suppose the male has decided to care. Then if the female deserts she will remate quickly and, since at this future time there is biparental care under π^*, the new mate will help her care for the brood. Thus, if her current mate cares she should desert $(D_f(C) > 0)$. (This is not strictly true at $t = 55$, where $D_f(C)$ is slightly negative – see Figure 7.4(b). Because the computations assume there are errors in decision making, as in Section 7.7, a small negative value of $D_f(C)$ means that there is still substantial desertion by females.) Suppose, alternatively, that the male has deserted. Then the female should care, since otherwise the brood dies $(D_f(D) < 0)$. The male should thus desert since $H_m(D, C) > H_m(C, D)$ (Figure 7.4(c)). This is case 2A of Table 7.3. The outcome is uniparental care by the female, (D, C). This case is analogous to the case given in Table 7.2: for any fixed decision of the female, the male would do better by caring but deserts in order to prevent her from deserting.

Now suppose that there is a period when there is uniparental care by the female and focus on the decision of a female just prior to this period. If the female deserts and finds a new mate then, since at this future time there is uniparental care by the female, her new mate will desert. This reduces the advantage of desertion to a female and so it is optimal to care even if the male also cares $(D_f(C) < 0)$. She will certainly care if the male deserts $(D_f(D) < 0)$. Thus the female will care whatever the male does and the male no longer has to desert in order to prevent the female from deserting. If the male deserts he has a low remating probability; furthermore, if he does remate he will have to desert this new brood in order to prevent the female deserting. It is therefore better to care $(H_m(C, C) > H_m(D, C)$, see Figure 7.4(c)) and the resultant pattern of care is biparental care (C, C). This is case 4B of Table 7.3.

The above arguments show that a period of biparental care is preceded by a period of uniparental care by the female, and a period of uniparental care by the female is preceded by a period of biparental care. The result is an oscillation between these modes of care.

7.9 General issues

Payoffs come from putting a decision within a context

The desertion game that we have just described is useful in highlighting our approach to modelling the payoffs that result from a given action. Figure 7.4 shows an oscillatory pattern of care and desertion over the breeding season. The decisions of the animals at any particular time can be understood in terms of the oscillating costs and benefits shown in the figure. These costs and benefits emerged from solving the desertion game; it is not likely that we would have assumed costs and benefits of this form *a priori*. Our approach is to put decisions into a life-history framework, specifying both an animal's ecology and the consequences of the animal's actions for its future. Costs of particular actions emerge from the specification of consequences, i.e. costs are derived quantities and are not part of the initial specification of the model.

In contrast to the approach based on consequences, some analyses of evolutionary games have started by assuming values for costs and benefits. For example, in the Hawk–Dove game the value V of a contested resource and the cost C of losing a fight are taken as given. To illustrate why such an approach may be misleading, we focus on a model for a contest between two animals for the possession of a territory analysed by Maynard Smith & Parker (1976). One animal is the current owner of the territory and the other is an intruder. Each animal can either play Hawk or Dove. Maynard Smith and Parker showed that the conditional strategy of playing Hawk when owner and Dove when intruder can be an ESS when the value V of winning the contest is less than the cost C of losing. They call this strategy 'bourgeois'. Grafen (1987) points out that V and C depend on the rules for fighting adopted by members of the population under consideration. If members of the population adopt the bourgeois strategy, then owners will tend to keep their territories and intruders will tend not to obtain territories. If a territory is needed in order to reproduce, then this will tend to make V relatively high and C relatively low. Thus, if the bourgeois strategy is widespread then, because V is likely to be greater than C, this strategy is unlikely to be stable. Intruders are unlikely to reproduce unless they obtain a territory and so they will fight rather than accept a conventional settlement. This is an example of what Grafen calls the 'desperado' effect. Mesterton-Gibbons (1992) put contests for territories into an ecological context by specifying parameters such as the ratio of animals to territories, and the probability per searching period that an animal finds a territory. He establishes regions in parameter space for which various strategies, such as bourgeois, are evolutionarily stable. Any particular ESS can be understood in terms of the values of V and C that hold at the ESS, but these values are derived rather than fundamental.

The importance of the future can also be seen in an analysis of repeated interactions between animals presented by Houston & McNamara (1991). Male animals are assumed to be competing with other males for access to females. In the model with a potentially unlimited number of contests, a proportion $1 - \theta$ of females encountered by the male are not contested by another male. The focal male mates with these females, increasing his reproductive success by V for each female mated. A proportion θ of females encountered are contested by other males. In an encounter each male can play Hawk or Dove. If one male plays Hawk and the other plays Dove, then the former wins and mates, increasing his reproductive value by V. If both play Hawk then a given animal wins with probability $1/2$ (gaining V) and loses with probability $1/2$, in which case it dies with probability z. If both play Dove then each has an expected gain of $V/2$. There is also a probability $1 - s$ of dying between encounters that is independent of fighting behaviour. Males continue to compete for females until they die. A male's strategy, π, is the probability that he will play Hawk as opposed to Dove. For some parameter values there can be three Nash equilibria for this model. An example is given in Figure 7.5. When the resident population strategy π is zero, the best response is $\hat{b}(\pi) = 1$, i.e. in a population of Doves the best response is to play Hawk. As π increases, the best response drops from unity to zero, giving an ESS π_1^*. So far, the best-response function resembles that of the standard Hawk–Dove game but, as the resident strategy continues to increase, a striking difference emerges. In the model based on repeated interactions, the best-response function rises from zero to unity. This rise is an example of the desperado effect. In a population in which an opponent is likely to play Hawk, if the probability θ that an encountered female will be contested is high and the probability of survival s is low then a male cannot afford to play Dove. As a result, $\pi_3^* = 1$ is an ESS.

The fact that for some parameter values there can be two possible ESSs is in marked contrast to results obtained in the standard Hawk–Dove game. The analysis of the standard game is based on the benefit V of winning a contest and the cost C of losing a contest. For given V and C there is a unique ESS. In our model we have not assumed a value for C. Instead we have found an animal's reproductive value in terms of its behaviour (as represented by π'), the behaviour adopted by other members of the population (as represented by π) and the parameters z, s and θ. Given π', π, z, s and θ, we can calculate the loss in reproductive value associated with fighting and hence calculate C. For values of z, s and θ that result in two possible ESSs, π_1^* and π_3^*, the associated costs C_1 and C_3 will be different. At each ESS, behaviour is consistent with the usual condition that $\pi^* = \min(V/C, 1)$: $C_1 > V$ and $\pi_1^* = V/C_1 < 1$, whereas $C_3 < V$ and $\pi_3^* = 1$.

For more on the desperado effect see Eshel & Sansone (1995).

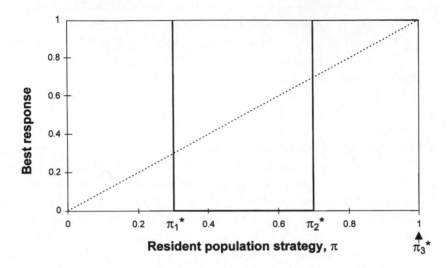

Figure 7.5. The best-response function in the repeated Hawk–Dove game of Houston & McNamara (1991); see Section 7.9. There are three Nash equilibria, π_1^*, π_2^* and π_3^*; of these only π_1^* and π_3^* are ESSs. The parameters are $z = 0.5$, $s = 0.95$, $\theta = 1$.

The importance of state

An organism following an evolutionarily stable strategy may make randomised decisions. Randomisation is often favoured because it is important not to give an opponent information about what one is going to do, since this information could be exploited by the opponent. For example, in the 'war of attrition' two animals contest a resource and the winner is the animal that is prepared to wait the longest. The ESS involves waiting for an exponentially distributed time (e.g. Maynard Smith 1982). The exponential distribution has the property that the probability of giving up in the next instant, given that one has not given up so far, does not depend on how long the game has been going on. This means that if an animal adopts the

ESS, then an opponent gains no additional information from the duration of the contest about whether the animal will give up. In contrast, if the resident population strategy was 'wait for a fixed time', a mutant strategy that waited for slightly longer would invade. Animals may, in certain circumstances, determine outcomes by a random process – e.g. certain forms of sex determination. In other cases, it may be possible to achieve essentially random decisions by means of a deterministic relationship between behaviour and a suitable state variable. Indeed, it has been argued that any mixed strategy can be viewed as a limiting case of a pure contingent strategy based on a state variable that has a negligible effect (Harsanyi 1973, Maynard Smith 1982). Thus, from a mathematical point of view, it makes no difference whether an animal is behaving randomly or following a pure contingent strategy based on an irrelevant state variable. But from the biological point of view, there will usually be state variables that are relevant in that they influence the consequences, and hence the payoffs, of adopting an action. Natural selection will favour organisms that base their decisions on such variables. Thus it is reasonable to expect organisms to adopt pure contingent strategies rather than randomised strategies.

We can illustrate the importance of a state variable in the context of the state-dependent Hawk–Dove game (Houston & McNamara 1988). In this game, the state variable is energy reserves. As we have already said, the ESS involves a critical level of reserves above which the animal plays Dove and below which it plays Hawk. Thus the ESS is a pure contingent strategy. The underlying state variable is, however, not irrelevant. When an animal has low reserves, the value of a contested food item is high and the animal should be prepared to fight for it. Playing Hawk is a strictly better option than playing Dove. Conversely, when reserves are high, avoiding death through injury in a fight is more important than getting food, and playing Dove is strictly better than playing Hawk. An individual following the ESS will spend some proportion q of its time playing Hawk and a corresponding proportion $1 - q$ playing Dove. It is important to note that an animal playing Hawk in each contest with probability q and Dove with probability $1 - q$ will have a much lower fitness than an animal following the ESS. What matters is not the overall frequency of performing an action, but the states in which the action is performed (see also Section 3.6).

The standard way of finding mixed ESSs is based on the fact that all pure strategies that are adopted under the ESS must have equal payoffs (Bishop & Cannings 1978; see also Section 7.3). If what we assume to be a mixed strategy is really a pure contingent strategy, then this equalisation need not hold, and the assumption that it does hold may lead to erroneous conclusions. For example, in the state-dependent Hawk–Dove game there is no equalisation, in that individuals that play Dove have higher reserves and hence higher reproductive value than those that play Hawk. Similar remarks apply to the state-dependent ideal free distribution model of McNamara & Houston (1990c); see also Section 6.7. In this model, members of a population choose between two patches that differ in terms of the distribution of food and the level of predation. When one patch is safe from predators and the other is not, then at the ESS animals with high reserves are on the safe patch and animals with low reserves are on the risky patch. In contrast to the standard analysis of ideal free distributions, in which there is no state variable (see Section 6.7), nothing is equalised across patches: animals in the dangerous patch have a higher rate of food intake, a higher predation risk and a lower reproductive value than animals on the safe patch. This lack of equalisation can also be found in the caller–satellite models of Lucas & Howard (1995) and Lucas *et al.* (1996). In these models a male can obtain a mate either by calling, or by remaining silent and intercepting females on their way to a calling male. A male decides between these options on the basis of its energy reserves. Calling males have higher reserves and achieve greater reproductive success than satellites. In these models there is the possibility of repeated switches between rôles as reserves fluctuate.

There are alternative patterns of male mating behaviour in which such switches do not occur; a male is committed to a given rôle in all its mating attempts. In such systems, the choice of rôle could be based on state variables such as body size, condition or competitive ability (e.g. Parker 1982, 1984, Gross 1996). In such cases the reproductive success of males need not be equalised across rôles. It may not, however, be easy to determine whether choice is state dependent. Male coho salmon (*Oncorhynchus kisutch*) are either large and have hooked jaws ("Hooks") or are small and cryptic ("Jacks"). The Hooks fight with each other for access to females, whereas the Jacks try to obtain matings by stealth. Gross (1985) found that each type of male had approximately equal lifetime success, which led him to suggest that Hooks and Jacks constitute a mixed ESS. Evidence now indicates that the decision about whether to be a Hook or a Jack is based on differences in growth rate early in life, with faster growing males becoming Jacks (Gross 1996). The modelling approaches of both Gross (1996) and Charnov (1993) assume that there is an environmental component to growth and that male salmon adopt a pure contingent strategy. The early growth advantage of Jacks gives them a higher lifetime reproductive success than Hooks (but this advantage may be small – see Charnov 1993).

In the above examples, an explanation that did not take state into account would have to be based on mixed strategies and would be erroneous. There are also some phenomena for which no explanation is possible unless a state variable is introduced. An example is provided by biparental desertion of young. Suppose that the young die when both parents desert. Thus if they desert the parents get nothing out of the breeding attempt and we might ask why they bred in the first place. The answer has to be that something has changed between the decision to breed and the decision to desert. Some changes might be environmental, e.g. deterioration in the food supply or disturbance by predators. There are also less obvious possibilities, based on the animals' obtaining information about each other. If members of a given sex differ in quality and this difference cannot be fully assessed when the decision to breed is taken, then it can be optimal to initiate breeding but subsequently to desert, if the mate turns out to be of poor quality (Webb *et al.*, in press).

The importance of the decision process

The original model of desertion proposed by Maynard Smith (1977) assumes that each parent makes a decision without knowledge of its partner's decision. We refer to this as a 'sealed envelope' decision procedure, because the players can be thought of as writing down their decision and placing it in a sealed envelope. This envelope is opened at the decision time, and commits each player to its choice of action, regardless of the behaviour of the other. Many other biological decisions (e.g. parental effort, aggression and vigilance) have been modelled in this way. This may be reasonable when the total duration of the interactions is essentially instantaneous. It is not reasonable when the outcome is the result of a decision process that extends over time. We showed in Section 7.8 that, when the desertion game is modelled as involving sequential rather than simultaneous choice, the ESS may be very different. We also argued that desertion may be the result of repeated interactions between a pair of animals in which information is exchanged. Similarly, the standard Hawk–Dove game is a sealed-envelope model of aggression. A more realistic model would involve a sequence of decisions concerning the level of aggression to adopt. Each decision in this sequence may have a sealed envelope component in that an individual will only have partial knowledge of what the opponent is doing, but it will not be a completely sealed envelope in that decisions are likely to be based on the opponent's previous behaviour.

To expose the limitations of the sealed envelope approach, we consider a game between two parents that are caring for dependent young (the 'parental effort' game). In the model of Houston & Davies (1985) each parent decides upon its level of parental effort u. If the male expends effort u_m and the female expends effort u_f then the immediate contribution to the reproductive success of each parent is $B(u_m+u_f)$. This function of total effort is increasing but decelerating (i.e. $B' > 0$, $B'' < 0$). If the male expends effort u_m its future reproductive success is $V_m(u_m)$ where $V' < 0$, $V'' < 0$. Thus, given that the female expends effort u_f the payoff to the male is

$$B(u_m + u_f) + V_m(u_m). \tag{7.35}$$

Similarly, given that the male expends effort u_m the payoff to the female is $B(u_m + u_f) + V_f(u_f)$ if she expends effort u_f.

Given that the female expends effort u_f the best effort for the male is the value of u_m that maximises expression (7.35). We denote this best effort by $\hat{r}_m(u_f)$. Similarly, the best effort for the female given that the male expends effort u_m is denoted by $\hat{r}_f(u_m)$. At the ESS found by Houston & Davies (1985) the effort of each parent is the best given that of its partner. Thus at the ESS the efforts of the male and female are u_m^* and u_f^* respectively, where

$$u_f^* = \hat{r}_f(u_m^*) \qquad \text{and} \qquad u_m^* = \hat{r}_m(u_f^*). \tag{7.36}$$

Houston and Davies gave a graphical method by which the efforts u_m^* and u_f^* can be found. This method is shown in Figure 7.6.

In the above model, when the population is at the ESS all males expend effort u_m^* and all females expend effort u_f^*. Thus the effort of an individual's partner is known in advance, and to model decisions as sealed envelope is reasonable since after 'the envelopes are opened' there is no reason for a parent to change its mind. All of this alters if individuals can differ in quality. If quality differences mean that individuals pay different costs for the same level of effort, individuals will differ in their optimal effort. If we assume that a parent does not know the quality of its partner *a priori*, it is worthwhile for a parent to adjust its effort once the level of effort of the partner is observed. In this way a parent can, for example, increase its own effort to partially compensate if its partner seems to be of poor quality. Thus if there is quality variation, we might expect both parents to adjust their efforts during the breeding season until efforts settle down to 'negotiated' levels. These negotiations, and the final levels, will depend on the rules that each parent uses to respond to the effort of its partner. Thus, when parents are not sure of the quality of their partners, we must seek evolutionarily stable response rules rather than just evolutionarily stable fixed levels of effort.

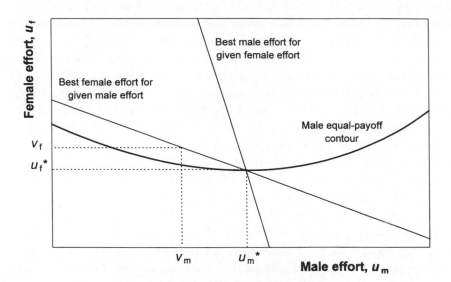

Figure 7.6. The parental effort game of Houston & Davies (1985). The best male effort, $\hat{r}_m(u_f)$, given female effort u_f and the best female effort, $\hat{r}_f(u_m)$, given male effort u_m are shown. The male and female efforts under the ESS found by Houston and Davies are u_m^* and u_f^* respectively (equation (7.36)). The solid curve shows the locus of combinations (u_m, u_f) of male and female efforts that give the male the same payoff as the ESS combination (u_m^*, u_f^*). Points above this curve give the male a higher payoff than points on the curve. Suppose the female responds to male effort u_m by changing her effort to $\hat{r}_f(u_m)$. Then it can be seen that if the male has used a fixed effort v_m then the female will adjust her effort to v_f and the male will achieve a greater payoff than if he had used fixed effort u_m^* (or had used response rule $\hat{r}_m$).

Consider first the response rule under which an individual always adjusts its effort so that it is the best effort for that individual given the effort expended by its partner. This may seem a sensible rule, but a resident population playing this strategy is invadable (McNamara *et al.*, unpub). This can be most easily seen when there are no quality differences between members of the same sex. In this special case, suppose that all population members adopt the rule. Then a male will respond to effort u_f of his partner by expending effort $\hat{r}_m(u_f)$, and a female will respond to effort u_m of her partner by expending effort $\hat{r}_f(u_m)$. As a result of repeated responses of parents to each other's effort, the level of efforts of the male and female will settle down to u_m^* and u_f^* respectively (cf. equation (7.36) and Figure 7.6).

Now consider a single mutant male within this population. Suppose the strategy of this male is not to respond to his partner's effort, but to play a fixed effort v_m, where $v_m < u_m^*$. Then the partner of this male will adjust her effort to $v_f = \hat{r}_f(v_m)$. As Figure 7.6 shows, provided that $u_m^* - v_m$ is not too large the male will then obtain a bigger payoff than had he adopted the resident male strategy. Thus, the population is invadable. This shows that in the parental effort game of Houston & Davies (1985), the functions $\hat{r}_m$ and $\hat{r}_f$ cannot be interpreted as evolutionarily stable response rules over the breeding season. Although the above argument has ignored quality variation amongst members of the same sex, the general conclusions still hold when quality variation is introduced. If an individual responds to its partner's effort by adjusting its own effort to the best effort given the effort of its partner, then the individual can be exploited by a lazy partner.

The approach to modelling evolutionarily stable response rules taken by McNamara *et al.* (unpub.) involves making two simplifying assumptions. First, they assume that the negotiation phase is short, so that the phase is cost-free and payoffs depend only on the efforts u_m and u_f that are the final outcomes of the negotiations. Second, they assume that a given response strategy by the male determines a function r_m such that the final effort of the male after negotiation is a function $u_m = r_m(q_m, u_f)$ of the male's quality q_m and the female's final effort u_f. Similarly, a given response strategy of the female determines a function r_f such that the female's negotiated final effort is $u_f = r_f(q_f, u_m)$. Under these assumptions, details of the negotiation phase are unimportant; all that matters are the response functions r_m and r_f that describe the final outcome. An ESS is then determined by a pair of response functions r_m^* and r_f^*, where the male response function is the best given that females have response function r_f^*, and vice versa. McNamara *et al.* found r_m^* and r_f^* for the case when the functions B, V_m and V_f take an especially simple form. They showed that

  (i) individuals should compensate for the low effort of their partner by increasing their own effort, but the rate at which their effort increases as their partner's effort decreases is less than if they were adopting the best effort given the effort of their partner.
 (ii) the final negotiated efforts are less than if each partner were behaving optimally given the fixed effort of their partner (i.e. less than that predicted by the model of Houston & Davies 1985).

These results hold provided that there is some quality variation amongst individuals. Furthermore, even in the limit of vanishingly small quality variation, the results are both qualitatively and quantitatively different from the case where there is no quality variation (i.e. the case considered by Houston & Davies 1985).

7.10 Future directions

Although there are various other aspects of biological game theory that deserve attention (e.g. n-person games and games with explicit spatial structure), we feel that the most fundamental issues are modelling the interactions between a pair of animals and understanding complex social interactions.

Although foraging behaviour may involve frequency dependence, it is possible to construct useful foraging models that do not adopt a frequency-dependent approach. In contrast, social interactions are fundamentally frequency dependent. The evolution of many aspects of social systems are not well understood.

In the previous section we argued that in the context of parental decisions, it is unrealistic to use a model based on a single 'sealed envelope' decision. We expect this conclusion to be valid in many contexts. Work on the evolution of co-operation has followed Axelrod & Hamilton (1981) in assuming repeated interactions, but typically of a simple form. In general we believe that in modelling interactions between two individuals, it is important to allow for differences in individuals. For example, in the parental effort game presented in Section 7.9, the existence of quality differences between individuals was crucial. Individuals may differ in relatively permanent attributes such as the ability to perform certain activities, or in relatively temporary states such as hunger. Leimar (1997a) emphasised the importance of individual differences in need and ability in the context of co-operation. Leimar (1997a, b) also argued that two other features are important. One is the fact that animals may make errors in their decisions. The other is that motivational states such as guilt and aggression are part of the behavioural mechanism for dealing effectively with social interactions, and that such states should be explicitly incorporated in models. 'This kind of representation provides a possible link between a game theoretical analysis and concepts from mechanistically oriented ethology, ...' (Leimar 1997b, p. 471.) We agree with Leimar's view, and feel that his state-based approach gives us a chance to understand some otherwise puzzling phenomena.

Within Leimar's framework, an individual is characterised by its states (some of which are mental states) and the transitions between them. A key feature of this approach is that there is a relatively small number of states and an organism's current state does not completely specify all relevant information from the past. This means that the organism has a certain lack of flexibility in its behaviour. In other words, the animal can be represented as an automaton or machine. (This approach has been used to investigate repeated games within economic game theory; see e.g. Osborne & Rubinstein 1994 and references therein.) This approach provides a possible explanation of the significance of bond formation between a pair of animals. If behaviour is infinitely flexible, in the sense that the choice of behaviour in the future is not constrained by behaviour in the past, then it is hard to see why repeated interactions between animals should result in the development of trust. If we regard a population of animals as a population of different automata, where a given automaton corresponds in some sense to a personality, then repeated interactions will give an animal some idea of the personality of its partner.

8

State-dependent life-history theory

8.1 Why base life-history theory on state?

In some species, individuals do not reproduce until they have grown to a large size (e.g. some trees); in contrast, other species mature rapidly (e.g. many insects and small mammals). Pacific salmon reproduce once and then die, whereas most mammals can have several episodes of reproduction over their lifetimes. Mice produce many small offspring at a reproductive bout, whereas wildebeest typically produce a single well-developed offspring. Life-history theory attempts to account for this diversity, and to explain the scheduling of reproductive behaviour over the life of any given individual.

Early models assumed that fecundity and survival are functions of an organism's age alone. These models have been valuable in establishing certain broad qualitative features of an optimal life history. For example, they have been used to examine whether an organism should breed once (semelparity) or many times (itcroparity) over its lifetime (Charnov & Schaffer 1973, Charlesworth 1994). Other applications include the effect of a decline in fecundity with age, or an increase in mortality with age, on the scheduling of reproductive effort over an organism's lifetime (e.g. Gadgil & Bossert 1970, Charlesworth & León 1976, Taylor 1991).

In assuming that fecundity and survival are functions of an organism's age alone, age is taken to be the only relevant state variable. Changes in fecundity and survival are presumably the result of changes in, among other things, an organism's physiology. If changes in physiology are tightly correlated with changes in age, then it is reasonable to take age as a state variable rather than explicitly describing the underlying physiological variables. The age-based approach, however, amounts to assuming that all individuals of the same age are equivalent. This is not a reasonable assumption for most organisms. There is abundant evidence that organisms of the same age differ in aspects of state that have a strong influence on their survival and reproductive ability (see McNamara & Houston 1996 for a review). Although some physiological variables may be tightly correlated with age, a realistic model may require state variables other than, or in addition to, age. For instance, considerable attention has been given to an organism's size as a relevant state variable. Models based on size have been used to investigate optimal patterns of growth and the size at which an animal starts to reproduce. For examples of size-based models see Kirkpatrick (1984), Kozlowski (1992), Perrin & Sibly (1993). Another relevant aspect of state might be quality or condition. Even when it is not clear what constitutes quality or condition there may be effects that suggest the importance of some such underlying state variable. The work of Gustafsson and his colleagues on the collared flycatcher (Gustafsson & Sutherland 1988, Gustafsson *et al.* 1994) is highly suggestive in this respect. They observed that if the number of young cared for by the mother was increased by manipulation, there was a decrease in both the maternal clutch size in future years and the subsequent clutch size of daughters surviving from the manipulated clutch. A possible interpretation of these findings is as follows: (i) an increase in clutch size increased maternal effort and hence decreased some aspect of maternal condition; (ii) this decrease in condition persisted into future breeding seasons and resulted in decreased clutch size in these seasons; and (iii) although total maternal effort increased, the effort per offspring decreased, which led to decreased condition of the daughters and hence to their lower clutch sizes at maturity.

In this chapter we describe a general framework for investigating optimal life histories. The framework is applicable to the special case of age-based models, but it can also handle other state variables. Many attempts to model life histories have assumed that all newborn individuals are equivalent. This is certainly true of the age-based approach, which assumes that all newborns have the same reproductive value. Our framework allows newborn individuals to differ in their reproductive value. As a result it can handle two important intergenerational effects. One is the ability of a parent to increase offspring quality by reducing their number (the number versus quality trade-off). The other is phenotypic correlations between mother and offspring (maternal effects).

8.2 What is fitness?

We begin by looking at the definition of fitness in the simplest possible case and then discuss how the definition needs to be modified to deal with more general and realistic situations. In particular, in Section 8.3 we discuss how to deal with state-structured populations.

We consider a population of a particular species. An annual census is made of this population. In each year the census time is just prior to the breeding season. We refer to the census times as $t = 0, 1, 2, \ldots$, where successive times are one year apart. In our basic model we assume that conditions (i) – (v) given below apply.

(i) The species is asexual.

(ii) There are non-overlapping generations with a generation time of one year. Thus an individual born just after time t reaches maturity by time $t + 1$, reproduces just after this time and dies before time $t + 2$.

(iii) At the census times all individuals of a given genotype are in the same state.

(iv) The physical environment does not fluctuate between years. Under this assumption there may be seasonal effects, but it is assumed that they are the same each year and that, say, the winters are equally harsh each year.

(v) There are no effects of density or frequency. In other words, the survival and reproduction of a population member is unaffected by the number and genotype of other population members.

Consider a cohort of individuals of the same genotype in this population. Each mature individual of the genotype present at time t reproduces immediately after t (and dies), and exactly λ of its offspring survive to maturity at time $t + 1$. Let $n(0)$ individuals be present at time 0; then, under these assumptions,

$$n(t) = n(0)\lambda^t \qquad (8.1)$$

individuals of the genotype are present at time t.

Now suppose the population is composed of a mixture of two genotypes, A and B, at time 0. Let individuals of these genotypes produce λ_A and λ_B surviving offspring respectively. Then the numbers of the genotypes at time t are $n_A(t) = n_A(0)\lambda_A^t$ and $n_B(t) = n_B(0)\lambda_B^t$. Thus

$$\frac{n_A(t)}{n_B(t)} = \frac{n_A(0)}{n_B(0)} \left(\frac{\lambda_A}{\lambda_B}\right)^t. \qquad (8.2)$$

It follows that if $\lambda_A > \lambda_B$, $n_A(t)/n_B(t) \to \infty$ and the population is eventually dominated by genotype A, while if $\lambda_A < \lambda_B$, $n_A(t)/n_B(t) \to 0$ and the population is eventually dominated by genotype B. In this way the population comes to be dominated by the genotype with the highest value of λ. Even small differences in λ will be significant given sufficient time. For example, if λ_A is 1% bigger than λ_B and genotypes A and B are initially present in equal numbers, then genotype A will comprise 73% of the population after 100 generations and 99.3% of the population after 500 generations.

The argument extends to a population initially composed of a mixture of many different genotypes. The genotype with the highest λ will eventually dominate the population. Thus if we take the number of offspring surviving to maturity, λ, as a measure of the fitness of a genotype, natural selection leads to a population in which fitness is maximised.

Here we have been concerned with population size at a sequence of discrete times one year apart. If instead we had been dealing with a continuum of times it would have been more natural to describe fitness by r where $e^r = \lambda$; r is called the intrinsic rate of natural increase. In terms of r, equation (8.1) becomes $n(t) = n(0)e^{rt}$. Maximisation of λ is equivalent to maximisation of r.

Clearly no population satisfies all the assumptions (i) – (v) and we now discuss the consequences of relaxing these assumptions.

Sexual reproduction.   In considering an asexual population, we are concerned with the growth in numbers of a genotype. When we consider a sexual population, the analogous concept is the growth in the numbers of an allele. Suppose there is no demographic difference between the sexes, in that survival and fecundity are the same for males and females (this would require a 1 : 1 sex ratio of offspring). Suppose also that, as assumed in condition (iii), all individuals of a given genotype are in the same state at the time of census. Then descendant numbers can simply be discounted by the relatedness to the focal individual. Fitness is the expected discounted number of descendants left in one year's time. If there are differences between the sexes then condition (iii) does not hold and the population is structured. We deal with structured populations in Section 8.3.

Overlapping generations.   In our basic model an individual that is mature at time t reproduces and immediately dies; it is only the surviving offspring that contribute to the population at time $t + 1$. How should fitness be modified if there is the possibility of parental survival until $t + 1$?

Suppose first that a parent that survives until $t+1$ is no different from any offspring surviving until this time. In other words, parent and offspring have equal future expectations of survival and fecundity. Then if the population is asexual the parent can just be treated as one of the offspring and the fitness measure is

$$\lambda \;=\; \text{expected number of surviving offspring}$$
$$+ \text{ probability that parent survives.}$$

If the population is diploid and reproduces sexually, and there are no demographic differences between males and females, then

$$\lambda \;=\; \tfrac{1}{2} \times \text{expected number of surviving offspring}$$
$$+ \text{ probability that parent survives.}$$

If an individual present at time t survives until time $t + 1$ we can regard the individual at time $t + 1$ as one of its own descendants. With this interpretation, both the above formulae express λ as the expected number of descendants left in one year's time, where descendant numbers are discounted by relatedness.

Both of the above expressions for λ assume that parent and offspring are indistinguishable at a census time. If this is not true, condition (iii) does not hold, and again the population is structured.

Density and frequency dependence. Assumption (v) will never hold. There cannot be unlimited population growth because an increase in population numbers will eventually lead to a shortage of key resources. In addition to density effects there may also be frequency effects: the resources available to an individual may depend on the types of other population members as well as their number.

To see how these effects modify the definition of fitness, consider a population which may be sexually reproducing and have overlapping generations but which still satisfies assumptions (iii) and (iv). Assume that population numbers are constant because of density and frequency effects; we refer to such a population as being at a dynamic equilibrium. Introduce a subpopulation of mutants into this population. The 'environment' experienced by these mutants is the result both of physical conditions and of the number and type of the resident population. As long as the number of mutants remains small compared to the size of the whole population and related individuals do not interact, we can take the environment experienced by the mutants as constant (there may still be seasonal variation). We can then define λ as the number of descendants that a mutant is expected to leave in one year's time. λ then equals the rate of growth of the mutant cohort, and mutant numbers will increase (i.e. the mutant can invade) if $\lambda > 1$.

If the endpoint of the evolutionary process is a population that is both dynamically and evolutionarily stable (and hence is an evolutionarily stable strategy (ESS), see Chapter 7), then the resident population has a growth rate of $\lambda^* = 1$ and any mutant has a growth rate into the resident population of $\lambda \leq 1$. Thus evolution leads to the maximisation of λ, but λ is not the growth rate under unlimited food as originally envisaged by Malthus, but the rate of invasion of a rare mutant into a resident population.

These results are equivalent to results on the lifetime production of off-spring. Suppose that all offspring are in the same state. Then in a population that is both evolutionarily and dynamically stable it can be shown that the lifetime number of offspring produced under the resident strategy is greater than under any mutant strategy. For a fuller discussion of this point see Mylius & Diekmann (1995) and the references therein. When not all offspring are in the same state a similar result holds but now the expected total reproductive value of offspring produced over a lifetime is maximised (see McNamara 1993a and Section 8.6.)

Demographic stochasticity.   Consider a cohort of genetically identical population members. Even if all individuals are in the same state at time t, the numbers of mature offspring left at time $t + 1$ by different individuals will differ. In particular, not all individuals will leave the mean number, λ, of descendants. Differences in descendant numbers occur because of differences in luck in obtaining food, avoiding predators etc. experienced by different individuals. This variation in survival and fecundity is referred to as demographic stochasticity.

Is λ still a good measure of fitness when there is variation about this mean? Certainly $\lambda \geq 1$ is necessary for a single mutant genotype to invade a resident population. But the probability that a line starting from a single individual can establish itself also depends, for example, on the variance in offspring number (Ewens 1969). Thus if mutation to a genotype is very rare then because of stochastic effects this genotype may not become established even if it has a higher value of λ than the resident population; and the probability of establishment will depend on λ and other parameters.

If, however, mutation to a genotype with $\lambda > 1$ is not rare then the genotype will eventually become established in the population. Thus, if a population is evolutionarily stable against common mutants we must have $\lambda^* = 1$ for the resident population and $\lambda \leq 1$ for all common mutants, as before.

For a genotype that becomes established, growth in genotype numbers can be taken to be essentially deterministic and given by $n(t + 1) = \lambda n(t)$, provided numbers are large enough to average over demographic stochasticity but are small compared with the population size.

Environmental fluctuations. The environment may fluctuate from year to year. This might arise because of the weather; for example, in some years winter temperature may be lower than others. But not all sources of fluctuation are caused by external factors. If density effects lead to unstable population dynamics with oscillations or chaos, then the resultant changes in population level between years produce changes in the 'environment' experienced by population members. When there are environmental fluctuations affecting essentially all population members, the simple fitness measure λ that we have introduced is not adequate. This topic is discussed in Chapter 10.

So far we have always assumed that condition (iii) holds. If there are differences between the sexes or if a surviving parent cannot be regarded as equivalent to its offspring at time $t + 1$ or if, for any other reason, not all descendants are in the same state at the census times, the definition of fitness becomes more complicated, as we now discuss.

8.3 Fitness for a state-structured population

From now on we consider a population in which organisms can be in a range of possible states at the annual census times. The species could be sexual or asexual, the generations may overlap and there may be density-dependent or frequency-dependent effects. Thus of the original assumptions (i) to (v), we have kept only assumption (iv). In this section we define the fitness of a rare mutant. In doing so we take the environment, which includes the number of animals in the resident population and their behaviour, as given.

In defining fitness for a mutant genotype in an asexual population we are concerned with the rate of increase of genotype numbers. Similarly, in a sexual population we are concerned with the rate of increase of allele numbers. Thus, as in the previous section, we can consider a subpopulation comprising individuals of the genotype (or possessing the focal allele) at time $t = 0$. At subsequent times we are interested in individuals in the population that are either initial population members or their offspring or their offspring's offspring and so on. We refer to this group of individuals as the cohort of descendants. We analyse growth in cohort numbers (suitably discounted to take relatedness into account) at subsequent times. Unlike the previous section, however, we now suppose that individuals of the same genotype can differ in their state, i.e. the population is state structured or, in the terminology of Caswell (1989), stage structured. The growth in cohort numbers in a one-year period may then be strongly dependent on the distribution of states over cohort members at the start of the year. Thus, to analyse growth we need to keep track of both cohort numbers and the composition (distribution of states) of the cohort. To do this we first define the population projection matrix.

As before, the population is censused annually at times $t = 0, 1, 2, \ldots$ one year apart. At a census time individuals can be classified as having one of a finite number of possible states. Here 'state' includes any measures that are important in influencing the future survival or fecundity of the organism. The state variable might be, for example, age, size, sex, parasite load, fat level or a combination of these or other physiological variables. External variables such as territory or mate quality might also be components of an organism's state.

Consider an individual in our focal cohort in state x at time t. The number and the states of descendants of this individual produced between t and $t+1$ that are alive at time $t + 1$ can be described by four quantities.

(i) *Survival of the focal organism until $t + 1$.* If the organism itself survives until time $t + 1$ it is included as its own 'descendant'. The probability of survival depends on the state at time t and is denoted by $S_{\mathrm{mat}}(x)$.

(ii) *State of focal organism at $t+1$.* If the focal organism survives until time $t + 1$ the probability that it is in state y at this time is denoted by p_{xy}.

(iii) *Number of offspring surviving until $t + 1$.* If the species is asexual and there is a single reproductive bout each year, then $N_{\mathrm{off}}(x)$ is the expected number of offspring produced by the focal organism between times t and $t+1$ that are alive at time $t+1$. If there is more than one generation between t and $t + 1$, $N_{\mathrm{off}}(x)$ can include the offspring of offspring etc. If instead the species reproduces sexually, $N_{\mathrm{off}}(x)$ needs to be discounted to take into account the relationship of offspring (or offspring of offspring etc.) to the focal organism.

(iv) *Offspring state at time $t + 1$.* We denote the probability that an offspring which survives until $t + 1$ is in state y at this time by b_{xy}.

Then the organism in state x at time t leaves a total of a_{xy} expected descendants in state y at time $t + 1$, where

$$a_{xy} = S_{\mathrm{mat}}(x)p_{xy} + N_{\mathrm{off}}(x)b_{xy} \tag{8.3}$$

The matrix $A = (a_{xy})$ of these numbers is called the population projection matrix. (Note that the matrix defined in this way is the transpose of the corresponding matrix defined in Caswell 1989.)

To motivate these abstract definitions we consider two specific examples.

Example 8.1. In this example the population is sexually reproducing, but there are no demographic differences between the sexes and the sex ratio of offspring is 1 : 1. All cohort members of the same age are in the same state. Thus the cohort population can be structured by age alone. Individuals born between times t and $t + 1$ that survive until time $t + 1$ are counted as age 1 at this latter time. A female that is age 1 at time t survives until time $t + 1$ with probability 0.5; if it survives it is counted as age 2 at this time. This female does not reproduce between time t and $t + 1$. A female that is age 2 at time t survives until time $t + 1$ with probability 0.5. On average one of the daughters produced by this female between t and $t + 1$ survives until time $t + 1$. A female that is age 3 at time t is dead by time $t + 1$. On average four of the daughters produced by this female between times t and $t + 1$ survive until time $t + 1$.

In this example there are three states, 1, 2 and 3, corresponding to the three age classes. Table 8.1(a) gives the projection matrix. (Again, note that this is the transpose of the usual Leslie matrix).

Table 8.1. Population projection matrices.

(a) Matrix for Example 8.1

Age of focal animal	Age of descendants one year later		
	1	2	3
1	0	0.5	0
2	1	0	0.5
3	4	0	0

(b) Matrix for Example 8.2

Quality of focal animal	Quality of descendants one year later	
	low	high
low	0.5045	0.2554
high	0.4851	0.7762

Example 8.2.   In an asexual population there are non-overlapping generations with a generation time of one year, so that all individuals have the same age, which we refer to as age 1, at all census times. Individuals do, however, differ in quality at the census times. For simplicity we assume they can be classified as either 'low-quality' or 'high-quality'. These terms describe the ability of an individual to provision its young. The amount of provisioning determines both the probability that the young survive until maturity and their quality if they survive. For example, if there were no parental care, quality might be equated with the resources available for egg production. The greater the provision per egg, the higher the young's probability of survival and expected quality at maturity. Alternatively, if an organism is caring for its young, quality might be equated with the ability to feed the young, and hence equated with territory quality or foraging ability.

Here we assume that low-quality individuals produce four offspring and high-quality individuals produce six offspring. Later in Example 8.4, we explore the consequences of allowing organisms to choose the number of offspring produced when there is a trade-off between offspring number and the amount of provisioning per offspring. Although the details of the various functions given in Example 8.4 do not concern us at the moment, they mean that each of the four offspring of a low-quality parent survives to maturity with probability 0.19 and that if an offspring survives it is low quality with probability 0.6639 and high quality with probability 0.3361. Although high-quality individuals produce more offspring they are also able to provide more resources per offspring. Each of these offspring survives with probability 0.2102 and is high quality with probability 0.6154 if it survives. In this example there are two states, 'low' and 'high'. Table 8.1(*b*) gives the projection matrix. To illustrate the derivation of its elements consider $a_{L,H}$. This is the expected number of high-quality surviving offspring left by a low-quality parent and is hence $4 \times 0.19 \times 0.3361 = 0.2554$.

Asymptotic growth rates

Given initial numbers in the cohort and the initial states of cohort members, the projection matrix can be used to look at numbers in the various states at later times. Let $n_x(t)$ be the number of cohort members in state x at census time t. Then for every state y at time $t + 1$

$$n_y(t + 1) = \sum_x n_x(t) a_{xy}. \tag{8.4}$$

To illustrate this formula, consider Example 8.1. A one-year-old at time $t + 1$ is an offspring of either a two- or a three-year-old at time t. Thus $n_1(t+1) = n_2(t) \times 1 + n_3(t) \times 4$. Clearly we also have $n_2(t+1) = n_1(t) \times 0.5$ and $n_3(t + 1) = n_2(t) \times 0.5$. This illustrates that equation (8.4) is a system of equations, one equation for each state y at time $t + 1$. Appendix 8.1 gives a more succinct expression for the equation system in vector notation.

Let $n(t)$ be the total cohort size at time t, that is

$$n(t) = \sum_x n_x(t). \tag{8.5}$$

Then

$$\lambda(t) = \frac{n(t + 1)}{n(t)} \tag{8.6}$$

represents the proportionate growth rate of the cohort between years t and $t + 1$. To look at the composition of the cohort at time t, for each x let

$$\rho_x(t) = \frac{n_x(t)}{n(t)} \tag{8.7}$$

be the proportion of the cohort in state x.

Now $\lambda(t)$ and the proportion of the cohort in the various states at time t depend on the initial composition of the cohort at time 0. However, if the population projection matrix satisfies certain technical conditions (for example *primitivity*; see Appendix 8.2) these quantities asymptote to limiting values as t increases, i.e.

$$\lambda(t) \to \lambda, \quad \text{say,} \tag{8.8}$$

and

$$\rho_x(t) \to \rho_x \text{ for each } x. \tag{8.9}$$

These limiting values do not depend on the initial composition of the cohort. This result is well known for an age-structured population, where the ρ_x define the stable age distribution (Leslie 1945; Charlesworth 1994), but the result applies to any state-structured population (Caswell 1989).

Table 8.2 illustrates convergence to the stable-state distribution for Example 8.2. To obtain the table the initial population has been taken as 10 000 individuals all of low quality. Subsequent entries have been calculated using equations (8.4) – (8.7). Since the population is initially composed entirely of low-quality individuals, the population size $n(t)$ drops. As the proportion of high-quality individuals $\rho_H(t)$ rises, the rate of population increase $\lambda(t)$ rises until it is above 1, and from then on the population size $n(t)$ increases.

Eventually the proportion of high-quality individuals settles down to a limiting value $\rho_H = 0.514$ and the rate of population increase tends to a limiting value $\lambda = 1.018$.

Table 8.2. Convergence to the stable state distribution for Example 8.2.

year t	$n_L(t)$	$n_H(t)$	$n(t)$	$\rho_L(t)$	$\rho_H(t)$	$\lambda(t)$
0	10000	0	10000	1.000	0.000	0.760
1	5045	2554	7598	0.664	0.336	0.928
2	3784	3271	7054	0.536	0.464	0.992
3	3495	3505	7000	0.499	0.501	1.011
4	3464	3613	7077	0.489	0.511	1.016
5	3500	3689	7189	0.487	0.513	1.017
10	3810	4031	7841	0.486	0.514	1.018
30	5405	5718	11123	0.486	0.514	1.018

Following the population forwards gives a robust and general method of calculating the limiting growth rate λ. Another method is to recognise λ as an eigenvalue of the projection matrix A (see Appendix 8.1) and use an appropriate method for finding eigenvalues. In fact, λ is the largest eigenvalue of A and is known as the Perron–Frobenius eigenvalue. The stable-state distribution ρ_x can also be recognised as the (left) eigenvector of A corresponding to the eigenvalue λ (Appendix 8.1). Not only does the growth rate per year settle down to a limiting value, but the stronger result

$$n(t) \sim K\lambda^t \tag{8.10}$$

also holds (this follows from results in Gantmacher 1959); here K is a constant which depends on the initial cohort size and composition.

Formula (8.10) can be used to motivate the interpretation of λ as a measure of fitness. Suppose that we are interested in characterising the ESS strategy π^* for a population. We assume that when the resident population strategy is π^*, not only is the population evolutionarily stable but it is also demographically stable in that there is a stable distribution of states. Consider first an asexual species. Let $n^*(t)$ be the total number of organisms with the genotype that codes for π^* at time t. By formula (8.10) we assume that $n^*(t) \sim K^*(\lambda^*)^t$ for large t, where λ^* is the Perron–Frobenius eigenvalue of the projection matrix for this genotype. Consider a mutant genotype within this population. Let $n(t)$ be the number of this genotype present at time t. Again by formula (8.10), $n(t) \sim K\lambda^t$, where λ is the eigenvalue for this genotype. Then

$$\frac{n(t)}{n^*(t)} \to \frac{K}{K^*}\left(\frac{\lambda}{\lambda^*}\right)^t.$$

If $\lambda > \lambda^*$ then the proportion of mutants in the population increases over time and the population is invadable. Thus a necessary condition for evolutionary stability is that $\lambda \leq \lambda^*$ for every possible mutant. Given these conditions we can take the Perron–Frobenius eigenvalue of the projection matrix associated with a genotype as a measure of the fitness of that genotype. At evolutionary and demographic stability the resident genotype maximises fitness, where the maximisation is taken over all possible mutants. Of course, if the resident population is not only evolutionarily stable but also dynamically stable (i.e. numbers are constant and the population is demographically stable) then $\lambda^* = 1$, and $\lambda \leq 1$ for every possible mutant.

To describe the analogous result for sexually reproducing populations it is first necessary to define what is meant by the fitness of a strategy. Consider a subpopulation, all members of which and their descendants follow some given life-history strategy. Then the fitness of the strategy is the asymptotic growth rate in the (discounted) number of individuals in this subpopulation. In other words, fitness is the Perron–Frobenius eigenvalue of the projection matrix under the strategy. Lande (1982) considered a species in which males and females of the same genotype have the same demographic parameters. He used quantitative genetic arguments to show that selection favours strategies (traits) that maximise this fitness measure. When there are differences between the sexes, the sex of an individual can be regarded as a component of the individual's state. The important work of Taylor (1990) is based on this approach. Using the results of Taylor and the optimisation results of McNamara (1991), Leimar (1996) showed that a necessary condition for a resident population strategy π^* to be an ESS is that all other strategies within this population have fitness no greater than π^*. The central point here is that Leimar equates the fitness of a strategy with the Perron–Frobenius eigenvalue of the resulting projection matrix.

8.4 Reproductive value

In the previous section we introduced a fitness measure for structured populations λ, defined as the asymptotic growth rate of a cohort of the genotype under study. Individuals of the genotype will differ in their ability to survive and reproduce. This does not mean, however, that individuals of the same genotype have different fitnesses. Fitness is not defined at the level of the individual, but is a measure assigned to a life-history strategy. It is, nevertheless, useful to have a measure comparing the abilities of different individuals following the same strategy to leave descendants far into the future. Here we describe a standard measure known as the reproductive value.

The expected number of offspring produced in the future by an organism in a given state is an obvious, simple measure of the reproductive potential of that state. There are two objections to adopting this simple measure universally. Traditional life-history theory for age-structured populations has emphasised that in a growing population it is important not only to produce many offspring but to produce them quickly. Thus one must discount offspring according to their time of production (Fisher 1958). If, however, we are considering an age-structured population that is evolutionarily and dynamically stable, the growth rate of the resident population is $\lambda^* = 1$. Under these circumstances there is no need to discount the future when calculating the reproductive value of a member of this resident population, and reproductive value is just the expected future number of offspring produced.

A more serious failing of just counting offspring number is that this procedure takes no account of the state of the offspring produced. Offspring state is important if it is affected by the action of the parent, e.g. when there is a trade-off between size and number of offspring or when the parent can control offspring dispersal and hence the local conditions experienced by the offspring after dispersal. Offspring state is also important if there are maternal effects.

To take into account population growth or differences in offspring state we must look at the descendants left far into the future. Focus on a particular genotype and consider an individual of this genotype currently in state x. Let $f_n(x)$ be the expected number of direct descendants left in n years' time by this individual. To evaluate $f_n(x)$ we note that the descendants left in n year's time are precisely the descendants left $n - 1$ years into the future by descendants present in one year's time. Since an individual in state x leaves a_{xy} expected descendants in state y next year we have

$$f_n(x) = \sum_y a_{xy} f_{n-1}(y). \tag{8.11}$$

Equation (8.11) is really a system of equations determining $f_n(x)$ for all possible x once $f_{n-1}(y)$ is known for all possible y. Since $f_0(y) = 1$ for all y one can find $f_1(y)$ for all y and hence $f_2(y)$ for all y, and so on.

To analyse the limiting behaviour in equation (8.11) choose some reference state, x_0, say. We look at the numbers of descendants left in n years' time by an individual in state x relative to the number left by an individual in the reference state x_0, i.e. we look at $V_n(x) = f_n(x)/f_n(x_0)$. Assuming that the population projection matrix for the genotype is primitive (Appendix 8.2) it can be shown (Gantmacher 1959) that $V_n(x)$ tends to a limiting value as one looks further into the future,

$$V_n(x) \longrightarrow V(x) \qquad \text{as } n \longrightarrow \infty. \tag{8.12}$$

$V(x)$ is the reproductive value of state x relative to state x_0 and represents the relative contributions of the two states to descendants far into the future. If x and y are any two states then their relative contribution to future descendants is $V(x)/V(y)$. This is independent of the reference state x_0 chosen, since a change of reference state merely multiplies the whole function V by a positive constant. The original concept of reproductive value was developed for age-structured populations by Fisher (1958). For discussion in the context of state-structured populations see Caswell (1989). For an interpretation of V as the right eigenvector of the projection matrix see Appendix 8.1.

To illustrate reproductive value consider Example 8.2 again. For this example, Table 8.1(*b*) shows that

$$\begin{aligned} f_n(\text{low}) &= 0.5045 f_{n-1}(\text{low}) &+& 0.2554 f_{n-1}(\text{high}) \\ f_n(\text{high}) &= 0.4851 f_{n-1}(\text{low}) &+& 0.7762 f_{n-1}(\text{high}). \end{aligned} \tag{8.13}$$

Table 8.3 gives numerical values of $f_n(x)$. The table takes the reference state to be 'high quality'. Thus $V_n(\text{low})$ is the ratio of the numbers of descendants left in n years' time by a low-quality individual relative to those left by a high-quality individual. As can be seen, the number of expected surviving offspring of low-quality individuals is only a fraction $V_1(\text{low}) = 0.602$ of the number of expected surviving offspring of high-quality individuals. Since the surviving offspring of a low-quality parent tend to be lower quality, and hence leave fewer offspring than the offspring of a high-quality parent, the corresponding ratio of grandchildren left, $V_2(\text{low})$, is less than 0.602. As n increases $V_n(\text{low})$ continues to decrease and asymptotes to the value $V(\text{low}) = 0.498$.

Table 8.3 also shows the ratio $f_n(\text{high})/f_{n-1}(\text{high})$ and the ratio $f_n(\text{low})/f_{n-1}(\text{low})$. As can be seen, both these ratios tend to λ as n increases. This result can be understood as follows. Since the asymptotic growth rate in descendant numbers is λ per year, an individual leaves on average λ times as many descendants n years into the future as it leaves

$n - 1$ years into the future, when n is large. For a general version of this result see Appendix 8.1.

Table 8.3. Expected numbers of descendants left n years in the future for Example 8.2. Here $V_n(\text{low}) = f_n(\text{low})/f_n(\text{high})$, $\tilde{\lambda}_n(\text{low}) = f_n(\text{low})/f_{n-1}(\text{low})$ and $\tilde{\lambda}_n(\text{high}) = f_n(\text{high})/f_{n-1}(\text{high})$; see Appendix 8.1.

n	$f_n(\text{low})$	$f_n(\text{high})$	$V_n(\text{low})$	$\tilde{\lambda}_n(\text{low})$	$\tilde{\lambda}_n(\text{high})$
0	1.000	1.000	1.000	—	—
1	0.760	1.261	0.602	0.760	1.261
2	0.705	1.348	0.523	0.928	1.068
3	0.700	1.388	0.504	0.993	1.030
4	0.708	1.417	0.499	1.011	1.021
5	0.719	1.443	0.498	1.016	1.018
10	0.784	1.576	0.498	1.018	1.018
30	1.112	2.235	0.498	1.018	1.018

8.5 Trade-offs

When there is a life-history trade-off, an organism has a range of possible reproductive behaviour, and can only enhance one component of its reproductive success at the expense of another component. In particular, its current reproductive success can only be increased at the expense of the future reproductive success of itself or its offspring. For example, increased reproductive effort in one breeding season may decrease the probability that the organism survives to breed again, or may decrease condition at the next breeding attempt and hence reduce reproductive success at this later time. In this section we present a modelling framework that can be used to express trade-offs. This framework is then used in Section 8.6 to characterise and find optimal life-history strategies.

As before, consider a population that is censused at times $t = 0, 1, 2, \ldots$ one year apart. An appropriate state variable (e.g. age, size, condition etc.) is used to characterise population members at the census times. At each census time t, each organism chooses a reproductive action that it will adopt between times t and $t + 1$. For example, a tree can be regarded as choosing the proportion of its available resources to allocate to growth as opposed to seed production. A bird can be regarded as choosing its laying date or clutch size. In this chapter we restrict attention to single actions such as choice of clutch size. More complex scenarios in which organisms make state-dependent decisions over the year are analysed in Chapter 9, which deals with annual routines.

The range of reproductive actions available to an individual will typically depend on its state. For example, the allocation of energy to growth and to seed production by a tree is constrained by the total energy available, which in turn may depend on the tree's size. In other organisms the production of offspring may be impossible before a certain level of maturity is reached.

Here reproductive actions are seen as affecting the number and states of descendants at the next census time. In Section 8.3 we considered an organism in state x at time t and we described its descendants at $t + 1$ in terms of four functions. There the organism followed a given life-history strategy. Here we wish to compare the consequences of performing different behaviours and extend the definitions of the four functions to allow them to depend also on the reproductive action chosen. In doing so, reproductive actions will be denoted by the symbol u.

(f1) *Survival of the focal organism until $t + 1$.* If the organism takes action u when in state x then it survives until the next census time, at $t + 1$, with probability $S_{\text{mat}}(x; u)$. The trade-off between current reproductive effort and survival can be modelled by an appropriate choice of the function S_{mat}. For example, if u is equated with reproductive effort, then S_{mat} would be chosen so that $S_{\text{mat}}(x; u)$ was a decreasing function of u for fixed x. Survival could also be allowed to depend on state, x, and S_{mat} could be chosen to represent how maternal survival depends on some interaction between x and u.

(f2) *State of the focal organism at $t + 1$.* If the organism in state x at time t survives until time $t + 1$, then the probability that its state at this time is y is denoted by $p_{xy}(u)$. The philosophy behind the state-dependent approach to modelling life histories is that state changes give a mechanistic basis for trade-offs. In particular, the effect of the reproductive action u chosen at time t on survival and fecundity from time $t + 1$ onwards is represented by the effect of u on the state at time $t + 1$. For example, if state is taken to be condition and u is reproductive effort, then the deleterious effect of effort on future state can be modelled by assuming that an increase in u decreases the probability $p_{xy}(u)$ of high states y and increases the probability $p_{xy}(u)$ of low states y. This may in turn affect condition in later years (after $t + 1$) and hence also affect survival and fecundity in those years.

(f3) *Number of offspring surviving until $t + 1$.*   The number of offspring produced between times t and $t + 1$ that survive until time $t + 1$ is denoted by $N_{\text{off}}(x; u)$. (Again x is the state of the parent at time t and u the reproductive action.) As before, N_{off} can include offspring of offspring etc., if there is more than one generation between t and $t + 1$. N_{off} needs to be discounted appropriately if the population is sexual (see Section 8.2).

(f4) *States of surviving offspring at time $t + 1$.*   The probability that an offspring which survives until time $t + 1$ is in state y at this time is $b_{xy}(u)$. The number versus quality trade-off can be modelled by appropriate choices of $N_{\text{off}}(x; u)$ and $b_{xy}(u)$. Maternal effects can be incorporated by allowing $b_{xy}(u)$ to depend explicitly on the maternal state x.

Given these four functions, the expected number of descendants left in state y at time $t + 1$ by an individual that is in state x at time t and chooses reproductive action u is

$$a_{xy}(u) = S_{\text{mat}}(x; u)p_{xy}(u) + N_{\text{off}}(x; u)b_{xy}(u). \tag{8.14}$$

Example 8.3. Reproduction versus growth.   To illustrate the trade-off between reproduction and growth we consider an organism that always has the possibility of increasing its size. The state variable is taken as the organism's size. If this is x at time t, then the organism has αx resources available, which it can partition between immediate reproduction and growth. If resources u are allocated to immediate reproduction, then $N_{\text{off}} = K_{\text{off}}u$ offspring survive until time $t + 1$, where K_{off} is a constant. All the surviving offspring are in the same state x_1 at this time. The remaining resources are devoted to growth, and, given that the organism survives until time t, its size at this time is a function $f(x, u)$ of its previous size and reproductive allocation. The probability that it survives until time $t + 1$ is also a function $S_{\text{mat}}(x; u)$ of x and u. An increase in u increases the number of offspring in state x_1, at the expense of decreasing both the parent's state $f(x, u)$ next year and the probability $S_{\text{mat}}(x; u)$ that it will survive to be in this state.

In this example the action u must be chosen in the range $0 \le u \le \alpha x$. Expected descendants produced are

$$a_{xy}(u) = K_{\text{off}}u \qquad \text{when} \quad y = x_1$$
$$a_{xy}(u) = S_{\text{mat}}(x; u) \qquad \text{when} \quad y = f(x, u)$$

with

$$a_{xy}(u) = 0 \quad \text{for all other } y.$$

Example 8.4. Offspring number versus quality. This example is an extension of Example 8.2 and is based on the ideas presented in McNamara & Houston (1992c). There are non-overlapping generations ($S_{\text{mat}} = 0$) with a generation time of one year. At a census time, an individual must decide how many offspring u to produce. As in Example 8.2 individuals are either of low or high quality. Low-quality individuals have 6 units of resource to put into provisioning of offspring, high-quality individuals have 12 units of resource to do so. Resources are shared equally amongst offspring produced. An offspring requires a resource r greater than 1 to have any chance of survival. This limits the number of offspring, u_L, produced by a low-quality individual to the range $u_L = 1, 2, \ldots, 5$, and limits the number, u_H, produced by a high-quality individual to the range $u_H = 1, 2, \ldots, 11$. If an offspring receives resource r ($r > 1$) it survives until maturity in one year's time with probability

$$S_{\text{off}}(r) = 0.25 \left(\frac{r-1}{r} \right)^{1/4}. \tag{8.15}$$

This function has been chosen to rise rapidly with r for r close to 1, and to tend to 0.25 as r becomes large. The upper limit of 0.25 can be thought of as being imposed by a factor such as predation that is independent of resource allocation. Under these assumptions the expected number of surviving offspring left by a low-quality individual that produces u_L offspring is

$$N_{\text{off}}(\text{low}; u_L) = u_L S_{\text{off}} \left(\frac{6}{u_L} \right).$$

This quantity increases with increasing u_L for $1 \leq u_L \leq 5$. Similarly, the number of surviving offspring left by a high-quality individual that produces u_H offspring is

$$N_{\text{off}}(\text{high}; u_H) = u_H S_{\text{off}} \left(\frac{12}{u_H} \right).$$

This quantity increases with u_H for u_H in the range $1 \leq u_H \leq 10$. .

If an offspring receiving resource r survives to maturity it is high quality with probability

$$h(r) = \frac{r^4}{r^4 + 10}. \tag{8.16}$$

This function is an increasing function of r. Thus an individual can increase

its expected number of surviving offspring by increasing the number of offspring produced, but this decreases the probability that each offspring will
be of high quality.

For this example the states are 'low' and 'high'. The number of offspring
produced corresponds to the action chosen. The expected numbers of descendants left by a low-quality individual that produces u_L offspring are

$$a_{L,L}(u_L) = u_L S_{\text{off}}\left(\frac{6}{u_L}\right)\left[1 - h\left(\frac{6}{u_L}\right)\right]$$

$$a_{L,H}(u_L) = u_L S_{\text{off}}\left(\frac{6}{u_L}\right) h\left(\frac{6}{u_L}\right).$$

(8.17)

Similarly, the expected numbers of descendants left by a high-quality individual that produces u_H offspring are

$$a_{H,L}(u_H) = u_H S_{\text{off}}\left(\frac{12}{u_H}\right)\left[1 - h\left(\frac{12}{u_H}\right)\right]$$

$$a_{H,H}(u_H) = u_H S_{\text{off}}\left(\frac{12}{u_H}\right) h\left(\frac{12}{u_H}\right).$$

(8.18)

8.6 Optimal strategies

A life-history strategy is a contingent rule determining behaviour. In our
framework, a strategy π is a rule that, for each state x, specifies the action,
$\pi(x)$, taken in that state. For instance, in Example 8.3 π is a function that
determines the allocation to reproduction, $\pi(x)$, of an organism of size x.
In this section, we discuss the relationship between fitness and the strategy
that an organism adopts. We begin by illustrating this relationship using
Example 8.4. We then use the example to make some general points about
fitness. Next, we look at various criteria that can be used to recognise
optimal strategies. Finally, we present a robust and powerful method by
which optimal strategies can be computed.

Let $A^{(\pi)}$ denote the population projection matrix of a cohort following the
life-history strategy π. Since π specifies the action $\pi(x)$ taken in state x, the
(x, y)th element of $A^{(\pi)}$ is $a_{xy}(\pi(x))$. The asymptotic growth rate of a cohort
following strategy π is $\lambda^{(\pi)}$, where $\lambda^{(\pi)}$ is the maximum eigenvalue of the
matrix $A^{(\pi)}$. In other words $\lambda^{(\pi)}$ is the fitness of strategy π. Since a genotype
codes for the life-history strategy used by individuals of the genotype, we
can define the fitness of a genotype as the fitness of the strategy that it
codes for. By the arguments of Section 8.3 natural selection leads to the

maximisation of $\lambda^{(\pi)}$ in the population. Thus we define a strategy π^* to be optimal if

$$\lambda^{(\pi^*)} = \max_\pi \lambda^{(\pi)}, \tag{8.19}$$

where the maximum is taken over all strategies that are possible. It is important to recall that in defining fitness we are concerned with the rate of invasion of a rare mutant strategy into a resident population. An optimal strategy π^* maximises this invasion rate and is hence a best response to the resident strategy. Thus in this context an optimal strategy is not necessarily an ESS strategy (see Sections 8.2 and 8.11).

To illustrate how fitness depends on strategy consider Example 8.4. In this example the states are 'low' and 'high'. A strategy π is specified by a pair of numbers (u_L, u_H), where $u_L = \pi(\text{low})$ is the number of offspring produced by a low-quality individual and $u_H = \pi(\text{high})$ is the number of offspring produced by a high-quality individual. The pair (u_L, u_H) determines a population projection matrix given by equations (8.17) and (8.18). The fitness of the strategy (u_L, u_H) is then the maximum eigenvalue of this matrix. We denote fitness by $\lambda^{(u_L, u_H)}$. Table 8.4 shows the fitnesses of all 55 possible strategies. It can be seen that the optimal strategy is $\pi^* = (3, 7)$; this strategy has fitness 1.026.

Table 8.4. Fitness $\lambda^{(u_L, u_H)}$ for each possible strategy (u_L, u_H) in Example 8.4.

u_H	Offspring produced by low-quality individual, u_L				
	1	2	3	4	5
1	0.245	0.245	0.250	0.505	0.662
2	0.476	0.478	0.480	0.523	0.664
3	0.681	0.688	0.694	0.705	0.727
4	0.833	0.854	0.866	0.873	0.870
5	0.906	0.952	0.974	0.976	0.957
6	0.904	0.985	1.019	1.018	0.983
7	0.860	0.978	1.026	1.023	0.978
8	0.808	0.958	1.017	1.015	0.966
9	0.760	0.933	1.001	1.002	0.952
10	0.714	0.900	0.975	0.982	0.936
11	0.654	0.841	0.922	0.941	0.907

Example 8.4 can be used to illustrate various general points.

Optimisation is in the long term. The strategy of always maximising the expected number of descendants present next year, and ignoring their type, will typically not be optimal when there are trade-offs. In the above example a parent maximises the expected number of descendants next year by producing $u_L = 5$ offspring when it is of low quality and $u_H = 10$ offspring when it is of high quality. This strategy is poor in the long term, however. A cohort following the strategy has a proportion $\rho_H = 0.172$ of individuals of high quality at the stable-state distribution, and the growth rate $\lambda^{(5,10)} = 0.936$ is less than 1. In contrast, the strategy given by $u_L = 4$ and $u_H = 8$ produces fewer descendants after one generation, but is better in the long term. A cohort following this strategy has a proportion $\rho_H = 0.3361$ of high-quality individuals at the stable-state distribution, and the growth rate is then $\lambda^{(4,8)} = 1.015$. McNamara & Houston (1992c) give an illustrative example, emphasising the long-term nature of the optimisation criterion; their example shows that even the strategy that maximises the expected numbers of grandchildren may be significantly inferior to the optimal strategy. A further example of this type of effect is given in Section 8.8.

Norms of reaction. A strategy specifies how an organism responds to its state. Here the "state" might be its own internal state or the external environmental state. Thus a strategy specifies an organism's norm of reaction to its environment. The optimal strategy is the optimal plastic response. This topic is discussed in Section 8.9.

Interactions between the effects of actions. A strategy is a rule specifying what action is to be taken in every possible state. The fitness of a strategy depends, in a non-linear way, on the whole suite of actions specified. The non-linearity means that the effects of actions are not additive in their contribution to fitness: the influence on fitness of the action chosen in a particular state depends on the actions specified under the rule in other states (McNamara 1993b). For instance, in Example 8.4 suppose high-quality individuals produce seven offspring; then, given this constraint, fitness is maximised if low-quality individuals producing three offspring (Table 8.4). If, instead, high-quality individuals produce three offspring, the best action for a low-quality individual is to produce five offspring.

Example 8.4 shows that the best action for an organism in one state depends on what the organism would have done if it had been in another state. This holds because fitness is defined on genotypes (and hence strategies), not on individuals. The focal individual may never experience other states, but its descendants are likely to do so, and the growth rate of the individual's line of descendants depends on the action these descendants take. In this example the action of a low-quality individual affects the proportions of its offspring which are of low and high quality. It is worth reducing total offspring number in order to increase the absolute number of high-quality offspring only if high-quality offspring are valuable compared with low-quality offspring. But the value of high-quality offspring depends on the action they take when they in turn reproduce.

A strategy π^* is optimal if, for every state x, the action $\pi^*(x)$ is the best action given the constraint that the actions taken in all other states y are given by $\pi^*(y)$; see below.

Although we have emphasised that there may be an interaction between actions taken in different strategies, this need not always hold. In certain simple cases the best action in a state is independent of the actions chosen in other states (e.g. see Section 8.9).

Selection acts on strategies. Fitness is defined on strategies rather than actions and it is at this level that natural selection operates. We can illustrate this by considering clutch size decisions in birds. There is a large literature on whether observed clutch sizes are optimal (see Roff 1992, Stearns 1992, and van der Werf 1992 for reviews). This question should really be phrased as whether the laying strategy of these birds is optimal. A strategy here is a rule specifying when to lay and how many eggs to lay as a function of the bird's physiological state, territory condition and the prevailing conditions of weather and food availability.

Recognising optimal strategies

As well as computing an optimal strategy from first principles, it may be desirable to check whether a given strategy is optimal. A possible optimal strategy may be suggested by intuition or by observation of the behaviour of a population. We now give conditions that allow us to determine whether a strategy π^* is optimal. The first step is to determine what actions are possible and what are the consequences of these actions in terms of the state and number of the descendants produced. Having done this, one extremely crude method is to calculate the fitness, $\lambda^{(\pi)}$, for every strategy π, and see whether $\lambda^{(\pi)}$ is maximised when $\pi = \pi^*$ (equation (8.19)). This is often not practical because the number of possible strategies may be extremely large. Here we present two alternative ways in which an optimal strategy can be characterised and hence recognised.

Local optimisation. McNamara (1993b) introduced the idea of neighbouring strategies in the space of life-history strategy. Strategies π^* and π are neighbours if the actions they specify agree for all but one state. We can then define π^* as a local optimum if no neighbouring strategy has higher fitness, i.e. $\lambda^{(\pi^*)} \geq \lambda^{(\pi)}$ for all neighbours π. McNamara (1993b) proved that if a strategy π^* has a primitive projection matrix (Appendix 8.2) then

$$\pi^* \text{ is locally optimal} \iff \pi^* \text{ is globally optimal,}$$
$$\text{i.e. } \pi^* \text{ satisfies equation (8.19).} \tag{8.20}$$

This result can be restated as saying that a strategy π^* is optimal if and only if, for every state x, the action $\pi^*(x)$ chosen in that state under π^* is the best action given the constraint that the actions taken in all other states y are given by $\pi^*(y)$. This is a surprising result given the interactions between actions taken in different states. A weaker result holds if π^* is not primitive (McNamara 1993b).

To illustrate this result we return to Example 8.4. We will use condition (8.20) to verify that the strategy $\pi^*(\text{low}) = 3$, $\pi^*(\text{high}) = 7$ is optimal. Since both low- and high-quality parents leave some high-quality offspring and some low-quality offspring under strategy π^*, π^* is a primitive strategy. As before, it is convenient to use the shorthand notation $(3,7)$ for π^*, and similarly for other strategies. π^* has two sorts of neighbour. Neighbours obtained by altering the action taken when quality is low are $(1,7), (2,7), (4,7)$ and $(5,7)$. It can be seen from Table 8.4 that these strategies all have lower fitness than π^*. Neighbours obtained by altering the action taken when high quality are $(3,1), (3,2), \ldots (3,11)$. Again these have lower fitness than π^*. Thus π^* is locally optimal. By condition (8.20) we infer that π^* is globally optimal, as can be verified by examining the other entries in the table.

If instead of looking at π^* we had chosen any suboptimal strategy $\tilde{\pi}$ we would have found at least one neighbour of $\tilde{\pi}$ with higher fitness.

Maximisation of reproductive value. The relative reproductive value of two states depends on the strategy used. Choose some fixed reference state x_0 and denote the reproductive value of state y relative to state x_0 under strategy π by $V^\pi(y)$. Given a strategy π, and hence the function V^π, we consider an organism (not necessarily following π) in state x at time t. Suppose the organism takes action u. Then it will leave $a_{xy}(u)$ expected descendants in state y at time $t + 1$. Assign each descendant in state y at time $t + 1$ a 'reward' $V^\pi(y)$. Then the expected total reward assigned to all descendants at time $t + 1$ is

$$\tilde{H}^\pi(x; u) = \sum_y a_{xy}(u) V^\pi(y). \tag{8.21}$$

We shall consider $\tilde{H}^\pi(x; u)$ as a function of the action u chosen, regarding x and π as fixed. For any two actions u_1 and u_2 the ratio $\tilde{H}^\pi(x; u_1)/\tilde{H}^\pi(x; u_2)$ is the expected number of descendants left far in the future by an organism choosing action u_1 at time t relative to that left by an organism choosing u_2, given that in both cases the behaviour of all descendants from time $t+1$ onwards is prescribed by strategy π.

First, let π be an optimal strategy π^*. Then, abbreviating $\tilde{H}^{\pi^*}(x; u)$ to $H^*(x; u)$ and $V^{\pi^*}(y)$ to $V^*(y)$, formula (8.21) becomes

$$H^*(x; u) = \sum_y a_{xy}(u)V^*(y). \tag{8.22}$$

This function of u describes the relative numbers of descendants left far into the future given that behaviour from time $t+1$ onwards is optimal. From this interpretation it is not surprising that the following criterion holds:

$$\begin{array}{ll} \text{A strategy } \pi & \text{For each state } x \text{ the action} \\ \text{is optimal} \iff & u = \pi(x) \text{ taken under } \pi \\ & \text{maximises } H^*(x; u). \end{array} \tag{8.23}$$

In other words, a strategy is optimal if and only if an organism following the strategy always maximises the expected reproductive value of descendants left in one year's time, i.e. it maximises

$$\sum_y a_{xy}(u)V^*(y), \tag{8.24}$$

where reproductive value V^* is evaluated under an optimal strategy.

Criterion (8.23) is conceptually interesting but is not very useful in determining whether a given strategy π is optimal. To apply it we require knowledge of V^* and hence we need to know at the outset which strategies are optimal. To avoid this circularity requires the following more powerful criterion:

$$\begin{array}{ll} \text{A primitive strategy} & \text{For each state } x \text{ the action} \\ \pi \text{ is optimal} \iff & u = \pi(x) \text{ taken under } \pi \\ & \text{maximises } \tilde{H}^\pi(x; u). \end{array} \tag{8.25}$$

In other words a primitive strategy π is optimal if and only if, for each state x at time t, the choice of action under π maximises the expected numbers of descendants left far into the future given that behaviour from time $t+1$ onwards is determined by π itself. Equivalently, a strategy π is optimal if and only if an organism following π always maximises the reproductive value of its descendants left in one year's time, where reproductive value is evaluated under π itself.

Characterisation of an optimal strategy in terms of the maximisation of reproductive value was given in the case of age-structured populations by Schaffer (1974a) and Taylor *et al.* (1974). Caswell (1982) extended these results to state-structured populations under rather restrictive conditions. The general criteria (8.23) and (8.25) were derived using an approach based on Markov decision processes by McNamara (1991, 1993b). For an interpretation of expression (8.24) in terms of changes in gene frequency, see Taylor (1990).

It may not be obvious how criterion (8.23) is related to the dynamic programming equation given in Chapter 3. This relationship is explained in Appendix 8.3.

To illustrate criterion (8.25) consider Example 8.4. Equations (8.17), (8.18) and (8.21) give

$$\tilde{H}^\pi(\text{low}; u_\text{L}) = u_\text{L} S_\text{off}\left(\frac{6}{u_\text{L}}\right)\left\{\left[1 - h\left(\frac{6}{u_\text{L}}\right)\right] V^\pi(\text{low}) + h\left(\frac{6}{u_\text{L}}\right)\right\} \quad (8.26)$$

$$\tilde{H}^\pi(\text{high}; u_\text{H}) = u_\text{H} S_\text{off}\left(\frac{12}{u_\text{H}}\right)\left\{\left[1 - h\left(\frac{12}{u_\text{H}}\right)\right] V^\pi(\text{low}) + h\left(\frac{12}{u_\text{H}}\right)\right\} \quad (8.27)$$

where the functions S_off and h are given by equations (8.15) and (8.16) respectively and where we have taken the state 'high' as our reference state, so that $V^\pi(\text{high}) = 1$. Let u_L^π be the value of u_L maximising $\tilde{H}^\pi(\text{low}; u_\text{L})$ and let u_H^π be the value of u_H maximising $\tilde{H}^\pi(\text{high}; u_\text{H})$. Criterion (8.25) states that π is optimal if and only if $u_\text{L}^\pi = \pi(\text{low})$ and $u_\text{H}^\pi = \pi(\text{high})$ both hold. Failure of one or both of these conditions implies that π is suboptimal. For example, when $\pi = (4, 4)$, $V^\pi(\text{low}) = 0.6928$ and it can be confirmed that $u_\text{L}^\pi = 5$ and $u_\text{H}^\pi = 9$. Thus criterion (8.25) fails, confirming that $\pi = (4, 4)$ is suboptimal. Conversely when $\pi = (3, 7)$, $V^\pi(\text{low}) = 0.4957$, $u_\text{L}^\pi = 3$ and $u_\text{H}^\pi = 7$, so that criterion (8.25) holds, confirming that $\pi = (3, 7)$ is the optimal strategy.

Maximisation of lifetime reproductive success

The element of a projection matrix counts the number of descendants in a given state, but does not distinguish between the focal animal and its offspring, given that they are in the same state. It is, however, of interest to keep track of the identity of an individual and hence follow it over the course of its lifetime. The projection matrix approach can be reformulated to do this, and hence to characterise what is maximised over the lifetime of an individual (McNamara 1993a). The results can be summarised a follows.

Suppose, as before, that we census a population at times $t = 0, 1, 2, \ldots$ one year apart. Individuals born between times t and $t+1$ are counted as having age 1 at time $t+1$, age 2 at time $t+2$ etc. In this chapter an organism's state x could include its age as a component. Here, however, it is convenient to separate out the age and regard an organism's state as a vector (x, s) where s is age and x contains all other components of state. For convenience, we refer to x as the residual state. A life-history strategy is now a rule π that specifies the action $\pi(x, s)$ taken for every combination of x and s.

Let π^* be an optimal strategy, i.e. π^* satisfies equation (8.19). Let $\lambda^* = \lambda^{(\pi^*)}$ be the fitness of this strategy and let $V^*(x, s)$ be the reproductive value of an organism in state (x, s) under the strategy.

Consider an organism following some given life-history strategy. Let the organism have age s and have residual state x at census time t. Let $b_{xy}^{(s)}$ be the expected number of offspring (suitably discounted by relatedness) produced by this organism between times t and $t + 1$ that have residual state y at time $t + 1$. Then since offspring have age 1 at time $t + 1$, their expected total reproductive value is

$$B(x, s) = \sum_y b_{xy}^{(s)} V^*(y, 1), \qquad (8.28)$$

where reproductive value is calculated using V^*. Now suppose the organism has residual state w at age 1. Given that an organism is alive at age s, its residual state at this time, $X(s)$, is a random variable whose distribution depends on both w and the life-history strategy. Thus given that an individual survives until the annual census time at age s, the expected reproductive value of offspring produced between ages s and $s + 1$ that survive to the annual census time one year later is

$$\overline{B}_w(s) = \mathbb{E}[B(X(s), s)]. \qquad (8.29)$$

Let $\ell_w(s)$ be the probability that the organism survives until the annual census time at age s. Then the expected discounted lifetime reproductive success under the given strategy is

$$\sum_{s=1}^{\infty} (\lambda^*)^{-s} \ell_w(s) \overline{B}_w(s). \qquad (8.30)$$

The functions $\ell_w(s)$ and $\overline{B}_w(s)$ depend on the residual state at age 1 and on the strategy used. A strategy is optimal if and only if it maximises expression (8.30) for every possible residual state w at age 1. This result was proved for age-based life-history strategies by Taylor *et al.* (1974) and

Schaffer (1974a). The general result, valid for any state-dependent life-history problem provided that there is a primitive optimal strategy, is given in McNamara (1993a).

The discount factor $(\lambda^*)^{-s}$ is present in expression (8.30) because in a growing population it is better to produce offspring sooner rather than later (see Fisher 1958). When a population is at evolutionary stability and is of demographically stable size, then $\lambda^* = 1$. Thus, within the population the resident ESS maximises

$$\sum_{s=1}^{\infty} \ell_w(s)\overline{B}_w(s) \tag{8.31}$$

for all w, where the maximisation is over all invading mutants. As a special case, suppose that all offspring that survive until their first annual census time are in the same state at this time. Then, by equation (8.28), $B(x, s)$ is proportional to the number of surviving offspring. It follows that expression (8.31) is maximised by maximising the expected number of surviving offspring produced over the lifetime (see e.g. Mylius & Diekmann 1995; also cf. Section 8.9).

Computing optimal strategies

The criteria (8.23), (8.25) allow us to decide whether a given strategy is optimal, but they are usually not useful in computing the optimal strategy, π^*, for a given life-history problem. Fortunately, there are straightforward ways of computing optimal strategies based on dynamic programming. The method explained below takes a target year in the distant future. Dynamic programming is then used to find the strategy that maximises the expected number of descendants left at this time. The behaviour specified by this strategy typically depends on both the organism's state and the time to go until the target year. As the time to go increases, the behaviour in a given state x tends to the limiting value $\pi^*(x)$. Thus, as the time to go increases, behaviour in any given state tends to that specified by the optimal life-history strategy π^* in that state.

The method of computing optimal strategies first finds the reproductive value $V^*(x)$ of all states x under the optimal strategy. The method used to find V^* is an extension of that used in Section 8.4 to find reproductive values under a given strategy.

Consider some target year far in the future. Set

$f_n(x)$ = maximum expected number of descendants left in the target year by an organism in state x with n years to go until the target year.

The maximum here is over all possible future reproductive behaviours of the organism and its descendants, and allows for behaviour to depend on both state and time to go until the target year. By definition we set $f_0(x) = 1$ for all states x. Given the function f_{n-1} we can find the function f_n by using the dynamic programming equation

$$f_n(x) = \max_u \sum_y a_{xy}(u) f_{n-1}(y), \tag{8.32}$$

where the maximum is taken over all reproductive behaviours u available in state x. Thus the sequence of functions $f_0, f_1, f_2, \ldots$ can be computed iteratively.

Now choose some reference state x_0 (any state will do). Set

$$V_n(x) = \frac{f_n(x)}{f_n(x_0)}. \tag{8.33}$$

Then $V_n(x)$ is the number of descendants left in the target year by an individual in state x relative to those left by an individual in state x_0 (when there are n years to go to the target year). Provided that certain technical conditions are met (see McNamara 1991 and the references therein), it can be shown that, for each state x, $V_n(x)$ tends to $V^*(x)$ as n tends to infinity, where $V^*(x)$ is the reproductive value of state x under the optimal strategy.

Having found V^* one can immediately deduce the optimal strategy. The reproductive decision chosen in state x under this strategy maximises expression (8.24).

In the next three sections we use the framework developed in this chapter to make some biological points about optimal life histories.

8.7 Apparent effects of age when age and state are correlated

In the traditional age-based approach to life-history theory, survival and fecundity are allowed to depend on an organism's age. This dependence then determines the optimal age at which to begin reproduction and the subsequent pattern of reproductive behaviour. Organisms in most natural populations differ in aspects of condition that strongly influence their survival and fecundity. Some condition measures are probably tightly correlated with age. Others, however, are not: individuals of the same age can differ markedly in their condition. Age-based theory ignores the latter condition measures. In doing so it ignores an important aspect of the organism's biology. In this section we describe an example given in McNamara & Houston (1996) in which age has no effect *per se*. Instead, survival and fecundity depend on condition. Unlike age, this condition variable can potentially increase or decrease over time depending on the behaviour of the organism. Despite the absence of explicit age-dependent effects, we demonstrate that there can be apparent effects of age because the condition-dependent behavioural strategy chosen by the organism means that age and condition are correlated.

Example 8.5. Apparent effects of age. The population is censused at the start of the breeding season. At this time, animals differ in condition, x. This variable lies in the range $0 \leq x \leq 1$ with high x denoting good condition. During the breeding season an animal chooses both the number of young to produce and the effort expended on provisioning any young. The number of young, c, is either 0 (no reproduction) or satisfies $c \geq 1$. Provisioning effort, u, satisfies $0 \leq u \leq 1$. When $c = 0$ the effort $u = 0$. When $c \geq 1$, the effort has to be positive if young are to have any chance of survival. Thus in this model an animal has to choose both c and u. A strategy π specifies how both c and u depend on x. If an animal in condition x at time t produces c offspring and expends effort u in provisioning them, descendant numbers at time $t + 1$ are as follows.

(f1) *Maternal survival.* $S_{\mathrm{mat}}(x; c, u) = 0.875(1 - u^2)$. Thus maternal survival does not depend on offspring number c, but decreases as provisioning effort u increases. In our basic model, for a given u survival is independent of condition x.

(f2) *Maternal condition.* If the animal survives, its mean condition in the following year is $x + 0.2 - 0.5u^2$. Thus increased effort decreases future condition, while for a given effort, condition at $t+1$ is positively correlated with condition at t.

(f3) *Number of surviving offspring.* The total amount of resources provided for the young is $R = 4ux$. Each of the c young receives resource $r = R/c$ and survives until time $t + 1$ with probability $S_{\mathrm{off}}(r) = \alpha r^2 / (r^2 + 0.02)$, where α is a parameter. Thus the expected number of offspring surviving until time $t+1$ is $N_{\mathrm{off}}(x; c, u) = cS_{\mathrm{off}}(r)$. Offspring that survive to this time are sexually mature and are classified as having age 1.

(f4) *Offspring condition.* All surviving offspring have mean condition 0.1875 at time $t + 1$.

In this model the state variable x can vary continuously between 0 and 1. To compute optimal strategies it is convenient instead to assume that x can only take a discrete set of values. The computations presented here are based on the nine condition states $x = 0, 0.125, 0.25, \ldots, 1$. The matrix elements p_{xy} and b_{xy} for these discrete states can be found from the four functions (f1) – (f4) by using the techniques of interpolation and smoothing given in Appendix 3.1.

With this choice of (f1) – (f4) there is a trade-off between current reproductive success (f3) and both maternal survival (f1) and future maternal condition (f2). In calculating the optimal strategy the parameter α is chosen so that the growth rate under an optimal strategy is 1. This is a crude way of incorporating density dependence acting on juveniles and ensures that the population size is stable. Since all offspring are in the same state, and the maximum growth rate is 1, maximisation of the number of descendants left far into the future is equivalent to maximisation of the expected number of offspring surviving to age 1 produced over a lifetime.

The optimal life-history strategy specifies that an animal should not breed when condition is below a certain critical level. When an animal's condition at age 1 is below the critical level it will not breed at this age, and as a result its condition at age 2 will be higher; see (f2). Condition continues to rise until it eventually exceeds the critical level and the animal reproduces. Because there is a stochastic component to the increase in condition the delay in breeding will not be the same for all animals. The result is variation in the age at first breeding (Figure 8.1(a)).

In this example, variation in the age at first breeding occurs because of variation in condition at a given age. This type of dependence on the age at first breeding on condition has been described in moose (Saether & Heim 1993), in bison (Green & Rothstein 1991) and in the northern elephant seal (*Mirounga angustirostris*) (Reiter & Le Beouf 1991). The work of Reiter and Le Boeuf on the elephant seal illustrates how adoption of a purely age-based approach leads people to ignore condition variables when they are clearly present and important. Some elephant seals in the study population breed at age 3, others at age 4. Reiter and Le Boeuf attempted to understand this variation by comparing the fitness of two age-based life histories, one in which females always start breeding at age 3 years and the other in which females always start to breed at age 4. They conclude that it is optimal to start breeding at age 4 at their study site, and are consequently unsure of why some females breed at age 3. But elephant-seal life-history strategies cannot be understood by a purely age-based approach; this approach ignores the fact that there is variation in the mass of females at weaning and that heavier females tend to breed earlier than lighter females.

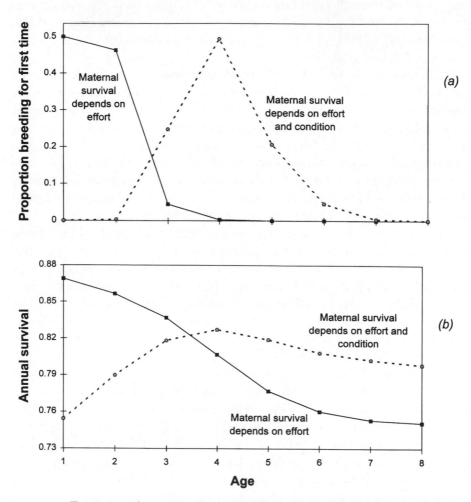

Figure 8.1. Apparent effects of age in Example 8.5. (*a*) Distribution of the age of first breeding, (*b*) the mean probability of survival until next year for individuals in each age class. Solid line, Example 8.5; $\alpha = 0.098$. Broken line, Example 8.5 modified so that $R = ux$ and maternal survival depends on both maternal condition and effort and is given by $S_{\text{mat}}(x; u) = 0.9(1 - u^2)(0.8 + 0.2x)$; $\alpha = 0.5755$.

In Example 8.5, condition tends to increase with age under an optimal strategy. Since offspring number increases with condition, mean offspring number increases with age. Effort also increases with condition and hence mean effort increases with age. This increase in effort results in a decrease in mean annual survival with age (Figure 8.1(b)). The increase in mean condition with age also results in an increase in mean reproductive value with age. Age-based life-history theory also predicts that under some circumstances reproductive effort increases with age (e.g. Charlesworth & León 1976). One reason for this is that a reduction in fecundity or an increase in mortality with age decreases reproductive value; as reproductive value decreases, effort put into current reproduction should increase (Clutton-Brock 1984, Roff 1992, Stearns 1992). This is not analogous to what is happening in Example 8.5, since in this example although mortality under the optimality strategy increases with age, mean reproductive value also increases with age. The age-based theory also predicts an increase in effort with age when fecundity increases with age (Charlesworth & León 1976; see also Taylor 1991). This is similar to what is happening in our Example 8.5.

In many species, not all the sexually mature individuals will participate in a given breeding season (see Bull & Shine 1979 for a review). We refer to this non-breeding as 'skipping reproduction'. A clear example is provided by the grey-headed albatross (*Diomedea chrysostoma*). This species has a breeding cycle of about 250 days. If a pair breeds successfully in one year, they do not breed in the next year. If a breeding attempt is a failure, the pair may breed in the following year (Prince *et al.* 1981). The black-browed albatross (*Diomedea melanophris*) has a similar body mass to the grey-headed albatross but has a slightly shorter breeding cycle and breeds every year. Such a pattern of behaviour cannot be understood in models based solely on age but has a natural interpretation in models based on state. Reproduction may sometimes cause such a severe deterioration in condition that the latter, in the year following breeding, may be below the level required for it to be optimal for the animal to attempt to breed. Thus state-dependent models can produce negative correlations between whether an animal has bred in the previous year and whether it breeds in the current year (e.g. Figure 8.2).

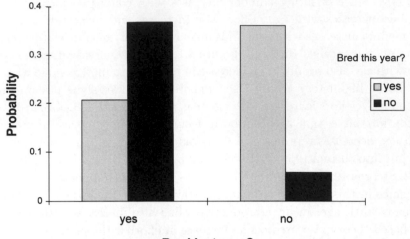

Figure 8.2. The dependence of whether an animal breeds this year on whether it bred last year. The analysis is restricted to animals of age 8 or more years that have previously bred at least once. The tendency to alternate arises because effort strongly affects condition. Example 8.5, modified so that $R = 4ux/(1 + 3x)$, maternal survival is $0.95(1 - u^2)$ and maternal condition next year is $x + 0.2 - 2u^2$; $\alpha = 0.3230$.

8.8 Intergenerational effects

In this section we show how our general framework can be used to analyse two important intergenerational effects, the number versus quality trade-off and maternal effects. When there is a trade-off between offspring number and quality, a mother can choose between, e.g. many low-quality offspring and a few high-quality offspring. For example, Fuchs (1982) manipulated the litter size of house mice and found that as litter size increased, the mother increased the supply of milk, but not enough to prevent a decrease in the milk obtained by each member of the litter. Females from large litters suffered a delay in the time until their own first litter, and a reduction in its size. When maternal effects are present, then within a genotype there are correlations between the phenotype of a mother and the phenotype of her offspring. In the current context, we take phenotype to be synonymous with state. An organism's state can change over its lifetime, and the state at a given time can include both internal and external variables. Thus an animal's dominance status may be a component of its state. In some primates and hyenas, a daughter inherits her mother's dominance status (e.g. Holekamp & Smale 1993). In the red squirrel, offspring may inherit their mother's territory (Price & Boutin 1993).

The age-based approach to life-history theory assumes that all offspring that survive to age 1 are equivalent. As a result, differences in offspring quality cannot be represented. The number–quality trade-off has been analysed by assuming some plausible relationship between the quality of offspring and their reproductive value (e.g. Smith & Fretwell 1974, Lloyd 1987). Similarly, the consequences of maternal effects can be analyzed in this way. The problem with this approach is that erroneous assumptions about the relationship between quality and reproductive value may result in erroneous conclusions. For example, the Trivers–Willard theory of parental investment is based on maternal effects, and takes the reproductive value of offspring to be the number of offspring that they themselves produce. This measure of reproductive value does not include the long-term consequences of quality differences. As Leimar (1996) showed, using the correct measure of reproductive values based on long-term consequences can change the conclusion of the model. We discuss this further below.

Our approach does not involve assuming a relationship between quality and reproductive value. Instead we specify the four functions S_{mat}, p_{xy}, N_{off} and b_{xy} of Section 8.5, which spell out the consequences of actions. From these functions and the optimisation procedure given in Section 8.6, we obtain both the optimal strategy and the dependence of reproductive value on state. We now illustrate this approach, using model II of McNamara & Houston (1996).

Example 8.6. Intergenerational effects. An asexual population is censused at the start of the breeding season each year. All individuals are mature at this time and reproduce. Having reproduced, they die. Offspring produced that survive until the next year are mature at this time, reproduce and die, and so on.

At the census times each individual is characterised by its quality x where $0 < x < 1$. An individual of quality x has total resources $4x$, which are divided equally between all offspring produced. The individual must decide on the number n of such offspring. If n offspring are produced each receives resource $r = 4x/n$ and survives to maturity with probability

$$S_{off}(r) = \alpha r^2/(r^2 + 0.2^2). \tag{8.34}$$

Thus the expected number of surviving offspring is $N_{off} = nS_{off}(r)$. The parameter α is chosen so that the growth rate under the optimal strategy is

$\lambda^* = 1$. Thus we are implicitly assuming density dependence acting on the juvenile phase.

If an offspring that receives resource r survives to maturity at the next census time its mean condition at this time is

$$0.5 + slope(x - 0.5) + \beta[r^2/(r^2 + 0.4^2) - 0.5], \tag{8.35}$$

(provided condition lies between 0 and 1). Here the parameter β (≥ 0) is the maximum amount by which a mother can increase the mean quality of her offspring by increasing the allocation of resources to them. The parameter *slope* is the slope of the regression of offspring quality on maternal quality for given r.

In this model there is a trade-off between offspring number and offspring quality when $\beta > 0$. There are maternal effects when *slope* > 0. For simplicity, equation (8.35) represents these intergenerational effects as acting additively.

When $\beta = 0$ the mother has no control over the state of her offspring and it is optimal for her to maximise the number of surviving offspring. When $\beta > 0$ she can increase their quality by decreasing their number, and it is optimal to maximise the sum of the reproductive values of offspring that survive to maturity. Of course, the measure of reproductive value used is that under an optimal strategy. The optimal strategy and the corresponding reproductive values can both be found simultaneously by the method presented in Section 8.6.

When *slope* $= 0$ reproductive value is proportional to the resources available to provision offspring and hence is proportional to x. This is no longer the case when *slope* > 0. When there are maternal effects, high-quality females have high reproductive value both because of their ability to produce many offspring and because daughters will themselves be of high quality and hence able to produce many high-quality offspring themselves; and so on down the generations. Under biologically plausible interpretations of the model *slope* will be less than 1. In these circumstances the effects of initial high quality will become weaker in successive generations. Nevertheless, the multiplicative nature of maternal effects means that the reproductive value of high-quality females relative to low-quality females can far exceed their relative abilities to produce or provision young (Figure 8.3(a)). A consequence in the current model is that increasing the strength of maternal effects leads to a reduction of offspring number and an increase in offspring quality under the optimal strategy (McNamara & Houston 1996). Figure 8.3(b) compares the strategy of maximising the expected number of surviving offspring (strategy π_1) with that of maximising the total reproductive values of surviving offspring (the optimal strategy, π^*) when the regression of offspring quality on maternal quality has *slope* = 0.5.

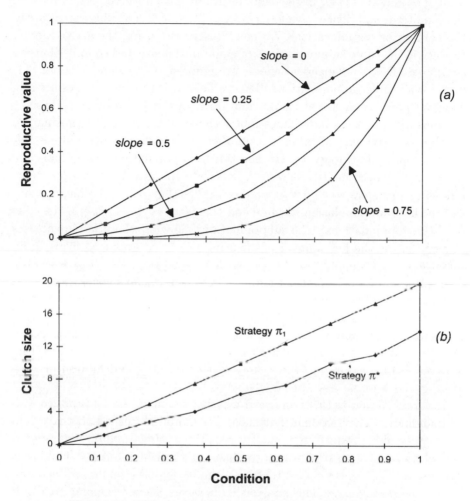

Figure 8.3. The consequences of intergenerational effects (Example 8.6). (*a*) Reproductive value as a function of condition for various strengths of maternal effects. When *slope* = 0 there are no maternal effects and reproductive value is proportional to the resource available to provision offspring. As *slope* increases maternal effects become stronger. (*b*) The number of offspring produced under the optimal life-history strategy, π^*, and the strategy of maximisation of the number of surviving offspring, π_1. $\beta = 0.25$ throughout; in (*b*) *slope* = 0.5.

Fitness is a measure of the number of descendants left far into the future, and the method used to calculate optimal strategies presented in Section 8.6 is based on maximisation of the number of such descendants. Other strategies may do better in the short term but have a worse performance in the long term and hence have lower fitness. Taylor (1985) illustrates this in the context of sex allocation. We now illustrate it using the above model. In Figure 8.4, three subpopulations of equal initial size and composition are compared. Density-dependent effects are ignored. (The same value of α is used for all three subpopulations. This value of α is the one that results in a constant population size at equilibrium if the optimal strategy is followed.) In one subpopulation, individuals follow the strategy π_1 of maximising the number of surviving offspring. In another subpopulation, individuals follow the optimal strategy π^*. In the third subpopulation, individuals use a strategy π_2 intermediate beween π_1 and π^*. Initially the subpopulations following strategies π_1 and π_2 increase, but as they do so the mean qualities in these subpopulations decline and this leads to a decrease in the sizes of both subpopulations. The subpopulation following π^* initially decreases in size, but as the low number of offspring produced under π^* leads to an increase in mean quality of this population, the population size levels off. This subpopulation is smaller than that following π_1 until generation 4 and is smaller than that following π_2 until generation 7. In the long term it is, however, much bigger. After 100 generations the sizes of the populations following π_1, π_2 and π^* are < 1, 186 and 8462 respectively. This example shows that strong maternal effects may mean that counting offspring is a very bad measure of fitness, and that counting grandchildren or even great-grandchildren may not be adequate.

Trivers & Willard (1973) analysed whether a female should produce sons or daughters. They assume that there are maternal effects, so that high-quality females produce high-quality offspring. High-quality sons are assumed to be able to produce more offspring than high-quality daughters. Trivers and Willard concluded that high-quality females should produce sons rather than daughters. This analysis is, however, flawed (Leimar 1996). It counts the number of grandchildren produced by the focal female, but fails to differentiate between high- and low-quality grandchildren. If, as assumed, there are maternal effects, high-quality daughters will tend to produce high-quality offspring, whereas high-quality sons will not (unless there are paternal effects). Thus if the focal high-quality female produces daughters, her grandchildren will tend to be of high quality, while if she produces sons,

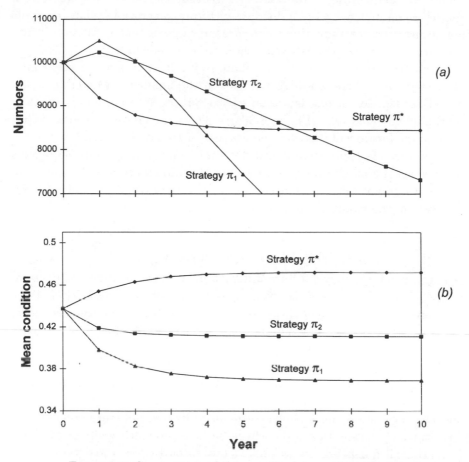

Figure 8.4. Comparision of the growth of descendant numbers under three strategies. Strategies π^* and π_1 are as in Figure 8.3(b). Under strategy π_2, clutch size is halfway between that under π_1 and that under π^*. Initially the three subpopulations following the three strategies each have 10 000 individuals, of which 5000 have condition 0.375 and 5000 have condition 0.5. (a) Subsequent sizes of each subpopulation. (b) Subsequent mean condition in each subpopulation. Example 8.6 is used, with $\beta = 0.25$, $slope = 0.5$ and $\alpha = 0.2402$.

her grandchildren will tend to be around the population average. Leimar (1996) shows that taking the quality of grandchildren into account can make it optimal for high-quality females to produce daughters. The analysis of Leimar works with the relative reproductive values of high-quality sons and daughters rather than looking at descendants left in future generations directly. Of course, the reproductive value of an individual is a measure of the descendants left by the individual in future generations. This example emphasises that reproductive value measures descendant numbers far into the future and does not just count children or grandchildren.

Populations are likely to be dominated by relatively few maternal lines especially when there is a high variance in the numbers of daughters produced per female. Strong maternal effects can produce large differences in reproductive value (e.g. Figure 8.3(a)). The numbers of descendants left far into the future will then vary enormously across individuals even though the numbers of offspring produced does not. Thus strong maternal effects will reduce the number of maternal lines that dominate a population and may have important consequences for the diversity of mitochondrial DNA observed in the population.

8.9 Phenotypic plasticity

A single genotype may produce different phenotypes in different environments. This phenomenon is referred to as phenotypic plasticity and the relationship between the environment and the resulting phenotype is known as the genotype's norm of reaction. Part of the reaction of a genotype to its environment may be the result of constraints imposed by the environment, while other aspects of the reaction may be adaptive (e.g. Schlichting 1986, Scheiner 1993, Gotthard & Nylin 1995). In the state-dependent approach to life histories that we have presented in this chapter, a strategy is a rule that specifies how an organism should respond to its internal state and external environment. Within this approach the set of reproductive actions available to an organism depends on its state. Thus, in particular, the external environment, as a component of an organism's state, restricts what options are available to the organism. An optimal strategy is the best plastic strategy within these constraints. Thus our framework provides a natural setting in which to discuss adaptive plasticity and the resultant optimal norms of reaction (Houston & McNamara 1992).

In Section 8.6 it was emphasised that fitness is a property of strategies, not of individuals or individual actions. Fitness is typically a non-linear function of the whole suite of actions prescribed in different states. In particular, it is easy to construct examples in which the best action to perform in one state depends on what the organism would have done were it in another state. Because of such effects, natural selection must be thought of as acting on a genotype's whole norm of reaction, not acting independently on the behaviours in different states.

These concepts can be illustrated by considering a population divided between habitats, where a proportion of the offspring produced in each habitat migrate to other habitats before maturity. Stearns & Koella (1986) analysed life-history decisions within such a setting. They treated each habitat in isolation, effectively assuming that all offspring lived in the same habitat as their parents. They then found the life-history strategy within a habitat that maximised the growth rate of descendants in that habitat. When there is migration between habitats this approach is wrong because it ignores the dependence of the best action in one habitat on the quality of other habitats and the actions taken in these habitats. The correct approach has been described by Houston & McNamara (1992). The analysis of Kawecki & Stearns (1993) is also correct, but only deals with a special case of the problem considered by Houston and McNamara. Kawecki and Stearns assumed that the probability that an offspring produced in one given habitat migrates to another given habitat is fixed and outside parental control. This means that there can be no parental control of the degree of dispersion.

There is a final twist to this analysis when the population as a whole is at a density-dependent equilibrium size, with some habitats acting as net sources and others as net sinks. If the population is evolutionarily stable, then the growth rate under an optimal strategy is $\lambda^* = 1$. Under these circumstances, the optimisation criterion used by Kawecki and Stearns reduces to that of maximisation of the expected number of offspring produced over an organism's lifetime (McNamara 1993a; see Section 8.6). Thus, given the assumptions of fixed offspring migration probabilities and of a dynamic and evolutionary equilibrium, the habitats can be treated in isolation. When migration probabilities are not fixed, habitats cannot be treated in isolation even when there is a dynamic and evolutionary equilibrium. It is then necessary to use the full analysis of Houston & McNamara (1992).

8.10 Empirical problems

It is clear from our description of the state-dependent approach to life-histories that a considerable amount of information is necessary if a state-dependent model is to be constructed. First, it is necessary to identify the relevant state variables. Then we need to establish the trade-offs. In other words, we need to determine the four functions S_{mat}, p_{xy}, N_{off} and b_{xy} of Section 8.5. In determining these functions, we have the general problem of establishing what would have happened had the organism adopted a behaviour other than the one that it chose. In the context of state-dependent models there is the additional problem of interactions between state and behaviour. For example, the effect of reproductive effort on maternal survival could depend on maternal condition. Such interactions might be crucial in determining the form of the optimal life-history strategy (Kisdi *et al.* 1998).

To illustrate some of the problems that can arise when state-dependent models are constructed, we now discuss a life-history model of Soay sheep. Clutton-Brock *et al.* (1996) used a long-term study of Soay sheep (*Ovis aries*) on St. Kilda to construct a model that predicted whether, in a given year, a female should not reproduce, should give birth to a single offspring or should give birth to twins. The study provided information about the lifetime reproductive success of marked individuals: some members of the population were caught once a year and weighed. An interesting feature of this population is that its size fluctuates with roughly a three-year cycle.

Given the information available on the sheep, Clutton-Brock *et al.* developed a model with the following state variables: weight (four classes); age (less than one year old, one–six years old, and over six years old); stage of the population cycle. (The years are classified according to the level of mortality. The years with the highest mortality are defined to be crash years.) Statistical analysis was then used to estimate the various functions needed for a life-history model. For example, Clutton Brock *et al.* needed to know the probability that a female survived to year $t + 1$ given her state and decision in year t. Generalised linear modelling was used to estimate this probability. The case of females older than one year (two categories combined) in non-crash years is particularly interesting. It can be seen from Figure 1(*d*) of Clutton-Brock *et al.* (1996) that survival probability is higher for females that reproduce than for females that do not reproduce. We do not believe that this result indicates a genuine advantage that results from breeding. It is more reasonable to suggest that there was an underlying difference in condition between females of which Clutton-Brock *et al.* were unaware and hence did not include in the female's state. Females in good condition may choose to breed and still survive better than females in poor condition that choose not to breed.

Given the limited information available on the state of the sheep, the only way in which Clutton-Brock *et al.* (1996) could investigate the importance of an unknown condition variable was by carrying out a sensitivity analysis. This analysis suggested that the optimal strategy was robust, in that it was not markedly changed if the difference in survival as a consequence of reproduction in non-crash years was removed.

Although we have highlighted the problem that arose with a particular component of the data, it is important to note that all the effects that were analysed by Clutton-Brock *et al.* are based on natural variation. The limitations of using unmanipulated data to estimate life-history costs are well known (e.g. Partridge & Harvey 1988; Lessells 1991). In the context of our approach, we can put the problem starkly. If there is a deterministic relationship between state and behaviour, then all animals that are in the same state will do the same thing. If animals that we take to be in the same state do not do the same thing, then either the rule that specifies behaviour is stochastic or we have not correctly identified the state. Even if we have correctly identified the state, we still need to establish the consequences of performing various actions when in a given state. It may be possible to estimate some of these consequences by performing manipulations. For example, Sinervo & DeNardo (1996) used both natural variation and a variety of manipulations to estimate the consequences for a lizard of various clutch masses.

State-dependent models can be useful in predicting the consequences of experimental manipulation. Once the model has been constructed and the optimal strategy has been found, we can of course run forward to find the expected behaviour of an organism following this strategy (Section 3.3). We can, however, also introduce a 'manipulation' in the model by, for example, imposing a state change on the organism and seeing how this changes expected behaviour under the optimal strategy. We illustrate this in Section 9.5 in the context of the influence of food supply on the timing of reproduction. In this case the effect of supplementary feeding is represented by imposing an increase in energy reserves on the model animal. McNamara & Houston (1996) modelled experiments in which clutch size is manipulated by imposing a change in clutch size on the model animal. They showed that results similar to those found empirically by Gustaffson *et al.* (1994) can be obtained.

8.11 Further issues

Two sexes. The worked examples that we have given have been based on either an asexual species or a species with no difference between the sexes. However, in many sexual species, the sexes have fundamentally different life-histories, and a model of optimal life history needs to include explicit differences between the sexes and allow for decision-making by each sex. The way to do this is to include sex as a component of state. Then, by allowing the decision to depend on its state, we are allowing decisions to depend on sex and other aspects of state, e.g. size, energy, reserves, territory quality. Taylor (1990) used the covariance formula of Price (1970) to establish a general framework in which the decisions of the sexes can be analyzed. The central idea is that reproductive value depends on the sex and hence on the state of an individual. The idea of different reproductive values for males and females has been used by a number of people to understand sex-ratio decisions (see Charnov 1982, Taylor 1985 for reviews). Given that the sex-ratio decision of the mother does not affect her in the future, she should choose the sex ratio so as to maximise the reproductive value of her offspring, where males and females may have different reproductive values. A similar principle applies to general problems of sex allocation, including sex change (Charnov 1982, 1993). Iwasa (1991) illustrated the application of dynamic programming to the problem of size-dependent sex change.

Frequency dependence. The optimality criterion that we have used (that of maximizing λ) takes the background environment as given. This background will, of course, depend not only on the physical environment but also on the behaviour of population members. In the terminology of Chapter 7, we have assumed a resident population strategy and have given a procedure for finding the best response. In this sense, we have spelt out one component of an ESS analysis. The full analysis requires us to find a life-history strategy which is the best response to itself. For a population-genetics perspective on this, see Day & Taylor (1996).

If an optimisation model is constructed without specifying the way in which the population behaviour affects the background (in other words, if we ignore what Mylius & Diekmann 1995 call 'environmental feedback'), then the parameters used in the optimisation model are unconstrained. By suitable choice of parameter values, it may be possible to obtain a wide range of effects from the model. If, however, we include the environmental feedback, then the condition that a population is at a density-dependent equilibrium will constrain the parameter values. As a result, some of the effects originally found may disappear. The implication is that models without environmental feedback may produce effects that are not biologically realistic. It is for this reason that the model described in Section 8.7 contains implicit density dependence on juveniles. Without this density dependence, the population might grow rapidly, favouring early reproduction, or might decline in size, favouring late reproduction.

Environmental feedback is an essential component of sex-allocation problems. The behaviour of the resident population determines the sex ratio and hence the reproductive values of males and females. Feedback is also essential if a realistic model of competition is to be achieved. Examples of models in this category include Abrams (1983, 1994), Maynard Smith & Brown (1986) and Kawecki (1993) (see also Chapter 7). Including feedback may lead to more than one ESS (Abrams 1983) or to a mixed ESS (Heino *et al.* 1997). For an example of the effect of environmental fluctuations on multiple ESS see Kisdi & Meszéna (1993).

When considering the invasion of a rare mutant into a resident population, we assume that the distribution of states of the cohort of descendants of the mutant settles down to a unique equilibrium distribution over time. This is a reasonable assumption provided that numbers in the cohort are small compared to the population as a whole. If, however, we are considering the distribution of states of members of the whole population, then this distribution may not settle down to a unique equilibrium distribution. In following the population as a whole forward in time, non-linear effects can result in chaotic or oscillatory behaviour of the distribution over states. Alternatively, the latter may settle down to a distribution which is not unique and which depends on initial conditions. What this means is that the physical environment and the resident population strategy do not uniquely determine the background experienced by population members. McNamara (1994a) considered a population with two age classes, in which older individuals outcompete young individuals for resources. A life-history strategy specifies the allocation of resources to survival versus reproduction as a function of age and total resource obtained. McNamara showed that if the population follows a particular life-history strategy, then it will settle down to one of two stable age distributions. At one distribution individuals allocate all their resources to reproduction and there is semelparity. At the other distribution some resources are devoted to survival and there is iteroparity. At each of these distributions, all animals are doing the best that they can, given the environment that they themselves create. This type of phenomenon may occur in a wide range of frequency-dependent systems, particularly in ones with the strong interactions that may characterise social behaviour. If it does, then two populations of a species might be found at what appears to be two totally different ESSs. It might be mistakenly assumed that either the physical environments are different or that the behavioural rules are different. The point of the example is that both populations might be following exactly the same 'high-level' rule but be trapped in different behaviours by the background that the behaviour generates.

Kin selection.   If young that are cared for by parents do not make any decisions, then they can be included in the state of the parents (Sections 2.2, 3.7 and 9.5). The problem becomes more complex if the young are making decisions while interacting with their parents. Problems like this are not specific to interactions between parents and young; they arise whenever there are two-way interactions between kin. The general theoretical framework for handling structured populations when there are interactions between kin is provided by Taylor (1990), and is further developed and illustrated by Taylor & Frank (1996).

Senescence.   Partridge & Barton (1996) defined aging or sensescence as a deterioration in state with age. Broadly speaking, there are two (not necessarily exclusive) explanations for aging. The first is that because selection pressure can be expected to decrease with age there will be a higher frequency of deleterious mutations whose effects occur late in life (e.g. Hamilton 1966, Partridge & Barton 1993, Charlesworth 1994). The second is that aging is part of an optimal life-history (e.g. Williams 1957, Kirkwood & Rose 1991, Partridge & Barton 1993, Charlesworth 1994, Abrams & Ludwig 1995). Within the second class of explanation, the 'disposable soma' model assumes that the level of resources allocated to repair of the body (as opposed to reproduction) determines the state of the body and hence the extent of aging (Kirkwood & Rose 1991, Abrams & Ludwig 1995; cf. our use of condition in Sections 4.8 and 9.5).

In Section 8.7 we described a model in which both condition and reproductive effort are found to increase with age under the optimal life-history strategy. A consequence of the increasing reproductive effort was an increase in mortality with age. McNamara & Houston (1996) referred to this as 'apparent sensescence' because although there is an increase in mortality there is no deterioration in state and hence no true senescence. Thus an observed increase in mortality with age cannot by itself be used to infer senescence.

Appendix 8.1 Vector notation for ρ and V

Let $n(t)$ be given by equation (8.5). Dividing both sides of equation (8.4) by $n(t)$ and writing $n_y(t+1)/n(t)$ as $[n(t+1)/n(t)][n_y(t+1)/n(t+1)]$ we obtain

$$\lambda(t)\rho_y(t+1) = \sum_x \rho_x(t)a_{xy},$$

where $\lambda(t)$ is given by equation (8.6) and $\rho_x(t)$ is given by equation (8.7). Thus letting t tend to infinity we find that

$$\lambda\rho_y = \sum_x \rho_x a_{xy} \qquad (A8.1.1)$$

for every state y. In vector notation this equation is $\lambda\rho = \rho A$ and shows that λ is an eigenvalue of the matrix A and ρ is a left eigenvector. λ is a particular eigenvalue of the matrix A known as the Perron–Frobenius eigenvalue. It is a positive real eigenvalue and is greater than or equal to the modulus of any other eigenvalue. When A is primitive, all other eigenvalues have strictly smaller moduli.

To obtain the equation determining $V(x)$, divide both sides of equation (8.11) by $f_{n-1}(x_0)$, where x_0 is a reference state, to obtain

$$\tilde{\lambda}_n V_n(x) = \sum_y a_{xy} V_{n-1}(y)$$

where $\tilde{\lambda}_n = f_n(x_0)/f_{n-1}(x_0)$ and $V_n(x) = f_n(x)/f_n(x_0)$. As the time to go increases, $V_n(x) \to V(x)$, $V_{n-1}(y) \to V(y)$ and $\tilde{\lambda}_n$ tends to a limiting value λ. Thus we have

$$\lambda V(x) = \sum_y a_{xy} V(y). \tag{A8.1.2}$$

Here λ does not depend on the reference state x_0 that is chosen. The reference state affects the scaling of V but not the relative value $V(x)/V(y)$ of any states x and y.

In vector notation, equation (A8.1.2) can be written as $\lambda \mathbf{V} = A\mathbf{V}$ and shows that V is the right eigenvector of the population projection matrix. λ is the Perron–Frobenius eigenvalue of A as before.

Appendix 8.2 Primitivity of the projection matrix

A population is said to be primitive if there is some time t such that, whatever the cohort composition at time 0 (i.e. whatever are the $n_x(0)$, provided not all of them are zero), there will be cohort members in every possible state at time t (i.e. $n_x(t) > 0$ for every x). (This definition is equivalent to the statement that there is a t such that all the elements of the matrix A^t are positive.)

The matrix of Example 8.2, given in Table 8.1(b), is obviously primitive. To see that the matrix of Example 8.1 is also primitive, first suppose that the initial cohort members are all of age 1. Then when $t = 1$ all cohort members are age 2. At $t = 2$, some are age 1 and some are age 3. At $t = 3$, some are age 1 and some are age 2. At $t = 4$ there are some cohort members in all three age classes, and this remains true for all later times. Similarly, it can be seen that if the initial cohort are all age 2, then there are cohort members in each age class for $t \geq 3$, while if the initial cohort are all age 3 then there are cohort members in each age class for $t \geq 5$. Thus, whatever the initial cohort composition, there are cohort members in each age class when $t = 5$.

There are two obvious circumstances under which primitivity fails. (i) There may be 'dead-end' states. An organism in such a state will never subsequently leave any surviving offspring. Post-reproductive states are dead-end states (provided that the individual is no longer caring for offspring), as are states for which the organism's condition is so poor as to preclude future reproduction. (ii) There may be non-overlapping generations with a fixed generation time, τ, of several years. In both cases, (i) and (ii), a simple modification removes the problem. In case (i) dead-end states are essentially equivalent to the organism's being dead. Thus we may remove such states from our state space and obtain a reduced state space such that all organisms have some possibility of leaving descendants in the future regardless of their state. In case (ii) we can take the population census every τ years instead of every year. Once these two modifications are made the projection matrix will typically be primitive.

For further discussion of primitivity see Caswell (1989).

Appendix 8.3 Comparison of the criterion for an optimal life-history strategy and the dynamic programming equations of Chapter 3

A life-history strategy is optimal if and only if any individual following the strategy always maximises expression (8.24). Let π^* be an optimal strategy. Then from the above criterion and equation (A8.1.2) we have

$$V^*(x) = (\lambda^*)^{-1} \max_u \sum_y a_{xy}(u) V^*(y), \qquad (A8.3.1)$$

where $\lambda^* = \lambda^{(\pi^*)}$ is the fitness of the optimal strategy. Not only do V^* and λ^* satisfy equation (A8.3.1) but, provided there is a primitive optimal strategy, any solution to equation (A8.3.1) for positive V^* and λ^* gives the reproductive value and fitness respectively under an optimal strategy. For a derivation and discussion of this equation for age-structured populations, see Schaffer (1974a) and Taylor *et al.* (1974). For its derivation and properties for a general state-structured population, see McNamara (1991, 1993b). Here we give an heuristic derivation of the equation from the dynamic programming equations of Chapter 3.

Suppose that the decision epochs $t = 0, 1, 2, \ldots$ in equations (3.1) and (3.3b) are taken to be one year apart. Following the terminology of these equations let $V(x, t)$ be the reproductive value of an individual in state x at time t given that its behaviour is optimal. Translating the components of equation (3.1) into the language of the present chapter, we have

$$B_{\text{off}}(x, t; u) = N_{\text{off}}(x; u) \sum_y b_{xy}(u) V(y, t+1)$$

and

$$S(x, t; u)\mathbb{E}_u[V(X', t+1)] = S_{\text{mat}}(x; u) \sum_y p_{xy}(u) V(y, t+1).$$

Thus equation (3.3b) becomes

$$V(x, t) = \max_u \sum_y a_{xy}(u) V(y, t+1), \qquad (A8.3.2)$$

where $a_{xy}(u)$ is given by equation (8.14).

Let $V(x, t)$ be taken as the number of descendants left in some target year T under the optimal strategy by an individual in state x in year t. Then for large $T - t$ we have

$$V(x, t) = \lambda^* V(x, t+1), \qquad (A8.3.3)$$

since λ^* is the growth rate under the optimal strategy. Thus we can write

$$V(x, t) = (\lambda^*)^{-t} V^*(x) \qquad (A8.3.4)$$

where $V^*(x) = V(x, 0)$. Similarly

$$V(y, t+1) = (\lambda^*)^{-(t+1)} V^*(y). \qquad (A8.3.5)$$

Using equations (A8.3.4) and (A8.3.5) to substitute for $V(x, t)$ and $V(y, t+1)$ in equation (A8.3.2) then gives equation (A8.3.1).

9

Routines

9.1 Periodicities in the environment

There are several periodicities in the environment that are biologically important. An obvious one is the day–night cycle. The foraging ability of animals that use vision to detect their food will be influenced by diurnal variation in light levels. The temperature difference between day and night may also be a significant factor. The tidal cycle determines food availability for wading birds, which feed when the tide is low, and animals such as barnacles and limpets, which feed when covered by the tide. Most temperate-latitude animals are affected by the annual cycle, which determines, among other things, food availability, temperature and day length. In this chapter we are concerned with the optimal behaviour of an organism when its environment is periodic.

If there is an important periodic environmental variable, then an optimal strategy will specify behaviour as a function of the organism's state and the time within the cycle. In this sense, optimal strategies are periodic. This does not imply, however, that the resulting behaviour will follow the same cycle. The state variables that characterise the organism either may not vary in a cyclic way or may have a cycle that is different from that of the strategy. For example, the grey-headed albatross typically breeds every other year (Prince *et al.* 1981). Presumably, there are some underlying physiological variables that determine whether the bird should attempt to breed (see Section 8.7). We refer to these variables as the bird's condition. We expect condition to decrease during reproduction and to increase afterwards. The optimal strategy specifies breeding at a particular time during a year if condition is above some critical level at this time. This critical level will depend strongly on the time of year. Given that there are no age-specific effects, the strategy will be the same from year to year, and will be cyclic

264

with a period of one year. Whether breeding occurs every year depends on how steeply condition declines during breeding and on how quickly it recovers. For the grey-headed albatross, a possible explanation is that condition cannot recover sufficiently quickly for breeding to occur in successive years; instead, breeding has a cycle of period two years.

In this chapter we present a general modelling framework that allows us to explore a range of issues concerning optimal behaviour in a periodic environment. As we have just discussed, a periodic environment need not result in periodic optimal behaviour, but if this were the case the period might not be the same as that of the environment. Our framework allows us to investigate this issue. Given that behaviour follows a regular cycle we can investigate the scheduling of activities over the cycle. For example, we might be concerned with predicting the number of reproductive bouts per year and when these bouts occur. If the organism performs other important activities such as migration, we might be interested in both the timing of reproduction and migration during the annual cycle.

We begin by discussing why the optimal scheduling of activities over a cycle is potentially problematic.

9.2 The problem of scheduling activities over a cycle

The optimal scheduling of activities over a cycle is a potentially difficult problem because whether an activity should be performed at a given time cannot be determined solely on the basis of whether the time is good or bad for performing the activity. We can distinguish two reasons for this.

(i) The performance of one activity may preclude the organism's performing another activity. In deciding whether to perform one activity now, it is necessary to consider whether there are good times to perform the other activity at other times within the cycle.

(ii) Performing an activity may influence an organism's state in the future and this may restrict the options that are available in the future and change their consequences.

These factors, which prevent a local analysis of the best decision, are common to dynamic optimisation problems in general. When dealing with cyclic problems we face the added difficulty that there is no final time from which to work back.

To provide a specific example, consider the annual routine of a bird in a seasonal environment. We might be interested in the timing of breeding, how many broods are produced, the timing of the moult and the behaviour adopted during the winter. As a consequence of finding the annual routine, we would also establish the patterns of energy reserves and energy expenditure together with the levels and causes of mortality through the year. The timing of breeding is a topic that has received considerable attention and has been discussed in terms of the time at which the parents are best able to feed dependent young (e.g. Lack 1950). There are, however, various other factors that may influence the timing of breeding and that link breeding with other activities during the year. Adults need to get into condition after the winter in order to start breeding (Perrins 1970, Drent & Daan 1980). They also need to moult at some time during the year. If breeding and moulting cannot be performed at the same time then the bird must decide which activity to perform first. The timing of the moult will influence the state of the feathers throughout the year, and hence the ability to survive the winter, build up condition and feed the young (Nilsson & Svensson 1996). The timing of breeding will affect the condition of the parent and young at the start of winter and hence affect their ability to survive the winter. Finally, the bird's foraging behaviour during winter will influence not only the probability of survival but also how quickly the bird can get into condition to breed in the spring.

It is clear from this example that the best action to choose in a particular state at a particular time of year depends on the behaviour adopted at all other times of the year. When the value of present actions depends on behaviour adopted in the future, the standard method of analysis is to use dynamic programming, working backwards from the future. It is not immediately obvious how to apply this technique here because there is no final time at which to start the backward induction process. If we knew the reproductive value as a function of state at some time during the cycle, we could take this time as final time, take the reproductive value as the terminal reward and work backwards over the cycle. Typically, however, we do not know *a priori* the reproductive value at any time of the year because it always depends on future behaviour. In this chapter we describe a technique for breaking into the cycle to establish both the reproductive value and the optimal strategy (and hence the optimal routine).

Previous work on annual routines (e.g. Iwasa & Cohen 1989, Holmgren & Hedenström 1995) has taken lifetime numbers of offspring as the fitness criterion. If, at independence, all young are equally valuable, then this is a reasonable optimisation criterion. There are two reasons why young might differ in value at independence. First, they may be in different states at a given time. It is important to allow for this possiblity when analysing the trade-off between the number and quality of offspring (Section 8.8). Second, they may be produced at different times of the year. It is well known in birds that young produced early in the season have a higher reproductive value than young produced later in the season (see Daan *et al.* 1989 for a review). This may be because young produced earlier have a better chance of getting into good condition before the winter. For small rodents, young produced early in the year have a chance to reproduce while conditions are still favourable, and hence these young have a higher reproductive value than young produced later in the year. The problem of dealing with the time at which young are produced does not arise in the models of Holmgren & Hedenström (1995) because in them reproduction occurs at a fixed time. If we wish to investigate the optimal time at which to reproduce, then Holmgren and Hedenström's criterion based on counting offspring will no longer suffice. The technique that we present automatically computes how the reproductive value of young depends on their state at independence and time of independence.

9.3 Daily routines

In this section we discuss routines that are driven by the day–night cycle. To keep the exposition simple, we confine ourselves to cases in which no young are produced. A general formalism for dealing with the production of young is presented in the next section. We start with a detailed analysis of a particular example.

Surviving a winter: daily routines of feeding and resting

Our example is based on a small bird trying to survive the winter. All days in the winter are taken to be identical in terms of the parameters that determine food availability, energy expenditure etc. Thus we are ignoring seasonal effects. Within the model, death occurs as a result of starvation or predation, and there is a energy–predation trade-off (cf. Chapter 6). The model can be used to investigate both foraging routines (McNamara *et al.* 1994) and regulation of body reserves (Houston & McNamara 1993).

A day in winter is taken to be a 24-hour period that starts at dawn and finishes at the following dawn. We refer to dawn as time of day 0 and dusk as time of day T. At each time t during the daylight period ($0 \leq t < T$) a bird can choose between resting or feeding. If it rests it finds no food but is safe from predators. If it feeds it finds a stochastic amount of food and risks being killed by a predator. Both activities are metabolically costly. The resting metabolics do not depend on body mass. Metabolic expenditure while feeding increases with body mass (see McNamara & Houston 1990b, Witter & Cuthill 1993, Houston *et al.* 1997). Since body mass increases with energy reserves, metabolic expenditure while feeding also increases with energy reserves. Overnight a bird rests. The metabolic energy loss overnight is taken to be a random variable with probability density function $\eta(e)$ specifying the probability that overnight expenditure is e. The bird dies of starvation if its energy reserves fall to zero at any time.

For this scenario we take a bird's state variable to be its level of energy reserves, x. If we knew the reproductive value $V(x, T)$ at dusk for all reserves x, then we could take this function as a terminal reward at time T and find the optimal strategy over a day by working backwards from T using dynamic programming. Unfortunately, we do not know the function $V(x, T)$. As a first approximation, we could take the terminal reward at dusk to be the probability that a bird survives to the following dawn. This terminal reward is given by

$$R_0(x) = \int_0^x \eta(e)de. \tag{9.1}$$

This function ignores the possible value of having more than the minimum level of reserves at the following dawn (McNamara & Houston 1982). The value of such reserves depends on the strategy adopted by the bird on subsequent days. To find the correct terminal reward, we need to embed a single day within a winter.

We now look at the strategy that maximises the probability that a bird survives the winter. Suppose that there are n full days to go after the present day and that a bird has energy reserves x at time of day t on this day. Let $f_n(x, t)$ denote the maximum probability that the bird survives the winter. Then we may compute $f_n(x, t)$ for all n, x and t as follows. First we set

$$f_0(x, T) = R_0(x) \tag{9.2}$$

since $R_0(x)$ is the probability that a bird with reserves x at dusk on the last day of the winter survives until the following dawn. Given $f_0(x, T)$ we can use this as a terminal reward at dusk and use dynamic programming to work backwards over the day. This yields $f_0(x, t)$ for all reserves x and times of day t. In particular this process yields the function of state $f_0(x, 0)$ that specifies how a bird's survival probability depends on its reserves at dawn on the beginning of the last day. The survival probability for a bird with reserves x at the previous dusk can then be computed from the formula

$$f_1(x, T) = \int_0^x \eta(e) f_0(x - e, 0) de$$

where $\eta(e)$ is the probability density function for the energy e expended overnight. Taking the function of reserves $f_1(x, T)$ as a terminal reward we can then use dynamic programming to find $f_1(x, t)$ for all x and t, and so on.

For fixed reserves x and time of day t, the sequence of survival probabilities $f_n(x, t)$ tends to zero as the number of days to go, n, tends to infinity. As in Chapter 8, however, one can rescale the functions f_n. Choose some reference level of reserves x_0 and set

$$V_n(x, T) = \frac{f_n(x, T)}{f_n(x_0, T)}.$$

Then $V_n(x, T)$ gives the survival probability of a bird with reserves x relative to a bird with reserves x_0, given that there are n days to go in winter. Strong backward convergence (Section 3.5) ensures that for each fixed x, $V_n(x, T)$ tends to a limit as n tends to infinity. We denote this limit by $V(x, T)$, i.e.

$$V(x, T) = \lim_{n \to \infty} V_n(x, T).$$

The convergence of V_n to V is illustrated in Figure 9.1.

The optimal strategy with n more days to go until the end of winter maximises the expected value of the terminal reward $f_n(x, T)$ at dusk on this day. The strategy obtained is, however, unaltered if this terminal reward function is replaced by another that is a positive multiple of $f_n(x, T)$. Since $V_n(x, T)$ is such a multiple, the optimal strategy with n days to go maximises the expected value of the terminal reward $V_n(x, T)$ at dusk on this day. Since $V_n(x, T)$ tends to a limit as n tends to infinity, the strategies on successive days settle down to a limiting strategy as n increases. Under this strategy the behaviour on each day maximises the expected value of $V(x, T)$ at dusk on that day. The function $V(x, T)$ is thus the terminal reward that we seek and specifies how reserves, x, at dusk contribute to future survival, given that there is a long time to go until the end of the winter. We can see that $V(x, T)$ is the correct terminal reward because it is self-consistent in the following sense. If $V(x, T)$ is taken as the terminal reward at dusk on a given day and we work backwards over the 24-hour period to find the reproductive value (relative to the same reference state x_0) at dusk on the previous day, then this reproductive value again equals $V(x, T)$.

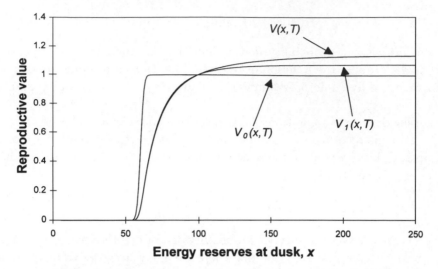

Figure 9.1. Convergence of the reproductive value of reserves at
dusk to its asymptotic limit. Let $V_n(x, T)$ be the reproductive
value of energy reserves x at dusk n days from the end of winter.
The figure shows $V_0(x, T)$, $V_1(x, T)$ and $V(x, T) = \lim_{n \to \infty} V_n(x, T)$.
The reference state x_0 is 100. The foraging model is taken from
McNamara *et al.* 1994a. (The model *foraging vs rest, continuous
foraging* with $G = 300$, $e = 15.5$, other parameters baseline).

We emphasise that the correct terminal reward $V(x, T)$ is not something
that is initially assumed; it emerges from the calculations at the same time
as the optimal strategy.

Although we have chosen to define f_0 via equation (9.2), strong backward
convergence (cf. Section 3.5) also implies that V is independent of the choice
of f_0. The choice that we have made assumes that all that is important for
the animal is surviving the winter. In reality, it may be valuable for a bird
to have high reserves at the end of winter in order to migrate or breed.
Although the form of f_0 will influence the strategy adopted near the end of
winter, the convergence result means that the daily strategy a long way from
the end of winter is independent of f_0 and hence of the details of behaviour
after the end of winter.

To analyse the routine that results from following the limiting strategy,
we consider a large group of birds that independently follow the strategy.
Near the start of winter, the energy reserves of group members will depend
on the reserves at the start of winter. Strong forward convergence (Section
3.5) means that the distribution of reserves for birds that are still alive will

become independent of the initial distribution as winter progresses. At a given time of day, the distribution of reserves of the remaining group members settles down to an equilibrium distribution that is the same on each day. As a consequence, the proportion of group members performing a given activity at this time of day also settles down to an equilibrium value. This is true for all times of day, and so we obtain a mean dailyroutine for the population as a whole. Figure 9.2 shows a limiting strategy, the corresponding equilibrium distribution of reserves over the day and the resulting mean daily routine of the population.

Although the mean population behaviour settles down to a daily routine, this does not logically imply that the behaviour of each individual in the population settles down to the same routine. A given individual will have good and bad luck in finding food and will hence spend different amounts of time foraging on different days.

Our model of foraging and resting is based on a trade-off between gaining energy and avoiding predation (see Chapter 6). We have, for simplicity, ignored seasonal effects; this is reasonable if the energy reserves of a bird change on a time scale that is rapid compared to the time scale of seasonal changes. In order for the strategy to settle down to depend only on the time of day and not on the time in winter, there must be either an upper limit on energy reserves or some cost that increases as energy reserves increase (Houston *et al.* 1997). From the model, we obtain two routines. One is the routine of expected behaviour (e.g. the proportion of birds foraging as opposed to resting). The other is the routine of expected levels of reserves throughout the day. In particular, Houston & McNamara (1993) used a model of this type to predict how fat levels at dusk will depend on environmental conditions.

The daily routine of foraging is driven by two factors: mass-dependent costs and stochasticity. An increase in metabolic expenditure with body mass (or an increase in predation risk with body mass, Houston & McNamara 1993, McNamara *et al.* 1994) favours a routine in which foraging is postponed until late in the day. For a given level of reserves at dusk, such a routine keeps mass low and hence avoids large mass-dependent costs (McNamara *et al.* 1994). Stochasticity means that if a bird postpones building up its reserves until late in the day, it cannot be sure of attaining a given level of reserves by dusk. Furthermore, if the bird allows its reserves to fall to a low level during the day, it risks dying from starvation if, through bad luck, it fails to find food. Thus stochasticity favours a routine in which the bird forages at the start of the day in order to build up reserves. Figure 9.2(*b*) illustrates the resultant average routine when both factors operate.

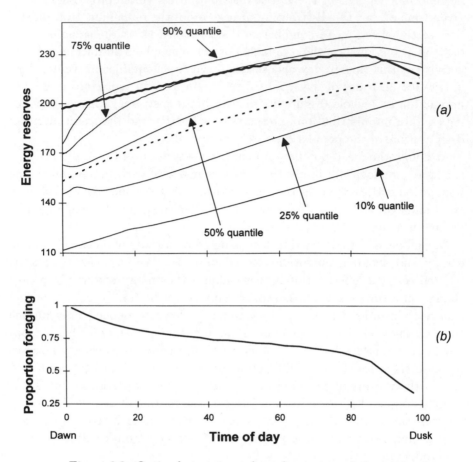

Figure 9.2. Optimal strategy and resultant mean daily routine in
the model used to generate Figure 9.1. (*a*) The critical level of
reserves below which it is optimal to forage (bold line), the mean
level of reserves over the day (broken line), and the 90%, 75%, 50%,
25% and 10% quantiles in reserves; the $x\%$ quantile is a level of
reserves such that $x\%$ of the population have lower reserves. (*b*)
The resultant proportion of birds foraging over the day (averaged
over each four successive time intervals).

Various modifications to this basic model have been made. McNamara *et
al.* (1994) allowed birds a range of foraging options that differ in both their
mean gain and their predation risk. This tends to result in a spreading out of
the routine, such that birds forage more evenly throughout the day (cf. the
risk-spreading theorem; see Houston *et al.* 1993 and Section 6.4). McNamara
et al. also considered the effects of interruptions to foraging caused by factors

such as snow cover. Incorporating such factors can significantly alter the routines. Bednekoff & Houston (1994) incorporated a second state variable that represents the amount of food in the gut. This may mean that foraging is not possible because the gut is full. An effect of this constraint is to spread foraging more evenly throughout the day.

Some other daily routines

As we have mentioned, carrying high levels of reserves is likely to be costly to a bird. One way in which it can avoid such costs is by storing food items in a cache or hoard rather than eating them. Hoarding food, however, has its own disadvantages in that caching and retrieving items takes time, and items that have been cached may be lost, either as a result of being removed by an other animal or because the hoarding animal forgets the location of the items. The optimal strategy of eating, hoarding, retrieving and resting over a day under constant environmental conditions has been modelled by McNamara *et al.* (1990) and Lucas & Walter (1991). McNamara *et al.* showed that if the hoard is likely to be lost overnight, birds tend to hoard at the start of the day and retrieve just before dusk. As a result of this pattern of behaviour, the reserves drop at the start of the day and rise sharply at dusk. This contrasts with the pattern obtained when the same bird is not allowed to hoard; typically, then, reserves rise throughout the day. Haftorn (1992) claimed that his data on the reserves of tit species failed to show this predicted difference between hoarding and non-hoarding species, but he appeared to be unaware that the prediction is based on a period when a bird of a hoarding species is actually hoarding, whereas his data are from a time of year when hoarding is not occurring.

Many songbirds sing intensely at dawn. This phenomenon, known as the dawn chorus, has been explained by assuming that dawn is a good time for singing or a bad time for foraging (see Kacelnik & Krebs 1983 for a review). In this chapter we have emphasised that it is not possible to isolate a particular time and predict behaviour at that time solely on the basis of conditions at the time in question. Such 'local' explanations ignore interdependencies; see (i) and (ii) in Section 9.2. To explore the logic of these interdependencies, McNamara *et al.* (1987) modelled the optimal routine for a male bird singing to attract a mate. During the day the bird can choose between singing and foraging; at night it has to rest. In the basic model, the foraging gain and the probability of attracting a mate while singing are independent of time of day. Even under these conditions, there may be pronounced dawn and dusk bursts of singing under the optimal strategy.

When there is stochasticity in overnight energy expenditure (e.g. because of fluctuations in temperature), the energy reserves of a bird at dusk are primarily determined by the worst possible conditions. On most nights, the bird does not experience the worst conditions and has excess reserves at dawn, and the result is a dawn chorus. Because of stochasticity in foraging, reserves for the night are built up before the end of the day, and typically this results in a burst of song at dusk. This period of song is analogous to the period of resting at dusk in the above model of surviving a winter (Figure 9.2).

Hutchinson *et al.* (1993) extended the analysis of McNamara *et al.* (1987) to consider the effect of different mating rules of females. In one model that they analyse, a female will not mate with a male unless the male has produced a long burst of continuous song. As a consequence, it pays a male to build up its reserves to a high level so that he can then sing continuously for a long time. This can result in the male's singing on alternate days. When this phenomenon occurs, there is an extended period of song at dawn. The male then spends the rest of that day and the whole of the next day building up reserves. He then sings again at the following dawn. The strategy of such a male has two components. If the male is not singing, the strategy specifies the combination of level of reserves and time of day for which song should commence. If the male is singing, the strategy specifies the combinations of reserves, length of the current song bout and time of day for which song should cease. Both components are the same on all days. As a result the mean behaviour of the population as a whole settles down to a routine that is the same on each day. Individuals within the population, however, tend to follow a two-day cycle.

So far, we have discussed daily routines in which the best action for an animal does not explicitly depend on the behaviour of other members of the population. Houston & McNamara (1987) and Iwasa & Obara (1989) presented analyses of daily routines based on competitive interactions between males, which are thus involved in a dynamic game; such games are discussed in Chapter 7. Houston & McNamara (1987) extended the work of McNamara *et al.* (1987) to allow the probability that a singing male attracts a mate to depend on whether other males are singing. Iwasa and Obara considered male butterflies searching for females and took the pattern of the emergence of females over the day as given. They established the evolutionarily stable pattern of male emergence over the daily cycle, given that each male is trying to maximise expected mating success subject to a constraint on his total daily searching time.

9.4 Annual routines

There have been several previous models of annual routines, but none of them has tackled the general problem that we have discussed earlier, in this chapter, in which reproduction is free to occur at any time of year. We now describe some of these models before presenting a general technique for finding optimal annual routines.

Katz (1974) analysed the annual feeding behaviour of the African weaver bird, *Quela quela*, as an illustration of a dynamic optimisation problem. The bird's state variable is its mass, x. The bird is assumed to starve if x falls below a critical level x_{min}. To avoid this, the constraint $x \geq x_{min}$ is imposed. The bird's decision is the proportion of each day that it spends foraging. Foraging results in a deterministic change in mass. Breeding occurs at a fixed time and imposes a constraint on the proportion of the day that can be spent foraging; it also increases the bird's energy expenditure. Given these costs of breeding and the pattern of seasonal changes in food availability, Katz showed that the bird's mass throughout the year can be predicted on the assumption that the total time spent foraging during the year is minimized. This criterion is reasonable in the context of the model if mortality increases with time spent foraging. Katz did not impose the condition that the mass at the end of a year is the same as the mass at the start of the year, but this boundary condition is satisfied by the optimal solution.

Holmgren & Hedenström (1995) investigated the timing of moult in birds. The state variables are a bird's geographical location and the state of its feathers. The behavioural options are: move north, move south, stay and do not moult, stay and moult. The consequences of these options are deterministic. There is a fixed breeding period, and the optimisation criterion is maximisation of the lifetime production of offspring. Holmgren and Hedenström found that their model is able to produce most of the observed annual cycles of moulting.

Iwasa & Cohen (1989) analysed the optimal annual routine of a perennial plant. A plant is characterized by two state variables; these are the size of the production part and the size of the storage part. The production part determines the plant's net daily production of resources through photosynthesis. This includes leaves, stems and roots. The storage part includes the allocation to reproduction in the current season and the storage of material for transferring to the production part in the next season. The plant is viewed as having two problems. (i) Within the growing season, how should it allocate the resources produced by photosynthesis to the production part as opposed to the storage part? (ii) At the end of the growing season, how should it allocate storage to reproduction as opposed to retaining it to promote production in the next season? In an elegant analysis, Iwasa and Cohen used Pontryagin's maximum principle to solve the first problem for arbitrary initial size of the storage part, and then used this solution to solve the second problem using dynamic programming.

In the annual routine model of Schaffer (1983) there is an arbitrary terminal reward at the end of the year. Because the reward chosen does not generate the same (rescaled) reward at the end of the previous year, the model does not 'close the loop'. The models of Katz (1974), Holmgren & Hedenström (1995) and Iwasa & Cohen (1989) deal with examples in which the expected total number of offspring produced over an individual's lifetime is maximised. We now present a modelling framework which 'closes the loop' and which can deal with offspring produced at different times and in different states.

Optimal annual routines

We take a year as starting at some convenient time point that we refer to as time of year 0. The year is divided up into decision epochs $t = 0, 1, 2, \ldots, T - 1$. Time T is exactly one year after time 0 and is also time of year 0 in the following year. At each of the decision epochs an individual is characterised by some appropriate state variables. When dealing with annual reproductive decisions in Section 8.5 it was implicitly assumed that offspring produced during a year were independent of the parent by the next annual census time. In dealing with the fine scale of behaviour over the year it would be very restrictive to assume that, say, offspring produced on day t were independent of their parents by day $t + 1$. Our approach to modelling young that are still dependent on a parent is not to take these young to be separate individuals at all, but to regard them and their own state as part of the state of the parent (see e.g. Section 2.2). When young do become independent, the state of the parent alters to take this into account and the young are then taken as descendants of the parent. It is for this reason that the definition of N_{off} given below counts only those young that become independent between t and $t + 1$ (and survive until $t + 1$).

At each of the decision epochs $t = 0, 1, 2, \ldots$, an organism has available a range of behavioural options. As usual the range of options available may depend on the individual's state. Options might be reproductive options, such as to produce young or to abandon young that are being cared for, or they might involve foraging decisions or options to migrate. An option may, however be more complex. In the example discussed in Section 9.5, an individual that is caring for young has to choose whether to continue to provision young and simultaneously has to choose a foraging intensity. These two decisions cannot be made in isolation from one another: if the individual decides to continue provisioning, it will have to forage at an intensity that gives sufficient food both to feed itself adequately and also feed the young.

The consequences when an individual in state x at time of year t chooses an option u can be denoted using a similar formalism to that of Section 8.5. The difference between the formalism of Section 8.5, which dealt with single annual decisions, and that given below is that we must now allow the consequences of actions to depend on the time of year.

We define

$$S_{\text{mat}}(x, t; u) = P(\text{focal organism survives until time of year } t+1)$$

$$p_{xy}^{(t)}(u) = P(\text{focal organism is in state } y \text{ at } t+1$$
$$\text{given it survives until this time})$$

$$N_{\text{off}}(x, t; u) = \mathbb{E}(\text{number of offspring that become independent}$$
$$\text{between time } t \text{ and } t+1 \text{ and are still alive at } t+1)$$

$$b_{xy}^{(t)}(u) = P(\text{an offspring which becomes independent between}$$
$$\text{time } t \text{ and } t+1 \text{ and which survives until } t+1$$
$$\text{is in state } y \text{ at this time}).$$

These consequences specify the number and state of descendants at time $t+1$ resulting from the action of the focal organism between t and $t+1$. As in Chapter 8, both young reaching independence between t and $t+1$ and the focal organism are taken to be descendants at $t+1$ if they survive until this time. From the above, an individual in state x at time t that chooses option u leaves

$$a_{xy}^{(t)}(u) = S_{\text{mat}}(x, t; u)p_{xy}^{(t)}(u) + N_{\text{off}}(x, t; u)b_{xy}^{(t)}(u)$$

descendants in state y at time $t+1$ (cf. equation (8.14)). As in Section 8.5 the number of surviving offspring, N_{off}, needs to be discounted by relatedness when there is sexual reproduction.

Under an optimal life-history strategy each individual maximises the number of descendants that it leaves far into the future. To find such a strategy we simply use dynamic programming to work backwards from the future. As in Section 8.6, consider a target year far into the future. To find the strategy that maximises the expected number of descendants left at the end of this target year, set

$f_n(x,t)$ = maximum number of descendants left at the end of the
target year by an individual in state x at time of year t,
n years back from the target year.

Here the target year itself is considered as 0 years back, the year before that
as 1 year back, and so on. Then we can take as a terminal reward

$$f_0(x,T) = 1 \quad \text{for all} \ x. \tag{9.3}$$

The dynamic programming equation is

$$f_n(x,t) = \max_u \sum_y a_{xy}^{(t)}(u) f_n(y, t+1) \tag{9.4}$$

for times of year t satisfying $0 \le t \le T-1$. Since the end of the nth year
back is the start of the $(n-1)$th year back we also have

$$f_n(x,T) = f_{n-1}(x,0). \tag{9.5}$$

These equations determine $f_n(x,t)$ for all states x, times of year t and
years to go n. They also determine the option u that achieves the maximum
on the right-hand side of equation (9.4). This option depends on x, t and n
in general but, as n increases, the option chosen at a given x and t settles
down to a limit that is independent of n. This limiting specification of
how the option chosen depends on state and time of year is the optimal
life-history strategy.

To define reproductive value choose a reference state x_0 at time of year T
and set

$$V_n(x,T) = \frac{f_n(x,T)}{f_n(x_0,T)}. \tag{9.6}$$

As n tends to infinity V_n tends to a limit that is independent of the choice
of f_0:

$$V(x,T) = \lim_{n \to \infty} V_n(x,T). \tag{9.7}$$

Then $V(x,T)$ is the reproductive value of being in state x at time of year
T relative to being in state x_0 at this time. An individual that follows the
optimal life-history strategy over a year maximises the expected value of
all descendants that it leaves at the end of the year, where the value of a
descendant in state x is taken to be $V(x,T)$. The function $V(x,T)$ is thus
the correct terminal reward to use at the end of the year. As with the
daily routine example of Section 9.3, V emerges from the calculation of the
optimal strategy rather than being assumed at the start.

For any time of year t set

$$V(x,t) = \lim_{n \to \infty} \frac{f_n(x,t)}{f_n(x_0,T)} \qquad (9.8)$$

where x_0 is the reference state at time of year T. Then $V(x,t)$ is the reproductive value of an individual in state x at time of year t relative to that of an individual in state x_0 at the end of that year. If an individual is following the optimal life-history strategy then its behaviour between times t and $t+1$ maximises the expected value of descendants alive at $t+1$, where the value of a descendant in state x at $t+1$ is $V(x,t+1)$; this follows from equations (9.4) and (9.8).

The function $V(x,t)$ measures reproductive value as a function of both state and time of year. In particular, it tells us how the value (to the parent) of offspring at independence depends on the state of these offspring and the time of year at which they are produced. In doing so it takes into account the survival and reproduction prospects of the offspring under the assumption that their behaviour is optimal. As with the reproductive value at the end of the year, $V(x,T)$, the reproductive value $V(x,t)$ emerges from the calculations rather than being assumed at the outset.

The fitness of the optimal strategy measures the asymptotic growth rate in the numbers of individuals following the strategy and is given by

$$\lambda^* = \lim_{n \to \infty} \frac{f_n(x,t)}{f_{n-1}(x,t)}. \qquad (9.9)$$

This limit does not depend on the values of x or t chosen or on the choice of the terminal reward f_0. Reproductive value $V(x,t)$ measures value relative to a reference individual at the end of the year. The dependence of $V(x,t)$ on t reflects both seasonal effects and the fact that in a population that is either growing or declining the time of production of young is important. For example, when $\lambda^* > 1$, and hence the number of organisms following the optimal strategy is growing, it is better to produce young now rather than in one year's time. Consequently, $V(x,0) > V(x,T)$.

If a population is at an ESS and is in density-dependent equilibrium then $\lambda^* = 1$. Under the optimal (i.e. ESS) strategy, the value of young produced is then independent of the year of production. In this case $V(x,0) = V(x,T)$ for all states x, and the change of $V(x,t)$ with t is purely due to seasonal effects.

9.5 An example based on the timing of reproduction

To illustrate the framework outlined in Section 9.4, we present an example
in which an animal can decide when and how often to reproduce. In this
example, the animal needs food to ensure that it does not starve, and if it is
caring for young, needs additional food to provision these young. Since the
environment is seasonal it must work harder to obtain food at certain times
of year than at other times. An increase in work rate increases metabolic
expenditure and may increase predation risk.

As before we refer to one of the state variables that characterise the animal
as 'condition'. This variable could be loosely thought of as the state of
the immune system of the animal. High metabolic expenditure decreases
condition and leads to increased mortality risk (cf. the discussion in Section
4.8). If we interpret condition as the state of the immune system, then the
source of mortality would be disease. When the animal does not work hard
its condition slowly recovers.

In this model, reproduction may directly decrease reserves and condition
and may further reduce condition because the animal must work harder to
provision young. Between reproductive bouts, reserves and condition have
a chance to recover. The timing and frequency of reproduction are driven
by the dynamics of reserves and condition and by the seasonal environment.

An additional feature of the model is that young animals have lower 'experience' and are hence poorer foragers than older animals. This allows us
to incorporate density dependence acting on juveniles and hence ensure that
$\lambda^* = 1$ (cf. Section 8.7).

Model details

For simplicity we consider an asexual population. Decision epochs during
the year are at times $t = 0, 1, 2, \ldots, 51$, which are one week apart. Time
0 is taken to be midwinter, so that time 26 is midsummer and time 52 is
midwinter of the following year.

At a decision epoch t, an animal that is not currently caring for a brood
must simultaneously choose its foraging intensity and whether to start a
brood. An animal that is caring for a brood must simultaneously choose its
foraging intensity and whether to abandon the brood.

The state of an animal at a decision epoch is given by a vector (x, y, e, a)
whose components are as follows:

$$
\begin{aligned}
x &= \text{energy reserves} \\
y &= \text{condition} \\
e &= \text{experience} \\
a &= \text{age of any brood being cared for}
\end{aligned}
$$

$(a = -1$ indicates that no brood is being cared for).

Here we first describe the dynamics of experience and the effect of experience on foraging ability.

Experience.   This variable e takes one of the values $0, 1, 2$. Newly independent young have experience $e = 0$. Thereafter experience tends to increase to its maximum value of 2 as follows. If experience is e at time epoch t, it is still e at $t + 1$ with probability $1 - p_{\exp}$ and is

$$
e' = \min(2, e + 1) \tag{9.10}
$$

at $t + 1$ with probability $p_{\exp}$. The figures are based on $p_{\exp} = 0.1$. For this value of $p_{\exp}$ it takes on average 20 weeks for newly independent young to reach full experience $(e = 2)$. Experience affects the foraging ability of an animal. Let γ be the mean gross energetic intake of an animal with full experience $(e = 2)$. Then under the same circumstances and foraging behaviour an animal with experience $e = 1$ has mean gross intake $\theta\gamma$ and an animal with experience $e = 0$ has mean gross intake $\theta^2\gamma$. Since θ is chosen to be less than 1 (see below), initially animals tend to improve at foraging as they get older.

Foraging intensity.   Foraging options are parametrised by a foraging intensity u, where $0 \leq u \leq 1$. Suppose that an animal has maximum experience $e = 2$ at time t and forages with intensity u between times t and $t + 1$. Let $\gamma(u, t)$ denote the mean gross energetic intake during this time interval. Then we can write the intake of an animal with experience e $(e = 0, 1, 2)$ and which forages with intensity u as $\theta^{2-e}\gamma(u, t)$. The figures that we present are based on assuming that the mean gross energetic intake is

$$
\gamma(u, t) = \left\{ 1 + \epsilon \sin \left[\left(\frac{t - 13}{26} \right) \pi \right] \right\} u, \tag{9.11}
$$

where ϵ is a constant. Thus for given t, mean intake increases linearly with u. For given u, mean intake varies sinusoidally with t, having a minimum when $t = 0$ (midwinter) and a maximum when $t = 26$ (midsummer). The parameter ϵ lies in the range $0 \leq \epsilon < 1$ and measures the degree to which seasons differ. The computations used to derive Figures 9.3 – 9.8 assume that actual intake is a random variable with mean $\theta^{2-e}\gamma(u, t)$. For ease of exposition, however, we ignore stochasticity in intake in describing the model, and give model details and the dynamic programming equations for the case where actual intake is always $\theta^{2-e}\gamma(u,t)$.

The metabolic expenditure between times t and $t+1$ is $c(u,t)$ when the foraging intensity is u. The figures are based on the function c given in Appendix 9.1. For this function, expenditure does not depend on the time of year and increases at an increasing rate with intensity u.

An animal that forages with intensity u between times t and $t+1$ is killed by a predator with probability $M(u,t)$. This probability increases with u at an increasing rate (Appendix 9.1).

Brood production and care.   The age of a brood is denoted by a and takes one of the integer values $a = -1, 0, 1, 2, \ldots, a_{\max}$. If $a = -1$ at time t, the animal is not caring for a brood at this time. If the animal decides to start a brood at time t, the brood age at time $t+1$ is then taken to be 0. Producing a brood directly depletes the animal's energy reserves by an amount Δ_{res} and decreases its condition by Δ_{cond}. For simplicity, we assume that the animal cannot choose the number of offspring in the brood, and that this number, n_{brood}, is fixed.

If the brood age at time t is in the range $a = 0, 1, 2, \ldots, a_{\max} - 1$, then the animal is caring for a brood of this age and must choose whether to abandon the brood or to continue to care for it. If the brood is abandoned then all brood members die. If the animal cares for the brood it must provide sufficient food to prevent starvation of brood members. We assume that if the parent animal dies between times t and $t+1$ or fails to achieve a gross energetic intake γ_{brood} during this time period then all brood members starve. If the parent's gross intake is γ_{brood} or more, all young survive until time $t + 1$. To achieve a gross intake of γ_{brood} an animal with experience e must forage with intensity greater than or equal to $u_{\mathrm{crit}}(e,t)$, where

$$\theta^{2-e}\gamma(u_{\mathrm{crit}},t) = \gamma_{\mathrm{brood}}. \qquad (9.12)$$

Thus if $u_{\mathrm{crit}}(e,t) > 1$ the animal is forced to abandon the brood, since brood members will starve even if the mother forages at the maximum intensity $u = 1$. If $u_{\mathrm{crit}}(e,t) \le 1$ and the mother forages with intensity u, where $u_{\mathrm{crit}}(e,t) \le u \le 1$, then the young survive and the mother herself receives a gross energetic intake of

$$\theta^{2-e}\gamma(u,t) - \gamma_{\mathrm{brood}}. \qquad (9.13)$$

Providing that the brood survives, its age increases by 1 each week until it reaches age $a = a_{\max}$. The young are then capable of foraging for themselves and are abandoned by their mother. Their experience is then $e = 0$. For simplicity, at independence reserves x and condition of young y are each taken to be 0.5.

If an animal abandons her young at time t then, whatever the age of young at abandonment, the animal cannot start a new brood before time $t + 1$. Her brood age at time $t + 1$ thus has the value $a = -1$.

Energy reserves. Reserves are denoted by x and are measured in units such that $0 \leq x \leq 1$. If reserves fall to $x = 0$ the animal dies of starvation. The upper limit $x = 1$ represents the upper fat-storage capacity of the animal. In defining the dynamics of reserves, it is convenient to use the truncation function chop(x) defined by

$$\text{chop}(x) = \max[0, \min(1, x)] \tag{9.14}$$

(cf. Mangel & Clark 1988). Suppose that at time of year t energy reserves are $x > 0$ and experience is e. Let the animal forage with intensity u between t and $t + 1$. We look at energy reserves at $t + 1$ given that the animal has not died of predation or disease by this time. We can identify three cases.

Suppose that the animal is not caring for a brood between times t and $t + 1$. This could be because it had no brood at time t and did not start a brood, or because it had a brood at time t but abandoned it at this time. Whatever the reason, reserves at time $t + 1$ are

$$x_{\text{nocare}} = \text{chop}[x + \theta^{2-e}\gamma(u, t) - c(u, t)]. \tag{9.15}$$

Suppose that the animal already has a brood at time t and continues to care for the brood. Then its foraging intensity u must lie in the range $u_{\text{crit}}(e, t) \leq u \leq 1$ and, from expression (9.13), reserves at $t + 1$ are

$$x_{\text{care}} = \text{chop}[x + \theta^{2-e}\gamma(u, t) - \gamma_{\text{brood}} - c(u, t)]. \tag{9.16}$$

Finally, suppose that the animal starts a brood at time t. Since the brood is not produced until time $t + 1$ there is no provisioning of the brood between t and $t + 1$. There is, however, a direct depletion of reserves by Δ_{res}. Thus reserves at $t + 1$ are

$$x_{\text{start}} = \text{chop}[x + \theta^{2-e}\gamma(u, t) - c(u, t) - \Delta_{\text{res}}]. \tag{9.17}$$

Condition. This variable is denoted y and lies in the range $0 \leq y \leq 1$. Suppose that an animal has condition y at time t. Let the metabolic expenditure of the animal between t and $t + 1$ be c, and consider the condition of the animal at $t + 1$ given that it is alive at this time.

If the animal does not start a brood at t, its condition at $t + 1$ is

$$y_{\text{nostart}} = \text{chop}[y + \alpha_0 - \alpha(c)]. \tag{9.18}$$

Here α_0 is the rate at which condition increases in the absence of metabolic expenditure and $\alpha(c)$ is the rate at which expenditure tends to decrease condition. We assume $\alpha(0) = 0$ and that $\alpha(c)$ increases at an increasing rate as c increases (Appendix 9.1).

Starting a brood may directly suppress condition. To allow for this we include a constant condition loss Δ_{cond}. Thus if the animal starts a brood at t, its condition at $t + 1$ is

$$y_{\text{start}} = \text{chop}[y + \alpha_0 - \alpha(c) - \Delta_{\text{cond}}]. \tag{9.19}$$

The probability that an animal dies of disease between times t and $t + 1$ is a function $D(y)$ of its condition y at time t. This probability, $D(y)$, increases as y decreases and is 1 when $y = 0$ (Appendix 9.1).

Density dependence. In this model the value of θ is not fixed but rescaled so that $\lambda^* = 1$. We are thus implicitly assuming that there is some form of density dependence acting on juveniles. Results are only presented for cases in which $\theta \leq 1$, since only these cases have a sensible biological interpretation.

The dynamic programming equations

In Section 9.4 we outlined the dynamic programming equations for a general annual routine model. Here we illustrate the general method by giving the specific equations that apply to the above model.

As in Section 9.4, choose some target year far into the future and let $f_n(x, y, e, a, t)$ be the maximum expected number of descendants left at the end of the target year by an individual whose state is (x, y, e, a) at time of year t, n years back from the target year. Suppose that the individual forages with intensity u between times t and $t + 1$. Then the probability that it is not killed by a predator or dies of disease by time $t + 1$ is

$$S(y, t; u) = [1 - M(u, t)][1 - D(y)]. \tag{9.20}$$

First suppose that at time t either the animal is not caring for a brood and does not start a brood or else abandons a brood. Then it leaves

$$H_{\text{nocare}}(u) = S(y, t; u)[(1 - p_{\text{exp}})f_n(x_{\text{nocare}}, y_{\text{nostart}}, e, -1, t + 1)$$

$$+ p_{\text{exp}}f_n(x_{\text{nocare}}, y_{\text{nostart}}, e', -1, t + 1)] \tag{9.21}$$

descendants at the end of the target year. Here x_{nocare} is given by equation

(9.15), y_{nostart} is given by equation (9.18) and e' is given by equation (9.10). The maximum payoff given no care is thus

$$H^*_{\text{nocare}} = \max_{0 \le u \le 1} H_{\text{nocare}}(u). \tag{9.22}$$

If, instead, the animal starts a brood at time t and forages with intensity u, then its payoff is

$$H_{\text{start}}(u) = S(y, t; u)[(1 - p_{\text{exp}}) f_n(x_{\text{start}}, y_{\text{start}}, e, 0, t + 1)$$

$$+ p_{\text{exp}} f_n(x_{\text{start}}, y_{\text{start}}, e', 0, t + 1)]. \tag{9.23}$$

Here x_{start} and y_{start} are given by equations (9.17) and (9.19) respectively. Thus given a brood is started the maximum payoff is

$$H^*_{\text{start}} = \max_{0 \le u \le 1} H_{\text{start}}(u). \tag{9.24}$$

If the animal is already caring for a brood of age a $(1 \le a \le a_{\text{max}} - 1)$ and continues to care, at time t, it must forage with intensity u satisfying $u \ge u_{\text{crit}}(e, t)$. Its payoff is then

$$H_{\text{care}}(u) = S(y, t; u)[(1 - p_{\text{exp}}) f_n(x_{\text{care}}, y_{\text{nostart}}, c, a + 1, t + 1)$$

$$+ p_{\text{exp}} f_n(x_{\text{care}}, y_{\text{nostart}}, e', a + 1, t + 1)], \tag{9.25}$$

where x_{care} and y_{nostart} are given by equations (9.16) and (9.18) respectively. Its maximum payoff is then

$$H^*_{\text{care}} = \max_{u_{\text{crit}} < u < 1} H_{\text{care}}(u), \tag{9.26}$$

where the maximum is now over values of the foraging intensity u such that the young do not starve. (Here we may formally take $H^*_{\text{care}} = 0$ if $u_{\text{crit}} > 1$.)

The dynamic programming equations are then as follows. For $x = 0$ we have

$$f_n(0, y, e, a, t) = 0. \tag{9.27}$$

For $x > 0$ and $a = -1$ we have

$$f_n(x, y, e, -1, t) = \max(H^*_{\text{nocare}}, H^*_{\text{start}}). \tag{9.28}$$

For $x > 0$ and $0 \le a \le a_{\text{max}} - 1$ then, since the brood dies if it is abandoned (no care), we have

$$f_n(x, y, e, a, t) = \max(H^*_{\text{nocare}}, H^*_{\text{care}}). \tag{9.29}$$

Finally, let $x > 0$ and $a = a_{\text{max}}$ at time t. Then brood members immediately become independent at this time and the parent is forced to abandon them.

Its only decision is then the choice of foraging intensity. The payoff to the focal individual now includes the contribution of the offspring to future descendants, which is $n_{\mathrm{brood}}f_n(0.5, 0.5, 0, -1, t)$ since n_{brood} offspring become independent at t. Thus

$$f_n(x, y, e, a_{\max}, t) = n_{\mathrm{brood}}f_n(0.5, 0.5, 0, -1, t) + H^*_{\mathrm{nocare}}. \qquad (9.30)$$

Equations (9.10) – (9.30) allow calculation of $f_n(x, y, e, a, t)$ for all x, y, e and a. Thus we can compute f_n, for all states, times of year and n, from these equations together with the terminal condition

$$f_0(x, y, e, a, t) = 1 \quad \text{for} \quad x > 0 \qquad (9.31)$$

and the wrap-around condition

$$f_n(x, y, e, a, 52) = f_{n-1}(x, y, e, a, 0). \qquad (9.32)$$

One slight complication is that young that are age $a_{\max}$ at time t are independent at time t, not at time $t + 1$. Thus equation (9.30) has a term involving f_n at time t on the right-hand side. In practice, this is not a problem because $f_n(x, y, e, a, t)$ is first computed for all x, y and e when $a = -1$.

Figures 9.3 – 9.8 are based on a computational grid of 17 values of x and 17 values of y (each running from 0 to 1 in steps of $1/16$). There is linear interpolation between grid points. Additional stochasticity is added to ensure biological realism and prevent grid effects (cf. Appendix 3.1).

Model output

We now illustrate the type of results that can be obtained from the above annual routine model. The idea of presenting these is to demonstrate the potential of such models. Thus we wish to illustrate the sorts of biological issues that can be addressed, rather than attending to any one of them in detail.

The model treats the reproductive bout very schematically and is not best suited to examining issues concerned with the details of parental care and desertion. The model does, however, include various costs of parental care, and can be used to examine how these costs and the seasonality of the environment combine to determine the timing and frequency of reproduction.

Figures 9.3 – 9.8 are based on the functions and parameters given above and in Appendix 9.1. The parameter ϵ, which controls the seasonality of the food supply, has value 0.3. Thus, by equation (9.11), at maximum foraging intensity ($u = 1$) the gross energetic intake rate $\gamma(1, t)$ of an animal of full experience ($e = 2$) is 0.7 in midwinter and 1.3 in midsummer. If this animal forages at intensity $u < 1$ its gross intake rate is proportionately less. The dynamics on condition are such that condition tends to decrease if $u > 0.5$ and increase if $u < 0.5$. In summer an animal (with $e = 2$) can both increase its reserves and its condition if it has no young. This is no longer true, however, if it is caring for young. Then, even at midsummer, an animal that tried to maintain its condition by foraging with intensity $u = 0.5$ would be losing on average 0.15 in reserves per week. Conversely, if it worked harder in order to maintain reserves, its condition would decrease. At midwinter, even without young, an animal that foraged with intensity 0.5 would be losing on average 0.05 in reserves per week. Thus, again, the animal would be unable to maintain both reserves and condition.

In order to ensure that $\lambda^* = 1$ we have taken $\theta = 0.839$. Thus, for given foraging intensity u, a newly independent animal with experience $e = 0$ would be $\theta^2 = 0.704$ times as efficient at foraging as a fully experienced animal, and an animal with intermediate experience of $e = 1$ would be $\theta = 0.839$ times as efficient.

The probability of death from disease increases as condition decreases. When condition $y = 1$ the probability is 0.004 per week, rising to 0.008 when $y = 0.5$ and 0.104 when $y = 0.25$. When $y = 0$ the animal always dies.

Reproductive value. Figure 9.3 shows how reproductive value at a given time of year depends on an animal's reserves and condition. It can be seen that, at both times of year illustrated, reproductive value increases more strongly with increasing condition than with increasing reserves. This is so partly because, in the scale at which we have represented reserves and condition, the probability of disease increases as condition decreases but the probability of immediate starvation only increases significantly when reserves are very low. But it is also a result of the fact that condition has a much-longer-term effect than reserves. Provided that the animal is not caring for young, it can rapidly increase reserves by increasing its foraging intensity, but can only produce a big increase in condition by a prolonged period of low intensity foraging.

A comparison of Figures 9.3(a) and 9.3(b) shows that reproductive value increases more strongly with reserves at the end of summer ($t = 39$) than at the end of winter ($t = 13$). Furthermore, the pattern of increase is different between the two times. At the end of winter, the reproductive value of an animal of condition $y = 0.5$ hardly depends on its reserves. The reason is that the animal is in too poor a condition to breed, and hence can concentrate on building up its reserves if they are low. The reproductive

value of an animal with condition $y = 1$ at this time increases more strongly
with reserves than does the reproductive value of an animal with condition
$y = 0.5$. The reason is that the animal with condition 1 will breed. Its level of

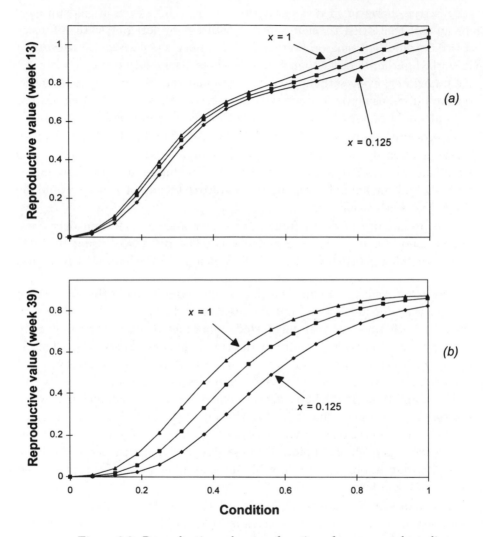

Figure 9.3. Reproductive value as a function of reserves and condition: (*a*) at the end of winter (week 13), and (*b*) at the beginning of
autumn (week 39). In each case the animal is not caring for young
($a = -1$) and has maximum experience ($e = 2$). Reproductive
value is shown as function of condition for three levels of reserves,
$x = 0.125$, $x = 0.5$ and $x = 1$.

reserves is thus important in determining whether it can successfully rear a brood, and whether it can avoid working very hard by allowing its reserves to decrease during care of the young. In contrast, at the beginning of autumn, having high reserves is more valuable to an animal in condition $y = 0.5$ than it is to an animal in condition $y = 1$.

Figure 9.4 shows how the reproductive value of young at independence depends on the time at which they become independent. Reproductive value peaks at the end of winter, at $t = 13$, because young that are independent at this time can gain experience and increase reserves and condition during the summer, when food is plentiful. The figure also shows the dependence of reproductive value on time of year in three other cases. In one case, an animal has the same reserves and condition as do young at independence, but has maximum experience ($e = 2$). In another case, an animal has maximum reserves and condition, but has the same experience as do young at independence. Finally, the reproductive value of an animal with maximum reserves, condition and experience is shown. Comparison of the four cases shows the influence on the reproductive value of the young of their lack of experience and their level of reserves and condition; it also shows the interaction between these factors.

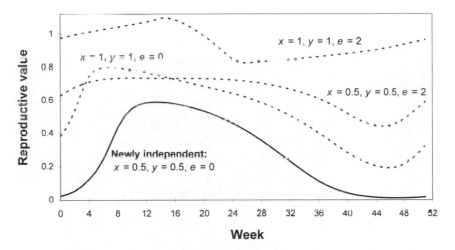

Figure 9.4. Reproductive value of newly independent young as a function of the time of year at which they become independent. Newly independent young have reserves $x = 0.5$, condition $y = 0.5$ and experience $e = 0$. For comparison, reproductive value as a function of time of year is shown for other combination of reserves, condition and experience. In all cases $a = -1$.

The optimal strategy.   It is never optimal for an animal of experience $e = 0$ to reproduce. An animal of experience $e = 1$ will start a reproductive bout between weeks 18 and 24 if reserves and condition are sufficiently high. However, since virtually no animals with experience $e = 1$ in week 18 will have survived the previous winter, this aspect of the optimal strategy is irrelevant to what would actually be observed in a population following the strategy. In practice, therefore, the only animals that reproduce have experience $e = 2$. For an animal of this experience it is never optimal to start a brood before time $t = 14$ nor after time $t = 25$. For each time in the range $14 \le t \le 25$ an animal with experience $e = 2$ will start a brood if its reserves and condition are sufficiently high (Table 9.1). As can be seen from this table, only animals with high reserves and in the top condition class start broods at $t = 14$. Later (e.g. $t = 22$), animals with lower reserves or condition will start a brood. However, by $t = 25$ it is again only animals with high reserves and top condition that start. In general, condition appears to have a stronger effect on the decision to start than do reserves.

Table 9.1. Critical levels of condition. For each combination of reserves and time, the table entry indicates the range in condition for which an animal with experience $e = 2$ will start a brood. If the table entry is '–', then the animal will not start a brood whatever its condition. If the table entry is an integer k, the animal starts a brood if and only if its condition is greater than or equal to $k/16$.

Reserve value × 16	Week													
	13	14	15	16	17	18	19	20	21	22	23	24	25	26
16	–	16	15	13	12	12	12	12	12	13	14	15	16	–
15	–	16	15	14	13	12	12	12	13	13	14	15	16	–
14	–	–	16	15	14	13	13	13	13	13	14	15	16	–
13	–	–	–	16	15	14	13	13	13	14	14	15	16	–
12	–	–	–	16	15	14	13	13	13	14	15	15	16	–
11	–	–	–	16	16	15	14	14	14	14	15	16	–	–
10	–	–	–	–	16	15	14	14	14	14	15	16	–	–
9	–	–	–	–	16	16	15	14	14	14	15	16	–	–
8	–	–	–	–	–	16	15	15	15	15	15	16	–	–
7	–	–	–	–	–	16	16	15	15	15	16	16	–	–
6	–	–	–	–	–	16	16	16	15	15	16	16	–	–
5	–	–	–	–	–	–	16	16	16	16	16	–	–	–
4	–	–	–	–	–	–	16	16	16	16	16	–	–	–
3	–	–	–	–	–	–	–	16	16	16	16	–	–	–
2	–	–	–	–	–	–	–	–	16	16	16	–	–	–
1	–	–	–	–	–	–	–	–	–	16	–	–	–	–

Given that under the optimal strategy all young produced are independent by time $t = 39$, this strategy can be understood in terms of the reproductive values illustrated by Figures 9.3(b) and 9.4. Focus on the period before time $t = 39$. Then an animal behaving optimally maximises the sum of its own reproductive value at $t = 39$ and the values of all offspring reaching independence. From Figure 9.4 it can be seen that it is better, in terms of the reproductive value of young, to produce these young as soon as possible after $t = 13$. However, if young are produced too early the parent has to forage at a very high intensity in order to provision the young and feed itself. Even so, it will lose reserves and may have to abandon the young prematurely in order to avoid its own starvation. It will also depress condition considerably by working so hard, and this will reduce its condition, and hence its reproductive value, at $t = 39$. The resulting optimal strategy is a compromise between these conflicting pressures.

Broods are abandoned if either the reserves are too low or the condition of the animal is too low. Critical levels of state variables depend strongly on brood age: broods of small age may be abandoned at intermediate levels of reserves and condition, while broods nearing independence are not abandoned unless the state variables are much lower.

Following a cohort forwards.   We start with a large subpopulation of newly independent young. For the purposes of illustration these young are all assumed to become independent at time $t = 24$. We then follow this cohort forward in time assuming all cohort members follow the optimal strategy. Given that there are large numbers in the cohort, we can find such quantities as the proportion of cohort members still alive at a time t by direct forward computation rather than simulation (see Section 3.3).

In the first winter after birth, there is high mortality because birds of low experience starve. Of the original cohort, 52.7% survive until time $t = 13$ at the end of this winter. Of these, 71.0% survive until $t = 13$, one year later. Thereafter, of those remaining at $t = 13$ in any given year, 71.6% survive until time $t = 13$ one year later. After the first winter, there is virtually no starvation. From their second complete year onwards, the pattern of mortality over the year is the same each year and is shown in Figure 9.5. As can be seen, the level of mortality from disease is well above that from predation. Mortality from predation peaks during care for young since parents are working at high foraging intensity at this time. Mortality from disease is highest at the end of the period of care and for a while afterwards. In this context, it should be noted that animals that start breeding at $t = 14$ are necessarily in top condition, whereas animals that start later at, say, $t = 18$ may be in much poorer condition (Table 9.1). Thus those animals that complete their period of care at $t = 25$ are liable to be in better condition on completion than those that complete at $t = 29$.

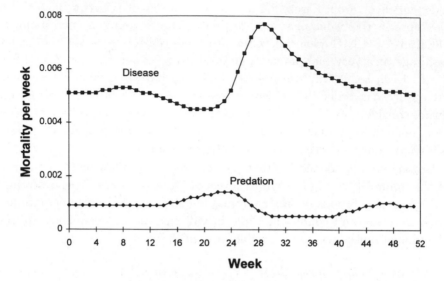

Figure 9.5. Sources of mortality over the year. A cohort of individuals is followed forward from birth. After several years the mortality of remaining cohort members over the year settles down to an equilibrium pattern that is the same each year and is independent of the state at birth and the time of birth of cohort members. The figure shows this equilibrium pattern. Mortality from disease in week t is defined as the proportion of those cohort members still alive at time t that die from disease by time $t + 1$. Mortality from predation and mortality from starvation are defined similarly; the latter is negligible and so is not shown.

Although there is virtually no starvation in this example, overwinter starvation will increase if the food supply in winter has a lower mean net gain or is more stochastic. Of course, such a change in the winter food supply could have consequences for both the optimal strategy and the cohort behaviour at other times of year.

As Table 9.1 shows, all broods are initiated between times $t = 14$ and $t = 25$. Although the table gives the strategy it does not tell us when, if at all within this period, cohort members start breeding. To find this out we need to combine knowledge of the strategy with knowledge of the distribution of states of cohort members (see Section 3.3). This analysis shows that, of those cohort members that survive to time $t = 13$ after their first winter, 70.3% start a brood that year and, of these, 85.8% manage to raise their young to independence. Of those cohort members that survive to time $t = 13$ in the following year, 84.4% start a brood and, of these, 88.3% manage to raise their young to independence. Subsequent years are similar to this second year. Not only is there less breeding in the first breeding season than subsequently, but animals tend to start breeding later (Figure 9.6). We emphasise that the strategy of cohort members is the same each year; it is the distribution of states of cohort members in the first breeding season that is different from the distribution of states in subsequent years. Owing to poor foraging success early on (before e rises to 2), animals tend to have lower reserves and be in poorer condition after their first winter than in subsequent years. They thus are more likely to start later if they start at all (cf. Table 9.1).

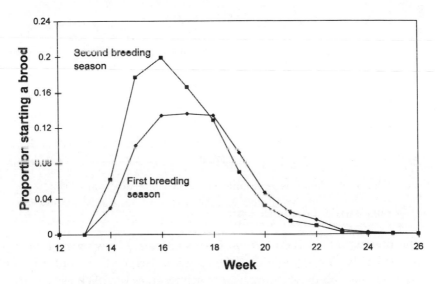

Figure 9.6. Timing of the breeding of cohort members. Cohort members all reach independence in week 24 of a year that we designate as 0. The figure shows the proportion starting a brood in each week of year 1 (first breeding season) and year 2 (second breeding season). For week t we show the proportion of those cohort members that are still alive at t that start a brood between times t and $t + 1$.

In this example, each animal produces at most one brood per year. By changing parameters in the model, it is easy to obtain cases in which there can be two or more reproductive bouts per year.

Figure 9.7 shows how the mean condition and reserves of cohort members change throughout the year once cohort behaviour has settled down to its equilibrium annual cycle. As can be seen, both mean reserves and condition decline during the breeding season and then subsequently recover. In late autumn reserves are very high, but then decrease throughout winter as animals behave to maintain condition at the expense of reserves.

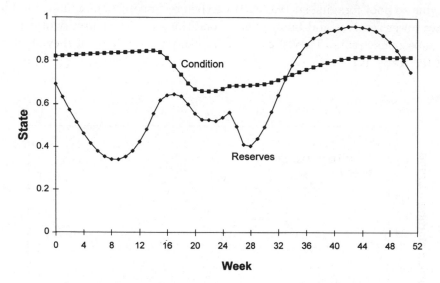

Figure 9.7. Mean reserves and condition as a function of time of year. As in the case of mortality (Figure 9.5), we give the equilibrium pattern for a birth cohort.

In the present model there are no disadvantages in having high energy reserves. It has been suggested that the metabolic expenditure of birds may increase with their body mass and hence with reserves (e.g. Witter & Cuthill 1993). Predation risk may also increase with reserves (e.g. Hedenström 1992). Such mass-dependent costs are likely to affect the annual pattern of changes in state.

Simulation results and within-individual correlations. In the above we followed a cohort forwards in time and recorded the proportion of remaining individuals that performed various activities. In doing so we did not keep track of the identity of cohort members, and are therefore unable to look at the correlation between the activities of an individual at one time and the activity of the same individual at a later time. For example, we noted that 70.3% of surviving cohort members started a brood after their first winter and 84.4% started a brood after their second winter. But this gives us no information on whether the animals that failed to breed after their second winter were mainly animals that failed to breed after their first winter. Specific questions on within-individual correlations can often be answered using an *ad hoc* device. For example, McNamara & Houston (1996) investigated the correlation between whether an individual breeds in one year and whether it breeds in the next by including a dummy variable that indicates whether the animal bred in the previous year. (This variable is only included when working forward to find expected behaviour; see Section 3.3). Such devices, however, are not a good way of surveying a wide range of characteristics that might be correlated.

To look at within-individual correlations in the model presented in this section we simulate the lives of a large number of animals. For each animal we choose a birth date using the distribution of birth dates for the whole population (which is found by forward iteration). We then simulate the change in state of the animal by a choice of random numbers drawn from the appropriate probability distribution for state changes. Mortality is similarly simulated. For each animal simulated we record the week of birth. A record is then kept of the details of each reproductive bout over the animal's lifetime. For each bout we record the year, and the week within the year, at which the bout starts. The state of the animal on starting and the outcome of the bout (either the young reached independence or died prematurely owing to abandonment by, or death of, the parent) are also recorded. Finally, we record the year and week of the animal's death and the cause of death. These data can then be analysed using normal statistical procedures. One outcome in this analysis is that there is a slight negative correlation between whether an animal breeds in a year and whether it bred in the previous year. Other issues that could be addressed include the dependence of the time of breeding in one year on the reproductive behaviour in the previous year and the relationship between sources of mortality and recent reproductive behaviour.

Manipulations.   Organisms may be experimentally manipulated in many ways. For example, the roosting environment of a bird may be altered by heating its nest box, or the bird may be given supplementary food or have the number of eggs in its nest changed. One can predict the effect of a manipulation by performing it on 'model organisms' in the computer. McNamara

& Houston (1996) used this approach to predict the effect of manipulation of a bird's current clutch size on the bird's subsequent clutch sizes. Here we demonstrate the effect of giving our model animal supplementary food prior to breeding.

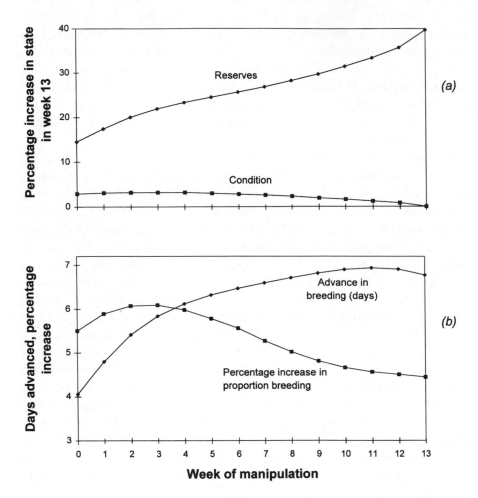

Figure 9.8. The effect of manipulating the food supply by increasing reserves by 0.5. The manipulation is carried out once at the start of a single week prior to the breeding season. Results are given in terms of the mean difference between manipulated and unmanipulated birds. (*a*) Effect of the week of manipulation on reserves and on condition in week 13. (*b*) Effect of the week of manipulation on the timing of reproduction and on the probability that reproduction occurs.

Our manipulation in the model involves focusing on a cohort of animals during the winter of their second year of life. To mimic giving the animals supplementary food during a particular week, the reserves of each animal at the start of that week are increased by 0.5 (up to the maximum value of 1). We assume that, before and after the manipulation, the environment is as in our model and that the animals follow the optimal strategy for this environment. Figure 9.8(a) shows the effect of the week in which the manipulation is carried out on the state of the animal in week 13. When animals are given food early, they subsequently forage at a lower intensity than they would otherwise do. As a result, they are able to increase their condition by week 13, although the increase in reserves at this time is much less than the amount of food given. The later the food is given, the less its effect on condition at week 13 and the greater its effect on reserves. The consequences for reproduction are shown in Figure 9.8(b). Under the optimal strategy, breeding never occurs before week 14, and the decision about whether to breed depends on reserves and condition (see Table 9.1). It can be seen from Figure 9.8(b) that manipulating food at about week 3 has the greatest effect on the proportion of animals that will breed, presumably largely because of the effect of manipulation on condition. In contrast, manipulation in week 11 has the greatest effect on the timing of reproduction, presumably largely because of the effect of manipulation on reserves.

9.6 Further application of the modelling framework

Solving the dynamic programming equations gives us an optimal strategy that specifies how behaviour depends on state and time of year. This strategy will typically follow the annual cycle, but this does not mean that the behaviour that results from following the strategy is cyclic. Even if behaviour is cyclic, it does not need to have a period of a year. Our framework thus enables us to investigate both the general question of which circumstances result in annual routines of behaviour and also the question of the particular circumstances that result in non-annual routines in some species (e.g. the king penguin (*Aptenodytes patagonicus*); see Olsson 1996 for details).

In the model of the previous section, our primary focus was on the timing of reproduction and how this was influenced by the long-term consequences of hard work. To incorporate such long-term consequences, we assumed that hard work decreased a state variable which we referred to as the condition. We interpreted condition as the state of the animal's immune system. Whether hard work does have long-term negative effects and whether these are mediated through the immune system are empirical questions (see

Apanius 1998 for a review). There is evidence in support of this in some organisms (e.g. Daan *et al.* 1996, Deerenberg *et al.* 1997), but it may not be relevant to all organisms, and furthermore it may not be necessary to include it in all models. What goes into a model reflects both the issues of interest and the biology that is thought to be relevant. Examples of problems that can be investigated with our general framework include the following.

Body mass. A simple model of changes in body mass throughout the year could be based on the assumption that changes in mass are solely caused by changes in energy reserves. In such a model the principal (perhaps the only) state variable would be energy reserves. If it is believed that reserves are determined by a food–predation trade-off (see Chapter 6) it would be necessary to specify how the mean and variability of the food supply depend on time of year and how reserves, activity and time of year determine the predation risk. It would also be necessary to specify energy requirements throughout the year for reproducing and non-reproducing animals. A more complicated model would allow for changes in other components of physiological state such as the skeleton and the digestive system (see Piersma & Lindström 1997 for a review).

The basic model, in which energy reserves are the only state variable, could be the starting point for a model of hibernation. The simplest model is one in which hibernation is represented as a foraging option with zero gross gain, low energy expenditure and a small risk of predation. A problem with such a model is that it assumes that the animal can switch essentially instantaneously from hibernation to non-hibernation. Furthermore, an animal in hibernation may not be aware of the state of its environment. One way to deal with these complications is to introduce a state variable that reflects the current degree of torpor. To deal with uncertainty about the environment at the time when the animal emerges from hibernation, it may also be necessary to introduce a state variable that characterises important features of the environment, such as temperature.

Migration. In trying to understand migration systems, the first question is why animals migrate at all. In addition, one can ask questions about the timing of migration and breeding, and about the foraging behaviour that is used to fuel migration. To apply our framework, the state of an animal must include its location. Another obvious component of state is energy reserves. An animal in a given location has the choice of staying and not breeding, of staying and breeding or of migrating to one of the possible alternative locations. (For some preliminary work on this see McNamara *et al.* 1998.) More sophisticated models might include components of the animal's physiology other than energy reserves (e.g., for birds, condition of the flight muscles) and might consider a range of foraging options that differ in terms of energetic gain and predation risk (cf. Weber *et al.* 1998).

Modelling details of the reproductive bout. In the model presented in Section 9.5, we were largely concerned with the timing of reproduction during the year. The reproductive bouts themselves were modelled schematically; a fixed number of offspring per bout was assumed and the level of parental care was fixed. The only sophistication was to include the age of the young. In fact, if we had assumed that young produced at time t were independent by time $t + 1$, then we could have dispensed with age as a state variable. If the fine details of reproductive bouts are of interest, a different approach is required. A basic component of the model would be a decision about the number of young to produce. The more young of a given quality that are produced, the more energy is required by the female for their production. If the young are fed after birth, then the number of young may influence the energy expenditure of a parent during this period of care. If care extends beyond the next time interval, then it is necessary to keep track of the number of surviving young, so this must be a component of the animal's state. It may also be necessary to include information on aspects of the state of the young, such as age or energy reserves. The decision about whether to breed is likely to be based on the animal's reserves, and because of stochastic variation in reserves prior to breeding, there will be variation in both timing of breeding and the number of young produced. Thus the model could be used to investigate the correlations between these aspects of reproduction (cf. Rowe *et al.* 1994).

In addition to variation in energy reserves, animals are likely to vary in features such as their foraging ability or territory quality (e.g. Daan *et al.* 1990). If, for example, territory quality is important, it must be included as a state variable. It would be necessary to specify how territory quality influences factors such as food availability and predation risk. It would also be necessary to specify how territory quality changes over time; in particular, how territory quality in one breeding season correlates with territory quality in the next season. In this model we could once again look at the correlation between the timing of reproduction and the number of young produced. Now, however, territory quality acts as an explanatory variable, and the dependence of offspring number on time of reproduction is liable to be different from the case in which there is no quality variation.

Environmental uncertainty. In the approach that we have described, the environment varies seasonally, but every year is the same. In reality, there is usually year-to-year variation in conditions at a given time of year. For example, the way in which temperature varies over the year may be different from year to year. Furthermore, the food supply at a given time of year may be dependent on previous temperatures. When there are year-to-year fluctuations affecting all population members, the measure of fitness that we have used in our framework is, strictly speaking, no longer appropriate. Instead, the fitness measure that should be used is either the geometric mean fitness or some generalisation of it (see Chapter 10 for details). If, however, year-to-year variations are small or the species is long lived, then it should be reasonable to use our framework as at least a first approximation.

If the dependence of temperature on time of year varies from year to year, then temperature could be included as a state variable within the framework. If food availability depends on the pattern of past temperature, then current food availability should also be included as a state variable. In the simplest case, it might be possible to model the environment using just these two state variables. In more complicated scenarios, information about previous temperatures might also have to be included. A crucial component of any model is the specification of how current temperature correlates with temperature in the future. This component is important because it determines the extent to which current temperature is a reliable predictor of future temperature and hence future food. An outcome of the model would be a prediction about how behaviour depends on time of year and on current environmental conditions. For example, the model could be used to identify the factors that determine the onset of reproduction. In particular the model could predict whether calendar date is the most important factor determining the onset of reproduction, and to what extent current conditions have an effect on the decision.

Appendix 9.1 Parameters for the timing-of-reproduction model

Figures 9.3 – 9.8 illustrate results from the timing-of-reproduction model of Section 9.5. These figures were obtained from the following functions and parameters.

Energy expenditure:

$$
\begin{aligned}
c(u, t) &= 0.3 + 0.4u^2 \\
\gamma_{\text{brood}} &= 0.4 \\
\Delta_{\text{res}} &= 0.2.
\end{aligned}
$$

Predation risk:

$$M(u, t) = 0.004u^2.$$

Change in condition:

$$
\begin{aligned}
\alpha_0 &= 0.0625 \\
\alpha(c) &= 100c^2/256 \\
\Delta_{\text{cond}} &= 0.1.
\end{aligned}
$$

Condition-dependent mortality:

$$D(y) = 0.004 + 0.996(1 - y)^8.$$

Seasonality:

$$\epsilon = 0.3.$$

Effect of experience:

$$\theta = 0.839.$$

Brood size:

$$n_{\text{brood}} = 1.$$

Age (weeks) at which brood become independent:

$$a_{\text{max}} = 10.$$

10

Life histories in fluctuating environments

10.1 Introduction

We can distinguish between two sources of stochasticity affecting an organism. This distinction is an artificially polarised one but is nevertheless useful. *Demographic* stochasticity is due to random events that affect each individual in a population largely independently of other population members. For example, at any given time the rate at which population members capture prey will differ. This difference is in part due to good or bad luck while searching for or pursuing prey. Here 'luck' is a source of demographic stochasticity. Similarly, whether an individual is captured by a predator depends in part on luck. Of course, the amount of food found and whether the individual is killed by a predator will also depend on the individual's state; nevertheless, even amongst individuals in the same state there will be variations.

In contrast to demographic stochasticity, *environmental* stochasticity involves fluctuations of the environment as a whole. At a given time all population members are subject to the same environmental conditions. This is in contrast to the good and bad luck of demographic stochasticity where, at any given time, different individuals are affected largely independently of one another. The weather is an obvious source of environmental stochasticity. Food availability may also fluctuate from year to year; for example, beech mast occurs sporadically, but when it does there is an abundant source of food for great-tit populations in the autumn of that year. Changes in population density also affect all population members. If fluctuations in density are brought about by unstable population dynamics these fluctuations cannot be regarded as due to stochastic effects. Nevertheless, the selection pressures on population members from, say, changes in food from year to year, are the same regardless of whether these changes are due to an external factor such as beech trees or due to variations in the density of the population under consideration. When there is environmental stochasticity and all population members are subject to identical environmental conditions at a given time, we will refer to the environment as a *fluctuating environment*.

To compare the effects of demographic and environmental stochasticity on the rate of spread of a gene, we introduce a simple example.

Example 10.1.   Consider an asexual species. The generations are non-overlapping, so that individuals born in year t reach maturity and reproduce one year later and then die. We focus on a particular genotype. Initially there are $N(0)$ individuals of this genotype, the number t years later is denoted by $N(t)$.

First, suppose that there is a demographic stochasticity but no environmental stochasticity. Assume that each individual of the genotype produces 0 offspring with probability 0.5 and 4 offspring with probability 0.5, independently of other individuals of the genotype. Thus the mean number of offspring per individual is 2. It is then easy to show that the expected number after t years is

$$\mathbb{E}[N(t)] = 2^t N(0). \tag{10.1}$$

The actual number after t years is a random variable. But if we assume that the initial number, $N(0)$, is large, the independence of different lines means that it is reasonable to assume that at least some genotype members have survived until time t and that their number, $N(t)$, can be expected to be close to $\mathbb{E}[N(t)]$. In other words the mean $\mathbb{E}[N(t)]$ is a good indicator of genotype numbers. By equation (10.1), $N(t)$ can be expected to increase rapidly, roughly doubling at each generation. Thus the genotype certainly has high fitness.

Now suppose that there is environmental stochasticity but not demographic stochasticity. Each year is 'good' with probability 0.5 and 'bad' with probability 0.5, independently of previous years. If good, each individual produces 4 offspring; if bad, each produces 0 offspring. For this scenario numbers quadruple in size each year until the first bad year, when all members of the genotype die:

$$P(N(t) = 4^t N(0)) = 2^{-t}$$

$$P(N(t) = 0) = 1 - 2^{-t}.$$

Averaging over environmental stochasticity we thus have

$$\mathbb{E}[N(t)] = 2^t N(0) \tag{10.2}$$

as before. But now $N(t) \to 0$ with probability 1. Thus $\mathbb{E}[N(t)]$ is not a good indicator of genotype numbers, and even though mean numbers are increasing rapidly, fitness of the genotype is, by any reasonable measure, very low.

Since mean numbers of descendants are not a good guide to fitness, how then should we define fitness in fluctuating environments? We consider this issue in the next section.

10.2 The definition of fitness

To give a general definition of fitness we first define it in a particular example and then discuss how the formulae can be generalized.

Example 10.2.   As in Example 10.1 we consider an asexual species with non-overlapping generations having a generation time of one year. Each year, the environment is good with probability 0.5 and poor with probability 0.5, independently of previous years. Suppose a population consists of two genotypes labelled A and B. Type A individuals born at time t leave

$$r_A \,(\text{good}) = 3$$

offspring at time $t + 1$ if the environment is good between times t and $t + 1$, and

$$r_A \,(\text{poor}) = 1$$

offspring if the environment is bad. Type B individuals are unaffected by the environmental state, always leaving 2 offspring, so that

$$r_B \,(\text{good}) = r_B \,(\text{poor}) = 2.$$

We are interested in whether the population is eventually dominated by genotype A or genotype B.

If the number of individuals of genotype B at time 0 is $N_B(0)$ then the number t years later is

$$N_B(t) = 2^t N_B(0). \tag{10.3}$$

Let $N_A(0)$ individuals of genotype A be present at time 0 and let $N_A(t)$ be the number present t years later. Unlike the case of genotype B, $N_A(t)$ is a random variable whose value depends on previous environmental states. To give some insight into its distribution we consider some simple cases. First, note that $N_A(1)$ is either $N_A(0)$ or $3N_A(0)$, each possibility occurring with probability 0.5. Thus $\mathbb{E}[N_A(1)] = 2N_A(0)$. Now consider $N_A(2)$. There are four equally likely combinations of environmental states in the previous two

years, poor–poor, poor–good, good–poor and good–good. These give values
of $N_A(2)$ equal to $N_A(0)$, $3N_A(0)$, $3N_A(0)$ and $9N_A(0)$ respectively. Thus the
mean population size is $\mathbb{E}[N_A(2)] = \frac{1}{4}(1 + 3 + 3 + 9)N_A(0) = 4N_A(0)$. Note,
however, that the distribution of $N_A(2)$ is skew and the median, $3N_A(0)$ is
less than the mean, $4N_A(0)$. Finally, consider $N_A(10)$. The possible values
of this random variable and their probability of occurrence are given in
Table 10.1. The mean is now $\mathbb{E}[N_A(10)] = 2^{10}N_A(0) = 1024N_A(0)$, while
the median of $N_A(10)$ is $243N_A(0)$. These examples illustrate the general
facts that

$$\mathbb{E}[N_A(t)] = 2^t N_A(0) \qquad (10.4)$$

for all t and the distribution of $N_A(t)$ is skewed, with the median less than
the mean. The disparity between the mean and the median grows as t
increases. Thus the mean, $2^t N_A(0)$, is not representative of the growth in
numbers of this genotype.

Table 10.1. Numbers of genotype A after 10 generations in Example 10.2.

Number of good years in previous 10	Value of $N_A(10)$	Probability of occurrence
0	1	0.001
1	3	0.010
2	9	0.044
3	27	0.117
4	81	0.205
5	243	0.246
6	729	0.205
7	2187	0.117
8	6561	0.044
9	19683	0.010
10	59049	0.001

To analyse growth further we follow the approach taken by Lewontin
& Cohen (1969) and take logarithms. Equation (10.3) gives $\log[N_B(t)] = \log[N_B(0)] + t \log 2$. Dividing by t then gives

$$\frac{1}{t} \log[N_B(t)] \to \log 2 = 0.6931 \qquad (10.5)$$

as t tends to infinity. To deal with genotype A, let S_t denote the environmental state between times t and $t + 1$. Then S_t is either 'good' or 'poor'

and so $P(S_t = \text{good}) = P(S_t = \text{poor}) = 0.5$. With this notation a type-A individual born at time t leaves $r_A(S_t)$ offspring at time $t+1$. Thus

$$\frac{N_A(t+1)}{N_A(t)} = r_A(S_t). \tag{10.6}$$

We now express $N_A(t)$ as

$$\frac{N_A(t)}{N_A(t-1)} \frac{N_A(t-1)}{N_A(t-2)} \frac{N_A(t-2)}{N_A(t-3)} \cdots \frac{N_A(1)}{N_A(0)} N_A(0), \tag{10.7}$$

so that

$$N_A(t) = r_A(S_{t-1}) r_A(S_{t-2}) r_A(S_{t-3}) \cdots r_A(S_0) N_A(0). \tag{10.8}$$

Taking logarithms and dividing by t gives

$$\frac{1}{t} \log[N_A(t)] = \frac{1}{t} \log[N_A(0)] + \frac{1}{t} \sum_{n=0}^{t-1} \log[r(S_n)]. \tag{10.9}$$

As t tends to infinity the first term on the right-hand side tends to 0, while the second tends to $\mathbb{E}\{\log[r(S_n)]\}$ by the strong law of large numbers. Thus

$$\frac{1}{t} \log[N_A(t)] \rightarrow \mathbb{E}\{\log[r(S_n)]\} \tag{10.10}$$

as t tends to infinity. To evaluate this limit note that if $S_n = \text{good}$, then $\log[r(S_n)] = \log 3$, while if $S_n = \text{poor}$, then $\log[r(S_n)] = \log 1 = 0$. Thus

$$\mathbb{E}\{\log[r(S_n)]\} = \tfrac{1}{2} \times 0 + \tfrac{1}{2} \times \log 3 = 0.5493,$$

and hence

$$\frac{1}{t} \log[N_A(t)] \rightarrow 0.5493 \tag{10.11}$$

as t tends to infinity.

We can now compare the growth in numbers of the two genotypes. Equations (10.3) and (10.4) show that the mean numbers of the two genotypes have the same growth equation, doubling each generation. However, equations (10.5) and (10.11) show that

$$\frac{1}{t} \log \left[\frac{N_B(t)}{N_A(t)} \right] \rightarrow 0.6931 - 0.5493 = 0.1438,$$

so that $\log[N_B(t)/N_A(t)]$ tends to infinity. It follows that $N_B(t)/N_A(t)$ tends to infinity. Thus the population will eventually be taken over by genotype B.

The general approach

To generalise the above approach assume that an asexual population is censused at times $t = 0, 1, 2, \ldots$ one year apart. Consider a particular genotype. Assume that at the census time all members of this genotype are in the same state. Let $N(t)$ be numbers of the genotype at census time t.

Denote the environmental state between times t and $t + 1$ by S_t. This might be a single random variable, such as the time at which an important food source becomes available or the mean overwinter temperature, or it might be a random vector specifying several quantities of importance to the organism. Here, for notational convenience, we present the results for the case when $S_0, S_1, S_2, \ldots$ are non-negative continuous random variables with common density function f, but directly analogous results still hold if the environmental states are random vectors or have a discrete distribution, or both.

Given that the environmental state between times t and $t+1$ is $S_t = s$, an individual of the focal genotype that is alive at time t leaves $r(s)$ descendants at time $t + 1$. We refer to $r(s)$ as the profile of the genotype. It specifies how well members of the genotype perform under different conditions.

Under weak assumptions on $S_1, S_2, \ldots$ the strong law of large numbers gives

$$\frac{1}{t} \sum_{n=0}^{t-1} \log[r(S_n)] \to g \tag{10.12}$$

as t tends to infinity, where

$$g = \int_0^\infty \log[r(s)] f(s) ds. \tag{10.13}$$

(Independence of the S_i suffices, but this result holds under weaker assumptions.) The arguments applied to genotype A in the previous example then show that

$$\frac{1}{t} \log[N(t)] \to g \tag{10.14}$$

as t tends to infinity.

Now suppose the population consists of several different genotypes. We can define the measure g for each. To see that g acts as a fitness measure, consider two genotypes, labelled 1 and 2, with profiles r_1 and r_2 and measures g_1 and g_2 given by equation (10.13). Let $N_1(t)$ and $N_2(t)$ denote the numbers of the genotypes at time t. By equation (10.14), $t^{-1} \log[N_1(t)] \to g_1$

and $t^{-1} \log[N_2(t)] \to g_2$. Thus $t^{-1} \log[N_1(t)/N_2(t)] \to g_1 - g_2$. Hence

$$\log\left[\frac{N_1(t)}{N_2(t)}\right] \to \begin{cases} \infty & \text{if} & g_1 > g_2 \\ -\infty & \text{if} & g_1 < g_2 \end{cases}$$

which implies

$$\frac{N_1(t)}{N_2(t)} \to \begin{cases} \infty & \text{if} & g_1 > g_2 \\ 0 & \text{if} & g_1 < g_2 \end{cases} \tag{10.15}$$

In other words, the genotype with the higher g-value will eventually dominate the other. Thus if we call g fitness, natural selection maximises fitness.

We will talk of g as measuring fitness here, but any increasing function of g will serve as well. In particular, since $e^{g_1} < e^{g_2}$ if and only if $g_1 < g_2$ we could take $G = e^g$ as a measure of fitness. G is usually referred to as the geometric mean fitness.

Other measures of fitness have been suggested. As above let $N(t)$ be the number of individuals of a given genotype at time t. Then, ignoring demographic stochasticity, $\log[N(t)]$ is a random walk whose step lengths are determined by the environmental states; g is the mean step length. Grey (1980) has considered other aspects of step length such as its variance. These other aspects are important in determining whether numbers of the genotype are likely to dip below some low level, and are hence important in assessing whether the genotype disappears when the rate at which genotype members are created by mutation from other genotypes is low. When mutation is not rare, it is the growth rate g of a mutant that is important, since if $g < 0$ the mutant will never become established while if $g > 0$ mutant numbers will eventually increase until the mutant is no longer rare.

Our definition of g needs some clarification. It is also valid more generally than would appear from the above derivation, but has its limitations, as we now describe.

Demographic stochasticity. For a given environmental state s, the number of descendants left by an individual is a random variable, owing to demographic stochasticity. In defining the profile $r(s)$ we have averaged over demographic stochasticity, so that it is as if each individual produces the mean number of descendants. This will be reasonable if mutation to the genotype under consideration is not rare (cf. Section 8.2). Demographic stochasticity will be important if we are considering the establishment probability of a rare mutant (Haccou & Iwasa 1996). Rare mutants will certainly die out if $g < 0$, but may still die out due to bad luck if $g > 0$.

Density and frequency dependence. In analysing $N(t)$ we have appeared to ignore frequency- and density-dependent effects. However, as in Chapter 8 we are really deriving conditions for evolutionary stability. If a population is evolutionarily stable then the genotype (or genotypes) present will have fitness $g^* = 0$ (corresponding to $e^{g^*} = 1$). The resident population, together with physical conditions, define the environment for population members. In particular, the density of this population will affect food, and fluctuations in population size will produce 'environmental' fluctuations. In this environment g defines the invasion rate of a rare mutant genotype. Thus at evolutionary stability $g \leq 0$ for all mutant genotypes different from those in the resident population. The definition of g therefore remains valid when there are density and frequency effects, but with an interpretation as the an invasion rate into a population. This interpretation is reviewed in Metz *et al.* (1992).

In this chapter we take the background determined by the environment and other population members as given and determine optimal strategies within these constraints. No full ESS analysis is attempted. For examples and analyses of ESSs in fluctuating environments see Bulmer (1984), Ellner (1985a, b), Kisdi & Meszóna (1993), Iwasa & Haccou (1994).

Sex. We have formulated g for asexual species. When there is sexual reproduction g defines the invasion rate of a rare mutant allele rather than a genotype. The simplest case to analyse is when there are no demographic differences between the sexes, in that survival and fecundity are the same for both males and females. Then for discrete non-overlapping generations $r(s)$ is the number of surviving offspring left in one year's time discounted by the relatedness of offspring to the focal organism.

Parental survival. We have defined $r(s)$ to be the expected number of descendants left in one year's time if the intervening environmental state is s. Here 'descendant' can mean the original organism if it is still alive or its offspring, or its offspring's offspring etc. For example, consider an asexual organism which may live for several years and which breeds once a year. Census the population just prior to breeding each year and assume that (i) all individuals of a genotype alive at the census time are in the same state as one another, and (ii) this state does not vary from year to year. In particular, we are assuming that offspring produced between times t and $t+1$ are mature by $t+1$ and in the same state as their parent, if both parent and young survive. Under these circumstances we need make no distinction between the parent and its offspring at time t. Given that an organism is alive at time t,

$$r(s) = P(\text{organism survives until } t+1 \mid S_t = s)$$
$$+ \mathbb{E} \left(\begin{array}{l} \text{Number of offspring produced by the organism between} \\ \text{times } t \text{ and } t+1 \text{ that are alive at } t+1 \mid S_t = s \end{array} \right).$$

If the organism were sexual but with no demographic differences between the sexes, the profile $r(s)$ would be given by

$$r(s) = P(\text{organism survives to } t+1 \mid S_t = s)$$
$$+ \tfrac{1}{2}\mathbb{E} \left(\begin{array}{l} \text{Number of offspring produced by the organism between} \\ t \text{ and } t+1 \text{ that are alive at } t+1 \mid S_t = s \end{array} \right).$$

We emphasise that the parental contribution is vital when defining fitness. Liou *et al.* (1993) examine the geometric mean fitness of different clutch sizes in birds. They point out that, when birds laying a particular clutch size produce no surviving offspring in some years, fitness is not well defined unless the parental contribution is included. They see inclusion of the parental contribution as one means of overcoming the problem. In taking this approach they are not employing the correct definition of fitness. The parental contribution is an integral part of the definition of fitness, not a convenient computational device. To illustrate the effect of the parental contribution suppose that there are two genotypes, A and B, and two environmental states, good and poor. Individuals of genotype A produce 3 expected surviving offspring in good years and 0.5 in bad years. Individuals of genotype B produce 1.25 expected surviving offspring every year. If individuals only reproduce once, at age 1, and then die the profiles are

$$r_A(\text{good}) = 3 \qquad r_A(\text{bad}) = 0.5$$

$$r_B(\text{good}) = 1.25 \qquad r_B(\text{bad}) = 1.25$$

and the fitnesses are

$$g_A = 0.2027 \qquad \text{and} \qquad g_B = 0.2231,$$

so that genotype B has higher fitness. If, however, individuals of both genotypes survive between breeding attempts with probability 0.5 irrespective of the environmental state, the profiles are

$$r_A(\text{good}) = 3.5 \qquad r_A(\text{bad}) = 1$$

and

$$r_B(\text{good}) = 1.75 \qquad r_B(\text{bad}) = 1.75,$$

which gives

$$g_A = 0.6264 \qquad \text{and} \qquad g_B = 0.5596,$$

so that genotype A has higher fitness.

For an analysis of the effect of parental survival on optimal clutch size see Haccou & McNamara (1998).

Structured populations. When the population is structured in a fluctuating environment the population projection matrix is a random matrix and the simple idea of geometric mean fitness G must be replaced by that of the dominant Liapunov exponent. The standard text on this subject is Tuljapurkar (1990). Optimisation for structured populations in fluctuating environments is discussed in Section 10.8.

10.3 Reduction of variability

For a particular genotype, the profile $r(s)$ is the mean number of descendants left by an individual, given environmental state $S = s$. Averaging over S, the overall mean number of descendants left is

$$R = \mathbb{E}[r(S)] = \int_0^\infty r(s)f(s)ds,$$

where f is the density function of environmental states. R is often called the arithmetic mean fitness.

Example 10.2 shows that two genotypes with the same R can have different fitnesses, g. In this example the genotype for which $r(S)$ is more variable has lower fitness. This is true generally. In this section we describe the effect of variability, and briefly review the types of adaptation that can reduce it.

By Jensen's inequality

$$g = \mathbb{E}\{\log[r(S)]\} \le \log\{\mathbb{E}[r(S)]\} = \log R,$$

with equality if and only if $r(S)$ is constant. Thus, if two genotypes have the same R, but one has a constant $r(S)$ (equal to R) and the other a variable $r(S)$, the constant-$r(S)$ genotype will have higher fitness. To quantify the effect of variability, set $\sigma^2 = \text{Var}[r(S)]$. Expanding $\log r$ in a Taylor series about $r = R$ in equation (10.13) gives

$$g \simeq \log R - \frac{1}{2}\left(\frac{\sigma}{R}\right)^2$$

for small σ^2 (Lewontin & Cohen 1969). The corresponding approximation to G is

$$G = e^g \simeq R\left[1 - \frac{1}{2}\left(\frac{\sigma}{R}\right)^2\right]. \tag{10.16}$$

These formulae show that, for given R, fitness declines linearly in $\sigma^2 = \mathrm{Var}[r(S)]$ for small σ^2.

Gillespie (1977) was one of the first to emphasise the advantage of reduced variance in a fluctuating environment, although many others have also analysed the effects of variance (see Seger & Brockmann 1987 for a review). Adaptations to increase fitness by reducing σ^2 can be described as falling into one of two categories (Seger & Brockmann 1987).

I. Reduction in the variance in the number of descendants left by an individual.

II. Spreading the risk over relatives by diversifying actions.

We now illustrate these mechanisms.

I Reduction in individual variance

Strategies that produce a high mean number of descendants (high R) will often be risky in that they are highly susceptible to environmental fluctuations. Reducing R may then increase fitness by reducing σ^2. To quantify this, suppose that behaviour over a year is determined by the choice of an action $u > 0$. For example, u might be the choice of clutch size in a year. Assume that R is a unimodal function $R(u)$ of u, having a maximum at $u = \tilde{u}$. Thus action $\tilde{u}$ maximises the mean number of descendants left, where the mean is formed by averaging over all possible environmental states. Let α be the coefficient of variation in descendant number, i.e. $\alpha^2 = \mathrm{Var}[r(S)]/R^2$. Assume that α is an increasing function $\alpha(u)$ of u. Equation (10.16) can then be written as

$$G(u) \simeq R(u)[1 - \tfrac{1}{2}\alpha^2(u)]. \tag{10.17}$$

Differentiating with respect to u and setting $G'(u^*) = 0$ we see that the action u^* that maximises fitness is given by

$$R'(u^*)[1 - \tfrac{1}{2}\alpha^2(u^*)] = \alpha(u^*)\alpha'(u^*)R(u^*). \tag{10.18}$$

By assumption, $\alpha'(u^*) > 0$. Also, the approximation in expression (10.16) assumes σ^2 is small, so that $1 - \tfrac{1}{2}\alpha^2(u^*) > 0$. Thus equation (10.18) shows that $R'(u^*) > 0$. But R is unimodal, having a maximum at $\tilde{u}$. Thus $u^* < \tilde{u}$. This shows that the action maximising fitness has a lower value than the action maximising expected numbers of descendants when the coefficient of variation α is increasing.

For many species, there is a a trade-off between the number of offspring and their size or quality (Smith & Fretwell 1974; see also Chapter 8). In a fluctuating environment, the success of a reproductive attempt will depend on the number of young produced, their quality and the environmental state. Obviously, if the environmental state is known, then there is no problem; each organism should maximise the total reproductive value of all offspring produced under the environmental conditions that occur. If, however, the reproductive decision has to be made before the environmental state is known, then the reproductive decision has to be analysed using the geometric mean fitness.

In many altricial bird species, reproductive success is limited by the rate at which parents deliver food to their young. This rate will depend on environmental conditions, which may not be known when the clutch is produced. In the production of seeds by plants, the resources given to each seed by the parent are known, but the growth of an individual seed will depend on its resources and on the environment conditions that it experiences. These conditions may not be known when resources are allocated

In a constant environment, the analysis presented by Smith & Fretwell (1974) shows that the same number of offspring should always be produced, and resources should be divided equally between them. Various workers have addressed the question of how these conclusions change if the environment is fluctuating. Lalonde (1991) investigated the best number of offspring to produce when each offspring receives the same allocation of resources. In the specific case that he considered, he showed numerically that the optimal number of offspring is less than the number that maximises the arithmetic mean fitness. Haccou & McNamara (1998) showed analytically that these results hold for a general class of models that include fixed parental survival. Under certain circumstances, randomising the number of offspring gives a higher fitness than producing any fixed number of offspring (e.g. Schultz 1991, Haccou & McNamara 1998). A criterion for deciding whether a fixed or a random strategy is optimal is given in Section 10.4. In addition to looking at the effect of a fluctuating environment on the optimal number of offspring, attention has also been given to whether resources should be divided equally between offspring (e.g. McGinley *et al.* 1987; see also work on brood reduction, e.g. Konarzewski 1993). It can be shown, however, that if offspring number is taken to be a continuous variable, a model in which allocation of resources to offspring is unequal is equivalent mathematically to a corresponding model in which the number of offspring is variable across breeding attempts but the allocation of resources to all young produced in an attempt is equal.

We now illustrate some of these issues concerning the trade-off between number and quality of offspring with an example based on clutch size.

Example 10.3. Clutch size in a fluctuating environment. We assume that there are no demographic differences between the sexes, in that survival and fecundity are the same for males and females. The sex ratio of offspring is then necessarily 1 : 1 and we can ignore males. For simplicity we also assume non-overlapping generations. At the beginning of the breeding season a female decides her clutch size u (> 0). This choice is made without knowledge of environmental conditions over the coming year. Clutch size u produces

$$r_u(s) = \tfrac{1}{2}ue^{-0.01su^2} \tag{10.19}$$

expected daughters when the environmental condition is s. Figure 10.1(a) illustrates the profiles $r_u(s)$ for a selection of clutch sizes u. Environmental condition declines as s increases. It can be seen from the figure that a large clutch produces more recruits than a small clutch in good years (low s), while the reverse is true in bad years (high s). It can be seen that the function $r_u(s)$ does not depend strongly on s when u is small, while $r_u(s)$ is highly dependent on s when u is large. Thus small clutches are less sensitive to environmental fluctuations than are large clutches.

We take the distribution of environmental conditions to be exponential with mean 1 so that $f(s) = e^{-s}$. For this distribution it is easily verified that the coefficient of variation $\alpha(u)$ is an increasing function of u. It can also be seen that $R(u) = 50u/(100+u^2)$. This is maximised at a clutch size of $\tilde{u} = 10$ and results in 2.5 surviving female offspring on average. Fitness is given by $g(u) = \log(\tfrac{1}{2}u) - 0.01u^2$ and is maximised at $u^* = \sqrt{50} = 7.07$. Thus as predicted $u^* < \tilde{u}$. Under the optimal strategy there are 2.36 surviving female offspring on average.

We have just referred to clutch size $u^* = 7.07$ as optimal. In fact, the above analysis shows that u^* is optimal amongst all non-randomised strategies. In Section 10.4 we will show that no randomised strategy can do better.

In the above analysis there are non-overlapping generations, and variance in offspring number is important. For relatively long-lived bird species, between-year variance in offspring number is less important since such variations tend to be averaged out over the lifetime of the bird. We would then expect maximisation of the geometric mean growth rate in descendant numbers to predict similar clutches to those predicted by maximisation of the arithmetic mean growth rate (e.g. Cooch & Ricklefs 1994, Haccou &

McNamara 1998). Thus for a long-lived species at density-dependent equilibrium, natural selection might be expected approximately to maximise the expected total lifetime offspring number.

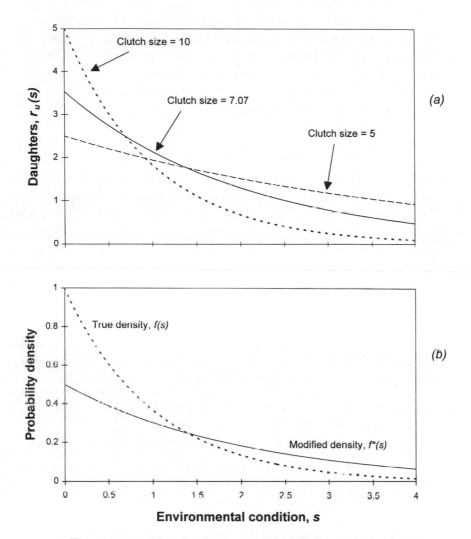

Figure 10.1. The clutch size model of Example 10.3. (*a*) The expected number of surviving daughters *r* as a function of environmental condition *s* for three values of clutch size *u*. (*b*) The true probability density of environmental states $f(s) = e^{-s}$ (broken line), and the modified distribution of environmental states $f^*(s)$ given by equation (10.28) (solid line).

Schaffer (1974b) looked at the variance in total lifetime offspring number of iteroparous species. He considers an example in which there is a trade-off between current reproduction and future survival and in which juvenile survival is more susceptible to adverse fluctuations than parental survival. In this model, within-individual variance in total lifetime offspring production is reduced by reducing the parental effort per year; the higher parental survival under the reduced effort increases the parents' lifespan and spreads the risk by allowing the parent more reproductive attempts. Similarly, others have invoked this type of risk-spreading to argue for the advantages of iteroparity over semelparity in fluctuating environments (e.g. Murphy 1968, Bulmer 1985, Orzack & Tuljapurkar 1989).

II Spreading the risk over relatives

Suppose an organism has the choice between two actions u_0 and u_1. For each λ in the range $0 \leq \lambda \leq 1$ we can define a strategy π_λ by the following rule. Toss a biased coin that gives a head with probability λ. If tails, then choose action u_0, if heads then choose action u_1. As many have pointed out it can be advantageous to allow such randomised strategies in a fluctuating environment (e.g. Cohen 1966, Cooper & Kaplan 1982). Cooper and Kaplan refer to these strategies as 'adaptive coin flipping'. The following example gives an illustration.

Example 10.4. Suppose that there are two environmental states, labelled s_0 and s_1, each occurring with probability 0.5. Let $d(u_i, s_j)$ denote the number of descendants left in one year's time if action u_i is chosen and the intervening environmental state is s_j. We suppose

$$d(u_0, s_0) = 3 \qquad d(u_0, s_1) = 1$$

$$d(u_1, s_0) = 1 \qquad d(u_1, s_1) = 3,$$

so that action u_0 is best when the environment state is s_0 and action u_1 is best when it is s_1. Consider the three strategies $\pi_0, \pi_{0.5}, \pi_1$. Strategy π_0 is always to choose action u_0 and has profile

$$r_{\pi_0}(s) = d(u_0, s) = \begin{cases} 3 & \text{if} & s = s_0 \\ 1 & \text{if} & s = s_1. \end{cases}$$

Similarly strategy π_1 is always to choose action u_1, so that

$$r_{\pi_1}(s) = d(u_1, s) = \begin{cases} 1 & \text{if} & s = s_0 \\ 3 & \text{if} & s = s_1. \end{cases}$$

Strategy $\pi_{0.5}$ chooses each action with equal probability and has profile

$$r_{\pi_{0.5}}(s) = 0.5r_{\pi_0}(s) + 0.5r_{\pi_1}(s) = 2 \qquad \text{for all } s.$$

Note first that all three strategies produce the same expected number of descendants, since $R(\pi_0) = R(\pi_{0.5}) = R(\pi_1) = 2$. It can be seen, however, that $r_{\pi_{0.5}}(S)$ has zero variance while the other profiles have a variance of 1. As a consequence $\pi_{0.5}$ has the highest fitness, with

$$g(\pi_0) = g(\pi_1) = 0.5\log 1 + 0.5\log 3 = 0.5493$$

and

$$g(\pi_{0.5}) = \log 2 = 0.6931.$$

In this example the actual number of descendants left by an individual is variable whichever strategy is used, and is 1 or 3 with equal probability. But in studying the effects of natural selection we are interested in how gene frequencies change, and the difference between $\pi_{0.5}$ and the other strategies becomes apparent once we consider the change in the number of genes coding for the various strategies. Suppose that a dominant gene codes for π_0. Then all individuals carrying the gene will produce the same number of descendants in a given year, 3 if $S = s_0$, 1 if $S = s_1$. Similarly if a gene codes for π_1 then the numbers of descendants left by all individuals carrying the gene are perfectly correlated. In contrast, consider a dominant gene coding for strategy $\pi_{0.5}$. Randomisation leads to reduced correlation between individuals carrying this gene, and in fact in this extreme example the correlation in the numbers of descendants left by different individuals carrying the gene is zero. This spread of the risk reduces between-year variability.

We now give two more examples in which adaptive coin flipping is optimal.

Example 10.5. *Cohen's model of seed germination.* Cohen's (1966) model can be described in our terms as follows. An annual plant dies at the end of summer leaving seeds. The population is censused at the beginning of spring each year. At these times the population is composed entirely of seeds. Each seed has two actions: u_0, remain dormant for another year; u_1, grow. The environmental state S in the growing season following germination has two values, 'good' and 'bad' where $P(S = \text{good}) = \beta$ and $P(S = \text{bad}) = 1 - \beta$.

A seed that remains dormant dies before the next census time with probability $1 - v$ independently of environmental conditions, so that the expected number of descendants left by this seed in one year's time is given as a function of the environmental conditions by

$$d(u_0, \text{good}) = d(u_0, \text{bad}) = v.$$

A seed that germinates produces on average y seeds in a good year and 0 seeds in a bad year, so that

$$d(u_1, \text{good}) = y \qquad d(u_1, \text{bad}) = 0.$$

A strategy is determined by the probability that a seed will germinate. If this probability is p then the expected number of descendant seeds at the next census time can be expressed in terms of the environmental state as

$$r_p(\text{good}) \;=\; py + (1-p)v$$

$$r_p(\text{bad}) \;=\; (1-p)v.$$

Assume $\beta y > 1$, since otherwise the seed population is bound to become extinct. The expected number of seeds present in one year is maximised by all seeds germinating, but this will not maximise fitness, which is given in terms of p by

$$g(p) = \beta \log[py + (1-p)v] + (1-\beta) \log[(1-p)v].$$

Differentiating with respect to p we see that $g(p)$ is maximised at $p = p^*$, where

$$p^* = \frac{\beta y - v}{y - v}.$$

Thus $0 < p^* < 1$.

There are now many variants on this model (e.g. Bulmer 1984, Ellner 1985a,b).

Example 10.6.   Diapause decisions.   Cohen (1970) introduced a model for the timing of entry into diapause in a fluctuating environment. Others have since considered this question (see Hanski 1988 for a review) and the related question of allocation to growth and reproduction in annual plants (Cohen 1971, 1976, King & Roughgarden 1982, Amir & Cohen 1990). Here we describe the work of McNamara (1994b), which provides a complete analytic solution of a simple version of the diapause problem.

Over the winter all mature individuals die and the only surviving individuals are eggs in diapause. At the end of winter, the eggs hatch and the resultant population goes through a series of discrete non-overlapping generations during summer. In each generation the eggs produced may either immediately enter diapause, in which case they remain in diapause until next year, or may attempt to grow to maturity and hence produce more eggs. The disadvantage of growth is that the individual will die if bad weather occurs. Eggs in diapause are unaffected by bad weather.

Define the decision times $n = 0, 1, 2, \ldots$ during a summer as follows. Decision time $n = 0$ is the time of emergence from overwinter diapause. The decision times $n = 1, \ldots, T$ are the times at which the nth generation reaches maturity.

Before bad weather occurs we can define the following.

$$
\begin{aligned}
D(n) &= \text{total number of eggs already in diapause before time } n \\
N_{\text{new}}(n) &= \text{number of new eggs produced at time } n \\
p(n) &= \text{proportion of these eggs entering diapause.}
\end{aligned}
$$

Then we assume that

$$D(n+1) = D(n) + p(n)N_{\text{new}}(n)$$

and

$$N_{\text{new}}(n+1) = \alpha[1 - p(n)]N_{\text{new}}(n),$$

where α (> 1) is the expected number of eggs produced at time $n + 1$ by an individual that starts growing at time n, given that the weather remains good until time $n + 1$. Let S be an integer-valued random variable, defined by $S = n$ if and only if the bad weather first occurs during the growth of generation n. Then S determines a 'season length', and the number of eggs already in diapause when bad weather first occurs is $D(S)$. We assume that bad weather always occurs before the maturity of the Tth generation at time T. An optimal strategy chooses $p(0), p(1), \ldots$ to maximise

$$\mathbb{E}[\log D(S)]. \tag{10.20}$$

McNamara (1994b) showed that the problem of finding the optimal strategy can be reduced to a dynamic programming problem in which the state variable at time n is the total proportion of all eggs present at this time that were already in diapause before this time,

$$\rho_D(n) = D(n)/[N_{\text{new}}(n) + D(n)].$$

Let $H(n) = P(S = n \mid S > n - 1)$ be the probability that bad weather first occurs during the growth of generation n, given that it has not previously occurred. Assume $H(1) \le H(2) \le \cdots \le H(T) = 1$ and set

$$c^*(n) = \min\left\{1, \left(\frac{1}{\alpha - 1}\right)\left[\frac{H(n)}{1 - H(n)}\right]\right\}. \tag{10.21}$$

Then the optimal strategy is to control the process $\{\rho_D(0), \rho_D(1), \ldots, \rho_D(T)\}$ in such a way that $\rho_D(n)$ is as close to $c^*(n)$ as possible.

Set $\tilde{n}$ as the first time at which $c^*(n) = 1$. The expected number of eggs in diapause overwinter is maximised if no eggs enter diapause before time $\tilde{n}$ and all eggs enter diapause at time $\tilde{n}$. In contrast, the optimal strategy maximising expression (10.20) involves a graded response in which a proportion $p^*(n)$ of new eggs produced at time n enter diapause;

$$p^*(n) = \frac{\alpha P(S = n+1) - P(S = n)}{\alpha P(S \geq n+1) - P(S \geq n)} \qquad (10.22)$$

for $n < \tilde{n}$ and $p(\tilde{n}) = 1$.

10.4 Optimal strategies

In all the examples presented so far in this chapter, the time of the annual census can be chosen so that all individuals of the same genotype are in the same state at this time of year. We now present a general framework that encompasses all such unstructured population problems. This framework will lead to a general criterion by which we can recognise whether a given strategy is optimal. It is often comparatively easy to find the best pure strategy for a problem. Having found this strategy, the criterion can be used to decide whether there exists a mixed strategy that does better. The criterion also suggests a general computational method for finding optimal strategies, whether pure or mixed.

The general model

As before there is a yearly census of the population. The population is unstructured. Thus there is some given state such that all individuals are in this state at the census time. At each census time each population member must choose an action from a set of available actions. (We take this set to be finite for notational convenience.) The choice is made without information about the environmental state in the coming year. If action u is chosen and the environmental state is s then the expected number of descendants left at the census time one year hence is $d(u, s)$. The environmental state in any year S is a random variable which, for simplicity of presentation, we assume to take non-negative real values and to be continuous with density function f.

In this setting a strategy π specifies the probability $P_\pi(u)$ that each pure action u will be chosen. For example, if a female bird must choose the number of eggs to lay, 'lay 3 eggs' and 'lay 5 eggs' are pure actions. The pure actions available to the female are thus $u = 0, 1, 2, \ldots, c_{\max}$ where $c_{\max}$

is the maximum clutch size. Then π defined by

$$P_\pi(\text{clutch size} = 3) = 1$$

is a pure strategy, while π given by

$$P_\pi(\text{clutch size} = 3) = P_\pi(\text{clutch size} = 5) = 0.5$$

is a mixed or randomised strategy.

The profile r_π of a strategy π specifies the expected number of descendants left under π when the environmental state is s, i.e.

$$r_\pi(s) = \sum_u d(u, s)P_\pi(u). \tag{10.23}$$

The fitness of π is given by

$$g(\pi) = \int_0^\infty \log[r_\pi(s)]f(s)ds. \tag{10.24}$$

We can then define an optimal strategy π^* as one satisfying

$$g(\pi^*) = g^* = \max_\pi g(\pi). \tag{10.25}$$

Optimal profiles

The profile function r_π of a strategy π specifies the performance of that strategy under the various possible environmental conditions. If there is a strategy π^* such that $r_{\pi^*}(s) > r_\pi(s)$ for all strategies π and all environmental conditions s, then clearly π^* is optimal. Typically, however, no strategy that satisfies this very strong requirement exists. Instead, if obviously inferior strategies are disregarded, for every remaining strategy π_0 one can find a strategy π_1 such that $r_{\pi_0}(s) < r_{\pi_1}(s)$ for some s, while $r_{\pi_0}(s) > r_{\pi_1}(s)$ for other s. In other words, π_0 is worse than π_1 under some environmental conditions but better than π_1 under other conditions. Figure 10.1(a) illustrates this phenomenon: large clutch sizes are superior to small clutch sizes when environmental conditions are good whereas the reverse is true when conditions are bad. Rather than being the best under all conditions, the optimal strategy is a compromise that does best in some average sense.

Although equations (10.24) and (10.25) formally define what is meant by an optimal strategy this definition is often not easy to work with. Instead, the following equivalent characterisation is easier to apply.

(C) Let π^* be a strategy with profile r_{π^*}. Then π^* is optimal if and only if

$$\int_0^\infty \frac{r_\pi(s)}{r_{\pi^*}(s)} f(s)ds \leq 1 \qquad (10.26)$$

for all strategies π.

A pure strategy is one that always chooses the same action. If strategy π_u chooses action u with probability 1, it has profile $r_{\pi_u}(s) = d(u, s)$. As equation (10.23) shows, the profile of any general strategy π can be expressed as a linear combination of these simple profiles. It is thus easy to see that condition (C) can be rewritten as follows.

(C') π^* is optimal if and only if

$$\int_0^\infty \frac{d(u, s)}{r_{\pi^*}(s)} f(s)ds \leq 1 \qquad (10.27)$$

for all pure actions u.

Since there must be equality in equation (10.26) when $\pi = \pi^*$, we also see that there must be equality in equation (10.27) for almost all actions u chosen under π^*. Conditions (C) and (C') were derived in special cases by Haccou & Iwasa (1995) and Sasaki & Ellner (1995) and established in the general case by McNamara (1995).

The profile function of an optimal strategy is unique in the sense that even if there is more than one optimal strategy, all must have the same profile function, r_{π^*}. Furthermore, any strategy π that is nearly optimal must have its profile r_π close to r_{π^*} (McNamara 1995). Given the profile r_{π^*} we can define a function of environmental conditions f^* by

$$f^*(s) = \frac{1}{K} \frac{f(s)}{r_{\pi^*}(s)}, \qquad (10.28)$$

where K is a normalising constant chosen so that

$$\int_0^\infty f^*(s)ds = 1.$$

Thus f^* acts as a modified density function on environmental conditions. We can now compare the criterion of maximisation of arithmetic mean fitness with that of maximisation of geometric mean fitness. The expected number of descendants left in one year's time by an individual following strategy π is (averaging over environmental conditions)

$$R(\pi) = \int_0^\infty r_\pi(s)f(s)ds.$$

The strategy maximising arithmetic mean fitness maximises this quantity. In contrast, let

$$R^*(\pi) = \int_0^\infty r_\pi(s) f^*(s) ds. \tag{10.29}$$

Then equation (10.26) shows that if a strategy maximises geometric mean fitness then it maximises R^*. In other words, maximisation of geometric mean fitness is equivalent to maximising the mean number of descendants left, this latter mean being an average formed using the modified distribution of environmental states f^* rather than the true distribution f (McNamara 1995). It can be seen from equation (10.28) that if the optimal strategy π^* does badly for an environmental condition s ($r_{\pi^*}(s)$ is small), the modified probability that this condition will occur is greater than its true value. Conversely, if the optimal strategy does well under condition s ($r_{\pi^*}(s)$ is large), the modified probability that s will occur is less than its true value. Thus in calculating R^* the impact of poor environmental conditions is emphasised. Figure 10.1(b) shows the true and the modified density of states for the clutch size problem of Example 10.3.

When are pure strategies optimal?

It is usually quite easy to find the best pure (i.e. non-randomised) strategy. For example, in Example 10.3 we found that the best pure strategy was to adopt a clutch of size $u^* = 7.07$. But how can we tell whether the best pure strategy is best amongst all strategies including randomised strategies? Equation (10.27) gives us such a criterion. Suppose that the best pure strategy is always to choose action u^*; then, by (C'), this pure strategy is optimal if and only if

$$\int_0^\infty \frac{d(u, s)}{d(u^*, s)} f(s) ds \le 1 \tag{10.30}$$

for all pure actions u (see Haccou & Iwasa 1995).

To illustrate this formula return to Example 10.3. Here we have

$$d(u, s) \equiv r_u(s) = \tfrac{1}{2} u e^{-0.01 s u^2}.$$

Thus for $u^* = \sqrt{50}$ we have

$$\int_0^\infty \frac{d(u, s)}{d(u^*, s)} f(s) ds = \frac{u}{\sqrt{50}} \int_0^\infty e^{-0.01 s (u^2 - 50)} e^{-s} ds$$

$$= \frac{u}{\sqrt{50}} \left[\frac{1}{0.01(u^2 - 50) + 1} \right].$$

Differentiating with respect to u we see that this is maximised when $u = u^*$. Thus

$$\int_0^\infty \frac{d(u,s)}{d(u^*,s)} f(s)ds \leq \int_0^\infty \frac{d(u^*,s)}{d(u^*,s)} f(s)ds = \int_0^\infty f(s)ds = 1.$$

Thus the strategy of always choosing a clutch of size $u^* = \sqrt{50}$ is optimal (i.e. the best strategy amongst all strategies, including randomised strategies).

If the best pure strategy for a problem, u^*, does not satisfy inequality (10.30) there is a mixed strategy that has higher fitness. The class of all mixed strategies is often very large, and a crude numerical search for the best strategy within the class may not be computationally feasible. Fortunately, criterion (C′) not only allows us to identify optimal strategies but also suggests a natural 'hill-climbing' method by which they can be found. This method is described in Appendix 10.1.

The analysis of this section has assumed that the organism must make a decision without information on the environmental state in the future. For analyses of models in which there is partial information, see Cohen (1967) and Haccou & Iwasa (1995).

10.5 Individual optimisation and implicit frequency dependence

Arguments in terms of adaptation are ultimately about the spread of genes in a population. Although the genetic view is the fundamental one, it is much easier in analysing behaviour or life histories to translate genetic arguments into phenotypic ones. For example, in a constant environment the genetic argument leads to the phenotypic principle that each individual should maximise its expected numbers of descendants in the distant future (Chapter 8). Thus the genetic argument is synonymous with individual optimisation. The big difference between constant and fluctuating environments is that in the latter the arguments about the spread of genes are no longer synonymous with individual optimisation.

Cooper & Kaplan (1982) illustrated the failure of individual optimisation with an example where each individual must choose between two actions S_0 and S_1. In their example, the strategy maximising geometric mean fitness, and hence maximising the rate of spread of genes coding for the strategy, is to 'coin-flip'. The expected number of descendants, averaged over environmental states, is less under S_1 than S_0. Thus some individuals of the optimal genotype end up taking an action that does not maximise the expected number of descendants produced. As Cooper and Kaplan say (p. 145):

It is as though each individual of the superior coin-flipping genotype were practising a form of 'coin-flipping altruism' by assuming the risk of getting stuck with the personally inferior strategy S_1. True, this is not the customary kind of altruism in which the altruist renders some tangible service to other individuals. It nonetheless represents a sacrifice of immediate individual fitness for the sake of the long term advantage of the genotype.

Delayed seed germination also illustrates the failure of individual optimisation. For example, in Cohen's model, outlined in Example 10.5, selection favours delayed germination, so that offspring are not all eliminated when there is a bad year. In this model each seed maximises its probability of growing to maturity by not delaying germination and growing immediately. The reason is that dormant seeds might die before the next year. Contrary to the suggestion of Westoby (1981) there is no parent–offspring conflict. The individual optimisation point of view is not valid in a fluctuating environment. As Ellner (1986) points out (p. 178):

Selection on embryo genotypes will favor the same delayed germination as is favored by selection on parent genotypes . . . for exactly the same reason. An allele influencing embryo behavior, present in multiple copies in a number of seeds, will have higher fitness by 'spreading the risk' over a number of risky opportunities for germination than by gambling everything on one opportunity. The calculations for the strategy maximizing allele fitness are exactly the same as for the strategy maximizing parental fitness, so there is no conflict.

Cohen's model has no competition between sibs. When the competition between sibs is more intense than between general conspecifics in the population, there may be a genuine parent offspring conflict (see Ellner 1986 and references therein).

Individual optimisation fails, but it is as though it has been replaced by a co-operative game involving individuals of the same genotype (or sharing some allele). For example, in the seed germination model, spreading the risk means that if most seed of the genotype grow now it is best for the others to delay germination, while if most delay germination it is best for the others to grow now. In this section we formalise the idea of a game-theoretic interpretation. The analysis is based on McNamara (1995).

Let π_1 and π_2 be strategies with profiles r_{π_1} and r_{π_2} respectively. Set

$$W(\pi_1, \pi_2) = \int_0^\infty \frac{r_{\pi_1}(s)}{r_{\pi_2}(s)} \, f(s)ds. \tag{10.31}$$

We will interpret $W(\pi_1, \pi_2)$ as the payoff to a single individual playing strategy π_1 in a game in which the resident population plays strategy π_2. Since the payoff is formed by taking an arithmetic average over environmental conditions, s, we can regard this game as one in a constant environment (see McNamara 1995 for more on this interpretation). Of course $W(\pi^*, \pi^*) = 1$ for any strategy π^*. Thus by inequality (10.26) a strategy π^* is optimal if and only if

$$W(\pi, \pi^*) \le W(\pi^*, \pi^*) \qquad \text{for all } \pi. \tag{10.32}$$

In the terminology of game theory, a strategy π^* is optimal if and only if it is a Nash-equilibrium strategy for the game (Section 7.3). If the optimal strategy is unique we can go further and show that condition (10.32) is equivalent to the condition that, for every $\pi \ne \pi^*$, either

$$W(\pi, \pi^*) < W(\pi^*, \pi^*) \tag{10.33}$$

or

$$W(\pi, \pi^*) = W(\pi^*, \pi^*) \qquad \text{and} \qquad W(\pi, \pi) < W(\pi^*, \pi). \tag{10.34}$$

These are now precisely the Maynard Smith (1982) conditions for π^* to be an ESS in a two-player game (see Section 7.4). In summary:

$$\left\{ \begin{array}{l} \pi^* \text{ maximises geometric} \\ \text{mean fitness in the} \\ \text{fluctuating environment} \end{array} \right\} \Longleftrightarrow \left\{ \begin{array}{l} \pi^* \text{ is an ESS in} \\ \text{the constant} \\ \text{environment game} \end{array} \right\}. \tag{10.35}$$

The ESS formalises the idea that individuals of the same genotype (or sharing the same allele for controlling behaviour) are involved in a co-operative game. Condition (10.32) is really saying that if almost all members of the genotype adopt strategy π^*, then the fitness of the genotype as a whole will not be improved by some genotype members changing to another strategy π. For a precise quantification of this statement, see McNamara (1995).

10.6 Phenotypic plasticity

So far we have been concerned with an individual that chooses a single action at the yearly census times, and have assumed that all individuals of the same genotype are in the same state when this action is chosen. In many cases an individual's state when making a decision is important. Furthermore, an organism may have to choose a sequence of actions during the year, where each action chosen depends on the organism's physiological state, its information about the environment and the time of year. Chapters 8 and 9 analyse these complex life-history decisions in the case when the environment is not subject to fluctuations from year to year. When there are environmental fluctuations the analysis becomes much more difficult, because the lack of individual optimisation means that the approach through dynamic programming set out in Chapters 8 and 9 is no longer directly applicable.

In this section we consider state-dependent decisions when there are no long-term consequences of these decisions. As in Section 10.4 we assume that there is a yearly time at which each organism must choose a single action. As before, this action must be chosen before environmental conditions for that year are known, but we now assume that, at the time the decision is made, individuals differ in their state. The state of an organism, its action and the ensuing environmental conditions combine to determine the expected number of descendants left in one year's time. Here we make the simplifying assumption that the distribution of states of surviving descendants is the same each year and is independent of parental state, parental action and environmental conditions. The extension of this model to the case where a sequence of state-dependent decisions is made during a year is dealt with in Section 10.7. The case where effects within one year persist into future years is treated in Section 10.8.

The following example, based on McNamara (1998), illustrates the scenario we are considering here.

Example 10.7. Dependence of clutch size on territory quality. Breeding territories for a bird species differ in quality. The quality of a territory is measured by a single number x, where $0 < x < 1$. Quality increases with increasing x. At the beginning of the breeding season each female obtains a territory. The assignment of females to territories is assumed to be random, with density function $p(x)$ describing the probability distribution of possible territory qualities obtained by a female.

The female decides how many eggs to lay on the basis of her territory quality. We denote clutch size by u and, in this schematic model, assume that u can be any positive real number rather than just an integer. Clutch size is decided before the environmental condition after hatching is known.

Environmental conditions are described by a single number s, where $0 < s < 1$. Condition deteriorates as s increases. We assume that environmental conditions in successive years are independent random variables, drawn from a distribution with density function $f(s)$ over $[0, 1]$.

Territory quality x and subsequent environmental condition s combine to determine the rate $e(x, s)$ at which food can be delivered to the nest. For the purposes of illustration we take

$$e(x, s) = 10x[1 + (0.5 - s)x]. \tag{10.36}$$

Thus the effect of increasing s is to reduce the delivery rate in all territories, while the effect of increasing x is to increase delivery rate whatever the value of s. Averaging over s, the delivery rate has mean $10x$ and variance proportional to x^4. Thus the coefficient of variation of delivery rate increases with increasing territory quality x, and high-quality territories (high x) are more susceptible to variations in environmental conditions.

Food delivered to the nest is shared equally by the brood. Thus each member of a brood of size u receives food at a rate e/u, where e is the delivery rate to the nest. The probability that an individual survives to maturity is a function $S_{\text{off}}(e/u)$ of the rate at which it receives food. The mean number of female descendants (averaging over demographic but not environmental stochasticity) left by a female that lays u eggs when her territory quality is x and when the environmental condition after hatching is s is then

$$\tilde{d}(x, u, s) = K + \frac{u}{2} S_{\text{off}} \left(\frac{e(x, s)}{u} \right), \tag{10.37}$$

where K is the probability that the parental bird survives and the factor $\frac{1}{2}$ is present because one-half of the surviving offspring are assumed to be female.

A strategy is a function π that specifies the clutch size, i.e. $u = \pi(x)$, for each possible territory quality x. A randomly selected female following strategy π leaves

$$r_\pi(s) = \int_0^1 \tilde{d}(x, \pi(x), s)p(x)dx \tag{10.38}$$

expected descendants when the environmental condition is s. Figure 10.2 illustrates the following three strategies.

(i) *Maximise arithmetic mean fitness.*   Suppose that s is a micro-environmental condition and that different females experience independent values of s. Then the allele (coding for strategies) that is selected maximises the arithmetic mean number of copies of itself left in the following year. That is, it maximises

$$\int_0^1 r_\pi(s)f(s)ds. \tag{10.39}$$

By equation (10.38) this is equivalent to each female's maximising her own expected number of descendants; i.e., a female with territory x should choose clutch size u to maximise

$$\int_0^1 \tilde{d}(x, u, s)f(s)ds. \tag{10.40}$$

(ii) Maximise geometric mean fitness. Suppose the same environmental condition s affects all females. Then the allele selected for will maximise the geometric number of copies of itself left next year. Equivalently it will maximise

$$\int_0^1 \log[r_\pi(s)]f(s)ds. \tag{10.41}$$

(iii) Individual maximisation of geometric mean descendant numbers. Maximisation of expression (10.41) is not equivalent to each female's maximising her geometric mean number of descendants. Under this latter strategy a female with territory x chooses u to maximise

$$\int_0^1 \log[\tilde{d}(x, u, s)]f(s)ds. \tag{10.42}$$

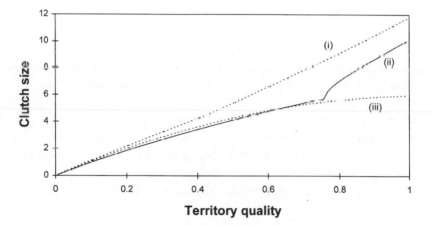

Figure 10.2. The model of clutch size as a function of territory quality (Example 10.7), showing the clutch size for the three strategies discussed in the text. Strategy *(i)* maximises the arithmetic mean number of descendants. Strategy *(ii)* is the optimal strategy, maximising the geometric mean number of descendants left by the genotype. Under strategy *(iii)* each individual maximises its geometric mean number of descendants. (From McNamara, 1998.)

When the same environmental conditions affect all females, maximisation of expression (10.41) gives the optimal strategy. Maximisation of (10.42) for fixed x does not give the optimal clutch size for a territory of quality x. The reason is that maximisation of (10.42) implicitly assumes that all members of the genotype have territory quality x. For illustration, suppose that all members of a genotype have territory quality $x = 1$. Then if genotype members lay a large clutch size their broods will be strongly affected if environmental conditions are poor, and the genotype as a whole will suffer a catastrophic failure. Thus, it is optimal for clutch size to be substantially reduced in order to guard against this eventuality (see Figure 10.2). In contrast suppose there is a spread in territory qualities amongst members of a genotype. Then a female of territory quality $x = 1$ that lays a large clutch will certainly be strongly affected by poor environmental conditions. But this will not lead to a catastrophic failure of the genotype as a whole if females with other territory qualities are less strongly affected. Because of this phenomenon, it is optimal in our example for females with high-quality territories to produce clutch sizes close to that predicted by maximisation of the arithmetic mean fitness (Figure 10.2).

This example emphasises that fitness is assigned to strategies. Individuals cannot be considered in isolation. Boyce & Perrins (1987) attempted to analyse the optimal clutch size of great tits in a fluctuating environment. The concept of a single best clutch size is only meaningful if all birds of the same genotype always produce the same clutch size irrespective of their own physiological state and territory quality. This is not true for great tits. Indeed, Boyce and Perrins attempt to 'correct' for the fact that a given female produces different clutch sizes in different years. Under such circumstances their attempt to ascribe fitness to a clutch size, rather than a strategy, is incorrect.

The strategy that maximises geometric mean fitness, presented in Figure 10.2, was computed using a formulation of the optimisation problem as a constant-environment game. This technique generalises the algorithm given in Appendix 10.1 and is described by McNamara (1998).

10.7 Dynamic optimisation over the year

So far, we have assumed that an organism takes a single instantaneous decision at the annual census time. This might be reasonable in modelling reproductive decisions in some species. In other species annual decisions on breeding extend over a period of days, weeks or months. Reproductive decisions that extend over a period may involve the organism's making a sequence of time- and state-dependent choices.

In a constant environment, given a suitable terminal reward at the end of a year, we can solve a dynamic optimisation problem by dynamic programming back from this terminal reward. In a fluctuating environment the situation is more complex, because of the failure of individual optimisation. Consider an individual in some given state, x, at some given time n during the year. As in a constant environment, the best action at time n depends on the actions of the individual in the future. When there are environmental fluctuations we cannot, however, just work backwards from the future. The reason is that the best action for the individual at time n also depends on the states and actions of related individuals at this time (see Section 10.6), and the distribution of possible states of relatives depends on their behaviour before time n. Thus we cannot determine the best choice of action at a time n until we have determined optimal behaviour at all times before n and all times after n.

Here we describe two approaches that can be used to deal with this complication. In both cases we are concerned with analysing behaviour over a period $[0, T]$ within a year, where time 0 is taken to be the census time at the start of the year and time T occurs before the next annual census time. We assume that the organism chooses an action at each of the times $n = 0, 1, 2, \ldots, T - 1$. Behaviour after time T and before the next annual census time is taken as given. We consider the subpopulation comprising individuals of a given genotype (or sharing a given gene). Some initial subpopulation is given at time 0. The fitness of the genotype is assumed to be a function of the number and distribution of states at time T of descendants of this initial subpopulation. Within this context the two approaches are as follows.

Control of the structure vector. Suppose that at each decision time individuals of the focal genotype can be classified as falling into one of K states labelled $1, 2, \ldots, K$. Let $N_x(n)$ be the number of genotype members in state x at time n. Let $\boldsymbol{N}(n) = (N_1(n), N_2(n), \ldots, N_K(n))$ be the vector of these numbers. Then we can regard the time evolution of the process $\boldsymbol{N}(0), \boldsymbol{N}(1), \ldots$ as being controlled by the collective actions of members of the focal genotype. Since genotype fitness is a function of the distribution of $\boldsymbol{N}(T)$, the dynamic optimisation problem can then be reduced to a problem of the optimal control of the vector process $\boldsymbol{N}(0), \boldsymbol{N}(1), \ldots$.

Let $\rho_x(n)$ be the proportion of those genotype members alive at time n that are in state x at this time. Let $\boldsymbol{\rho}(n) = (\rho_1(n), \rho_2(n), \ldots, \rho_K(n))$ be the vector of these proportions. We will refer to $\rho(n)$ as the structure vector for the genotype at time n. It can be shown that the problem of optimal control of the vector process $N(0), N(1), \ldots$ can always be mathematically transformed into a problem of the optimal control of $\rho(0), \rho(1), \ldots$ (see e.g. Section 10.8).

The above concepts are illustrated by the timing of entry into diapause (Example 10.6). In this example the decisions are made immediately after new eggs are produced, when the eggs decide whether to go into diapause or to grow immediately. At a decision time all relevant genotype members are in the form of eggs, because adults die after egg production. Eggs present can be classified as already in diapause or otherwise. The vector describing genotype numbers and their states is thus $N(n) = (D(n), N_{\mathrm{new}}(n))$, where $D(n)$ is the number of eggs already in diapause and $N_{\mathrm{new}}(n)$ is the number of new eggs that have just been produced. Since the summer ends before time T, the number of eggs in diapause at T equals the number of eggs, $D(S)$, in diapause when summer ends. Expression (10.20) thus shows that the genotype as a whole receives reward $\log D(T)$ at final time T, and the objective of the genotype is to maximise the expected value of this reward through the collective action of genotype members.

For this problem the population structure vector at time n is $\rho(n) = (\rho_D(n), 1 - \rho_D(n))$, where $\rho_D(n)$ is the proportion of eggs already in diapause at n. Eggs not in diapause at this time can either enter diapause or grow. The optimal collective action of these eggs depends only on $\rho_D(n)$ and n (McNamara 1994b; see Example 10.6).

Dynamic optimisation as a constant environment dynamic game.   In Section 10.5 we described how a simple problem of optimisation in a fluctuating environment could be transformed into a constant-environment game. Here we outline the approach of McNamara *et al.* (1995), which extends the methods of Section 10.5 to a simple case in which dynamic decisions are made.

As before there is an annual census of the population, but now behaviour is analysed during the period of T days following the census time. The states of different individuals of the same genotype may differ at the start of the period, but the distribution of states is the same each year. We let the random variable $X(0)$ be the state of a randomly selected organism and, as in Example 10.7, assume $P(X(0) = x) = p(x)$. The environmental state during the T-day period is assumed to be predictable and the same each year. The environmental state after the end of the T-day period is unpredictable and is a random variable with density $f(s)$ as before. During the T-day period the organism must choose a sequence of actions without knowledge of the environmental state after the period. A deterministic strategy over the interval is a rule $\pi(x, n)$ that determines the action taken in each state x and for each day n, where $0 \leq n \leq T - 1$. The initial state $X(0)$ and strategy π determine the possible states of the organism at the end of the period and their probability of occurrence. If the organism is in state x at the end of the period and the ensuing environmental state is s then the organism leaves $D(x, s)$ expected descendants at the census time next year. The distribution of state of these descendants is independent of both x and s.

In this model consider an individual in state x at the census time in year t. If this individual follows strategy π during the time interval of length T then it leaves $\mathbb{E}_\pi[D(X(T), s) \mid X(0) = x]$ expected offspring at the census time in year $t + 1$, when the intervening environmental state is s. Let

$$d(\pi, s) = \sum_x p(x) \mathbb{E}_\pi[D(X(T), s) \mid X(0) = x], \qquad (10.43)$$

be the expected number of descendants averaged over all states at the census time in year t. If the set of pure actions is taken to be the set of deterministic state-dependent strategies π over the interval then the problem of finding an optimal dynamic strategy over the interval of length T is equivalent to the model of Section 10.4 (McNamara *et al.* 1995). Let π be a randomised strategy, in which the pure deterministic strategies $\pi_1, \pi_2, \pi_3, \ldots$ are chosen with probabilities $\theta_1, \theta_2, \theta_3, \ldots$ respectively. Then the profile of π is

$$r_\pi(s) = \sum \theta_i d(\pi_i, s).$$

Let the function R_π be given by

$$R_\pi(x) = \int_0^\infty \frac{D(x, s)}{r_\pi(s)} f(s) ds. \qquad (10.44)$$

We refer to R_π as the terminal reward at T generated by strategy π. Then, if we regard π as a resident population strategy we can regard the strategy maximising

$$\mathbb{E}_{\pi'}[R_\pi(X(T))] \qquad (10.45)$$

over all strategies π' as the best mutant response to π. This best mutant response can be found by straightforward dynamic programming. McNamara *et al.* (1995) show that a strategy π^* is optimal if and only if it is a best mutant response to itself.

The two approaches to dynamic optimisation that we have described have been illustrated by especially simple cases. In the diapause problem the optimal action of an individual egg at decision time n depends on the current proportion of eggs already in diapause, $\rho_D(n)$. Before summer has ended, however, $\rho_D(n)$ is a deterministic function of n. This means that the optimal action of an egg at decision time n can be given as a function of n alone. Thus the optimal strategy can be implemented by the collective action of members of a genotype: each member responds only to its own state and the time of season. Similarly, in the model of McNamara *et al.* (1995) the optimal strategy is realisable if individuals respond only to their own state and time. In more complex dynamic optimisation problems the structure vector $\rho(n)$ need not be a deterministic function of time. Optimal strategies may then not be realisable by individuals' responding to their own current state and time, but may require individuals to have explicit knowledge of the states of relatives, or failing this a memory of previous environmental conditions. This topic is discussed further in the next section.

10.8 Optimal life histories for structured populations

In this section we consider state-dependent life histories in fluctuating environments. We are concerned with the following general model. A census is made of the population at times $t = 0, 1, 2, \ldots$ one year apart. At the census times individuals can be classified by state, where, as usual, 'state' could be age, size, territory quality etc., or some combination of these. At a census time each individual must choose an action from some set of possible actions. The action chosen at census time t is made before the environmental state $S(t)$ between t and $t + 1$ is known. The individual's state x, the action chosen u and the environmental condition s determine the expected number of descendants $a_{xy}(u, s)$ left in state y at time $t + 1$. Environmental states $S(0), S(1), S(2), \ldots$ in successive years are assumed to be independent, identically distributed, non-negative random variables with common density function f.

This topic combines the state-dependent approach of Chapter 8 with the fluctuating-environment approach of this chapter. Population projection matrices are now random matrices, and fitness no longer has the simple interpretation in terms of descendant numbers given in Chapter 8. In this section we outline the approach of McNamara (1997) to this topic. Since there is no individual optimisation this approach employs the 'state' of the genotype rather than individual states. The genotypic 'state' used is the structure vector, ρ, which specifies the proportion of genotype members in each of the individual states. The genotype is regarded as obtaining an immediate contribution $m(\rho)$ to its reproductive success when its structure vector is ρ. The fitness of the genotype is then the long-term average annual reproductive success obtained by the genotype. Rather than assigning reproductive value to individuals, the approach of McNamara (1997) assigns reproductive value to structure vectors. Reproductive value can then be decomposed into an immediate contribution to reproductive success and a future contribution. This decomposition leads to an iterative method of simultaneously calculating reproductive value and fitness and to a criterion by which an optimal strategy can be characterised.

To give an idea of the type of problem that might arise for a structured population in a fluctuating environment we give a verbal description of an example given in McNamara (1997). Here we just use the example to motivate various ideas. For a detailed technical specification of the model, and a detailed analysis, see the original paper.

Example 10.8. Offspring number in a structured population. A population is asexual with non-overlapping generations. At census time t all individuals are mature. They reproduce at this time, care for the young between times t and $t + 1$ and die before time $t + 1$. Offspring that survive until $t + 1$ are mature at this time and they in turn reproduce, care and die and so on. During reproduction each individual must decide whether to produce one or two offspring. This decision is made before environmental conditions during the year are known.

At the census time individuals differ in their ability to provision young. This might be, for example, because individuals vary in their ability to catch prey or because they are territorial and territories vary in their quality. We classify individuals into two ability classes and assume that an individual is either low-quality or high-quality. The quality of a parent and the ensuing environmental conditions determine the total resources delivered to young. Resources are divided between the young. The probability that an offspring survives to maturity increases with the resource it receives during care. The probability that a surviving offspring is high-quality at maturity also increases with the resource it receives.

The idea behind this model is that a parent faces two trade-offs in deciding on the number of young to produce. If it produces one young this young has a high probability of surviving to maturity. If two young are produced, both are liable to survive if environmental conditions are reasonable, but both will probably die if conditions are poor. There is thus a trade-off between ensuring at least one offspring survives by producing only one offspring and maximising the mean number of surviving offspring by producing two offspring. Since the quality of offspring tends to increase with the resource that they receive, the second trade-off is between the mean number of surviving offspring and offspring quality. Of course, since a high-quality parent is better able to provision young than a low-quality parent, the high-quality parent is in a better position to produce two offspring.

The definition of fitness

We now develop a general definition of the fitness of a genotype. In order to define fitness, we focus on the subpopulation comprising members of the focal genotype within a population. Fitness will be defined as an appropriate rate of growth in number of members of this genotype. Before defining fitness various concepts need to be introduced.

Strategies. A strategy is a rule specifying how to choose behaviour as a function of state. Since we must allow strategies to be randomised, a strategy specifies, for each possible state, the probability with which each action is chosen in that state.

In Example 10.8 above a strategy is specified by two numbers p_L and p_H. Under this strategy, a low-quality individual produces one offspring with probability $1 - p_L$ and two offspring with probability p_L, and a high-quality individual produces one offspring with probability $1 - p_H$ and two offspring with probability p_H.

Later we will extend the definition of a strategy to include more complex rules for choosing actions.

Population structure vector. We are concerned with a subpopulation of genetically identical individuals (or individuals sharing some focal gene). To analyse how the size of this subpopulation changes with time we must keep track of the composition (by state) of the subpopulation. To do this, define $\rho_x(t)$ to be the proportion of individuals in state x at time t. We will refer to the vector $\boldsymbol{\rho}(t) = (\rho_1(t), \rho_2(t), \ldots)$ of these proportions as the structure vector for the genotype at time t.

In Example 10.8 $\boldsymbol{\rho}(t) = (\rho_L(t), \rho_H(t))$, where $\rho_L(t)$ and $\rho_H(t)$ are the proportions of the focal genotype which are low- and high-quality respectively at time t. Of course $\rho_L(t) + \rho_H(t) = 1$.

It is instructive to compare how the structure vector $\boldsymbol{\rho}(t)$ changes with time in a fluctuating environment with how it changes in a constant environment. In both cases we assume that numbers are large, so that demographic stochasticity can be ignored. In a constant environment $\boldsymbol{\rho}(t + 1)$ is a deterministic function of $\boldsymbol{\rho}(t)$. This can be seen from equations (8.4), (8.5) and (8.7). In a similar way to that described in Section 8.3, the deterministic sequence of structure vectors $\boldsymbol{\rho}(0), \boldsymbol{\rho}(1), \boldsymbol{\rho}(2), \ldots$ converges to a limiting vector ρ, which is the stable distribution of states for the population under study (cf. equation (8.9)). In contrast, in a fluctuating environment $\rho(t+1)$ depends on both $\rho(t)$ and the environmental state between t and $t+1$, $S(t)$. Since $S(t)$ is a random variable, the sequence $\rho(0), \rho(1), \rho(2), \ldots$ is now a stochastic process (which is Markov if $S(0), S(1), S(2), \ldots$ are independent). We would not now expect this sequence to converge. Although it does not converge, given appropriate regularity conditions the sequence will exhibit long-term stationary behaviour.

Profile of a strategy.   Let the focal genotype have structure vector ρ at time t, and let s be the environmental state between t and $t+1$. We then define $r(\rho, s)$ to be the expected total number of descendants left at time $t+1$ by a genotype member selected at random at time t.

Growth in population size and fitness.   Let $N(t)$ be the number of members of the focal genotype present at time t. Since $r(\rho(t), S(t))$ is the expected number of descendants left at time $t+1$ by a genotype member selected at random at time t we have

$$\frac{N(t+1)}{N(t)} = r(\rho(t), S(t)), \tag{10.46}$$

assuming that the population is large enough that demographic stochasticity can be ignored. This equation is the analogue of equation (10.6) except that r now depends on $\rho(t)$ as well as $S(t)$. Following (10.7) we write

$$N(t) = \frac{N(t)}{N(t-1)} \frac{N(t-1)}{N(t-2)} \frac{N(t-2)}{N(t-3)} \cdots \frac{N(1)}{N(0)} N(0). \tag{10.47}$$

Thus taking logarithms and using equation (10.46) we have

$$\frac{1}{t} \log[N(t)] = \frac{1}{t} \log[N(0)] + \frac{1}{t} \sum_{n=0}^{t-1} \log[r(\rho(n), S(n))], \tag{10.48}$$

the analogue of equation (10.9). Now, exactly as in Section 10.2 we can argue that the fitness of the genotype can be taken to be

$$g = \lim_{t \to \infty} \frac{1}{t} \log[N(t)] \tag{10.49}$$

(cf. Tuljapurkar 1990). Since the first term on the right-hand side of equation (10.48) tends to zero as t tends to infinity,

$$g = \lim_{t \to \infty} \frac{1}{t} \sum_{n=0}^{t-1} \log[r(\rho(t), S(t))]. \tag{10.50}$$

The right-hand side of this equation is a long-term average over both $\rho(t)$

and $S(t)$. Since $S(t)$ is independent of $\rho(t)$ we can average over $S(t)$ first, defining $m(\rho)$ to be the average value of $\log[r(\rho, S(t))]$ averaged over $S(t)$ while holding ρ fixed. Thus

$$m(\rho) = \int_0^\infty \log[r(\rho, s)]f(s)ds; \qquad (10.51)$$

then g is the long-term average of $m(\rho(t))$. By equations (10.46) and (10.51) we can express $m(\rho)$ as

$$\mathbb{E}\left\{\log\left[\frac{N(t+1)}{N(t)}\right]\middle| \rho(t) = \rho\right\}. \qquad (10.52)$$

Thus $m(\rho(t))$ is a measure of how the collective action of genotype members at time t determines the change in their number between times t and $t+1$. We shall regard $m(\rho(t))$ as the immediate contribution to the reproductive success of the genotype at time t. The fitness of the genotype is then equal to its long-term average yearly reproductive success.

Reproductive value

In a constant environment, the reproductive value of an individual is a relative measure that compares the number of descendants left far into the future by this individual with the number left by a reference individual (Section 8.4). In a fluctuating environment the value, to the genotype, of one individual depends on the states of other genotype members. To reflect this we will define reproductive value in a fluctuating environment as a measure $V(\rho)$ of the value to the genotype of its structure vector, ρ. Again reproductive value will be defined as a relative measure. Compare two populations, each composed of members of the focal genotype. One population has structure vector ρ at time 0, the other has the reference structure vector $\rho^{(1)}$. At time 0 the two populations have equal size. Let the sizes of these populations after t years be $N(t)$ and $N^{(1)}(t)$ respectively. We define the reproductive value of structure vector ρ (relative to the reference vector $\rho^{(1)}$) by

$$V(\rho) = \lim_{t\to\infty} \mathbb{E}\left\{\log\left[\frac{N(t)}{N^{(1)}(t)}\right]\right\} \qquad (10.53)$$

(McNamara 1997).

It can be shown that reproductive value satisfies the equation

$$V(\rho(t)) = m(\rho(t)) - g + \mathbb{E}[V(\rho(t+1))] \qquad (10.54)$$

(McNamara 1997). This equation decomposes the reproductive value of the structure vector at time t into two components. The term $m(\rho(t)) - g$

depends on the growth between t and $t + 1$ and is the difference between the reproductive success in that year and the mean yearly success. The term $\mathbb{E}[V(\rho(t + 1))]$ measures how the structure vector at time $t + 1$ affects future success. Thus, as with a constant environment, reproductive value can be decomposed as a sum of immediate reproductive success and future reproductive success, except that now reproductive success and reproductive value are functions of the distribution of states of genotype members, rather than of individual states.

Equation (10.54) provides the basis of a computational procedure that can be used to compute simultaneously both fitness, g, and the function $V(\rho)$. This procedure is outlined in Appendix 10.2.

Optimal strategies

A strategy is a rule specifying behaviour. We have emphasised that this rule may be probabilistic, in that the rule may specify the probability of each behavioural action. To examine optimal strategies we need to differentiate between two sorts of strategy. Under an *individual-based* strategy, the probability distribution of possible actions employed by an organism in a particular state depends only on that state. Under a *population-based* strategy, the probability distribution may depend on both the state of the individual and the structure vector ρ for the genotype (or strategy) under consideration. It can be seen that the class of individual-based strategies forms a subset of the wider class of population-based strategies.

In Example 10.8, an individual-based strategy is specified by two numbers p_L and p_H, where p_L is the probability that a low-quality individual produces two offspring and p_H is the probability a high-quality individual produces two offspring. A population-based strategy for this example is specified by two functions $p_L(\rho)$ and $p_H(\rho)$. Here $p_L(\rho)$ is the probability that a low-quality individual produces two offspring when the structure vector is ρ. The function $p_H(\rho)$ is similarly defined.

Let the fitness of strategy π be denoted by g^π. Let $m^\pi(\rho)$ denote the immediate contribution to the fitness of the genotype coding for π when this genotype has structure vector ρ. McNamara (1997) showed that a strategy π^* is optimal amongst the class of population-based strategies if and only if, for each structure vector $\rho(t)$, the expression

$$m^\pi(\rho(t)) + \mathbb{E}^\pi[V^{\pi^*}(\rho(t + 1))] \tag{10.55}$$

is maximised when $\pi = \pi^*$. Here $\mathbb{E}^\pi$ denotes the expected value under strategy π. This equation shows that an optimal population-based strategy always maximises the sum of the immediate contribution to the reproductive success of the genotype and the future reproductive success. McNamara (1997) gave an iterative method for calculating this optimal strategy.

Implementation of a population-based strategy requires genotype members to know the current structure vector for the genotype at each decision time. In other words they must know the distribution of states of relatives. It may not be reasonable to assume such knowledge. If individuals must base their behaviour on their own state alone, it is necessary to seek an optimal strategy within the restricted class of individual-based strategies. In general, no single individual-based strategy will maximise expression (10.55) for every structure vector $\rho(t)$; instead, the optimal individual-based strategy maximises the mean value of expression (10.55) averaged over states $\rho(t)$ using the equilibrium distribution for ρ under π^* (McNamara 1997).

10.9 Further problems

We have presented methods for dealing with a number of optimisation problems, but these methods do not cover all possible cases of biological interest. More work needs to be done to develop techniques that address the following issues.

The general dynamic optimisation problem. The optimisation problem considered by McNamara *et al.* (1995) involves a special case in which decisions are made before environmental fluctuations occur. In many circumstances, organisms will make a sequence of state-dependent decisions while the environment is fluctuating. The method based on control of the structure vector (outlined in Section 10.7) can be used to find an optimal strategy for this more general problem. The drawback with this approach is that it yields a population-based strategy in which individuals base their decisions not only on their own state but also on that of related individuals. We have argued that individual-based strategies should be found. At present, no simple technique exists for finding such strategies.

Simplifying the state-dependent life-history problem. In principle, we know how to find both individual-based and population-based strategies when there are annual decisions in a structured population (Section 10.8). There is, however, a practical problem. If members of the population can be in one of n discrete states at the annual census time, then the structure vector for the population lies in an $(n-1)$-dimensional real space. Unless n is very small (e.g. $n = 2$), the problem of controlling the structure vector is both analytically and computationally intractable. There are two obvious potential ways to tackle this issue, neither of which has been explored. One method is to convert the optimisation problem into a constant-environment game. The other method is to assume that environmental fluctuations are not large, and to develop an optimisation criterion based on a suitable approximation.

Appendix 10.1 Computing optimal strategies for unstructured populations

Assume the general model of Section 10.4. Choose some initial strategy π_0 and construct a sequence of strategies $\pi_0, \pi_1, \pi_2, \ldots$ iteratively, as follows. Given strategy π_n define

$$W_n(u) = \int_0^\infty \frac{d(u, s)}{r_{\pi_n}(s)} f(s) ds$$

for each action u. Let u_{n+1} satisfy

$$W_n(u_{n+1}) = \max_u W_n(u).$$

Then one of the following holds.

(i) $W_n(u_{n+1}) = 1$. In this case, inequality (10.27) shows that the strategy π_n is optimal and the iterative procedure stops.
(ii) $W_n(u_{n+1}) > 1$. In this case, for $0 \leq \lambda \leq 1$ let the strategy $\pi_n(\lambda)$ be the randomised strategy that chooses π_n with probability $1 - \lambda$ and the pure action u_{n+1} with probability λ. Let λ^*_{n+1} be the value of λ that maximises the fitness $g(\pi_n(\lambda))$ of $\pi_n(\lambda)$. Set $\pi_{n+1} = \pi_n(\lambda^*_{n+1})$.

The procedure that we have outlined yields a sequence $\pi_0, \pi_1, \pi_2, \ldots$ of strategies that have strictly increasing fitness. It can be shown that, under suitable regularity conditions, the sequence either terminates at the optimal strategy after a finite number of iterations or else tends to the optimal strategy as the number of iterations tends to infinity.

Appendix 10.2 Computing fitness and reproductive value for state-structured populations

Here we give an iterative procedure for computing the solution of equation (10.54). This equation defines V up to an additive constant. We thus choose some reference structure vector $\rho^{(1)}$ and by convention set $V(\rho^{(1)}) = 0$. Given this condition, equation (10.54) defines g and the function V uniquely.

To start the iterative procedure set $V_0(\rho) = 0$ for all ρ. The sequence $V_0, V_1, V_2, \ldots$ of functions is then calculated iteratively as follows.

Suppose $V_{n-1}(\rho)$ has been calculated for every ρ.

(i) For each $\rho(t)$ define $f_n(\rho(t))$ by

$$f_n(\rho(t)) = m(\rho(t)) + \mathbb{E}[V_{n-1}(\rho(t+1))].$$

(ii) Define the real number g_n by

$$g_n = f_n(\rho^{(1)})$$

where $\rho^{(1)}$ is the reference state.

(iii) For each ρ set

$$V_n(\rho) = f_n(\rho) - g_n.$$

Using this procedure it can be shown that

$$g_n \to g \quad \text{as} \quad n \to \infty,$$

and that for each ρ

$$V_n(\rho) \to V(\rho) \quad \text{as} \quad n \to \infty,$$

where g and V satisfy equation (10.54).

References

Abrahams, M.V., and Dill, L.M. (1989). A determination of the energetic equivalence of the risk of predation. *Ecology*, **70**, 999–1007.

Abrams, P.A. (1982). Functional responses of optimal foragers. *American Naturalist*, **120**, 382–90.

Abrams, P.A. (1983). Life-history strategies of optimal foragers. *Theoretical Population Biology*, **24**, 22–38.

Abrams, P.A. (1987). The functional responses of adaptive consumers of two resources. *Theoretical Population Biology*, **32**, 262–88.

Abrams, P.A. (1989). Decreasing functional responses as a result of adaptive consumer behavior. *Evolutionary Ecology*, **3**, 95–114.

Abrams, P.A. (1990). The effects of adaptive behaviour on the type-2 functional response. *Ecology*, **71**, 877–85.

Abrams, P.A. (1991). Life history and the relationship between food availability and foraging effort. *Ecology*, **72**, 1242–52.

Abrams, P.A. (1992). Adaptive foraging by predators as a cause of predator–prey cycles. *Evolutionary Ecology*, **6**, 56–72.

Abrams, P.A. (1993a). Why predation rate should not be proportioned to predator density. *Ecology*, **74**, 726–33.

Abrams, P.A. (1993b). Optimal traits when there are several costs: the interaction of mortality and energy costs in determining foraging behavior. *Behavioral Ecology*, **4**, 246–53.

Abrams, P.A. (1994). Evolutionarily stable growth rates in size-structured populations under size-related competition. *Theoretical Population Biology*, **46**, 78–95

Abrams, P.A. (1997). Evolutionary responses of foraging-related traits in unstable predator–prey system. *Evolutionary Ecology*, **11**, 673–86.

Abrams, P.A., and Ludwig, D. (1995). Optimality theory, Gompertz' law, and the disposable soma theory of senescence. *Evolution*, **49**, 1055–66.

Abrams, P.A., and Matsuda, H. (1997a). Prey adaptation as a cause of predator–prey cycles. *Evolution*, **51**, 1742–50.

Abrams, P.A., and Matsuda, H. (1997b). Fitness minimization and dynamic instability as a consequence of predator–prey coevolution. *Evolutionary Ecology*, **11**, 1–20.

Amir, S., and Cohen, D. (1990). Optimal reproductive efforts and the timing of reproduction of annual plants in randomly varying environments. *Journal of Theoretical Biology*, **147**, 17–42.

Apanius V. (1998). Stress and immune defense. *Advances in the Study of Behavior*, **27**, 133–53.

Axelrod, R., and Hamilton, W.D. (1981). The evolution of cooperation. *Science*, **211**, 1390–96.

Ball, M.A., and Parker, G.A. (1996). Sperm competition games – external fertilization and adaptive infertility. *Journal of Theoretical Biology*, **180**, 141–50.

Ball, M.A., and Parker, G.A. (1997). Sperm competition games: inter- and intra-species results of a continuous external fertilization model. *Journal of Theoretical Biology*, **186**, 459–66.

Barnard, C.J., Brown, C.A.J., Houston, A.I., and McNamara, J.M. (1985). Risk-sensitive foraging in common shrews: an interruption model and the effects of mean and variance in reward rate. *Behavioral Ecology and Sociobiology*, **18**, 139–46.

Bednekoff, P.A. (1996). Risk-sensitive foraging, fitness, and life histories: where does reproduction fit into the big picture? *American Zoologist*, **36**, 471–83.

Bednekoff, P.A. (1997). Mutualism among safe, selfish sentinels. *American Naturalist*, **150**, 373–92.

Bednekoff, P.A., and Houston, A.I. (1994). Avian daily foraging patterns: effects of digestive constraints and variability. *Evolutionary Ecology*, **8**, 36–52.

Bishop, D.T., and Cannings, C. (1978). A generalized war of attrition. *Journal of Theoretical Biology*, **70**, 85–124.

Bolton, M., Houston, D., and Monaghan, P. (1992). Nutritional constraints on egg formation in the lesser black-backed gull: an experimental study. *Journal of Animal Ecology*, **61**, 521–32.

Boyce, M.S., and Perrins, C.M. (1987). Optimizing great tit clutch size in a fluctuating environment. *Ecology*, **68**, 142–53.

Brown, J.S. (1988). Patch use as an indicator of habitat preference, predation risk, and competition. *Behavioral Ecology and Sociobiology*, **22**, 37–47.

Brown, J.S. (1992). Patch use under predation risk: I. Models and predictions. *Annales Zoologici Fennici*, **29**, 301–9.

Bull, J.J., and Shire, R. (1979). Iteroparous animals that skip opportunities for reproduction. *American Naturalist*, **114**, 296–303.

Bulmer, M.G. (1984). Delayed germination of seeds: Cohen's model revisited. *Theoretical Population Biology*, **26**, 367–77.

Bulmer, M.G. (1985). Selection for iteroparity in a variable environment. *American Naturalist*, **126**, 63–71.

Byrd, J.W., Houston, A.I., and Sozou, P.D. (1991). Ydenberg's model of fledging time – a comment. *Ecology*, **72**, 1893–6.

Cade, W. (1975). Acoustically orienting parasitoids: fly phonotaxis to cricket song. *Science*, **190**, 1312–13.

Caraco, T. (1979). Time budgeting and group size: a theory. *Ecology*, **60**, 611–17.

Caraco, T. (1980). On foraging time allocation in a stochastic environment. *Ecology*, **61**, 119–28.

Caraco, T., and Chasin, M. (1984). Foraging preferences: response to reward skew. *Animal Behaviour*, **32**, 76–85.

Carlson, A. (1983). Maximizing energy delivery to dependent young: a field experiment with red-backed shrikes (*Lanius collurio*). *Journal of Animal Ecology*, **52**, 697–704.

Caswell, H. (1982). Optimal life histories and the maximization of reproductive value: a general theroem for complex life cycles. *Ecology*, **63**, 1218–22.

Caswell, H. (1989). *Matrix Population Models.* Sunderland *Ma*: Sinauer.

Charlesworth, B. (1994). *Evolution in Age-structured Populations.* Cambridge: Cambridge University Press.

Charlesworth, B., and León, J.A. (1976). The relation of reproductive effort to age. *American Naturalist*, **110**, 449–59.

Charnov, E.L. (1976a). Optimal foraging: attack strategy of a mantid. *American Naturalist*, **110**, 141–51.

Charnov, E.L. (1976b). Optimal foraging: the marginal value theorem. *Theoretical Population Biology*, **9**, 129–36.

Charnov, E.L. (1982). *The Theory of Sex Allocation.* Princeton: Princeton University Press.

Charnov, E.L. (1993). *Life History Invariants.* Oxford: Oxford University Press.

Charnov, E.L., and Krebs, J.R. (1974). On clutch size and fitness. *Ibis*, **116**, 217–19.

Charnov, E.L., and Schaffer, W.M. (1973). Life-history consequences of natural selection: Cole's result revisited. *American Naturalist*, **107**, 791–3.

Christiansen, F.B. (1991). On conditions for evolutionary stability for a continuously varying character. *American Naturalist*, **138**, 37–50.

Clark, C.W., and Ekman, J. (1995). Dominant and subordinate fattening strategies: a dynamic game. *Oikos*, **72**, 205–12.

Clark, C.W., and Mangel, M. (1986). The evolutionary advantages of group foraging. *Theoretical Population Biology*, **30**, 45–75.

Clark, C.W., and Ydenberg, R.C. (1990). The risks of parenthood II. Parent–offspring conflict. *Evolutionary Ecology*, **4**, 312–25.

Clutton-Brock, T.H. (1984). Reproductive effort and terminal investment in iteroparous animals. *American Naturalist*, **123**, 219–29.

Clutton-Brock, T.H. (1991). *The Evolution of Parental Care.* Princeton: Princeton University Press.

Clutton-Brock, T.H., Stephenson, I.R., Marrow, P., MacColl, A.D., Houston, A.I., and McNamara, J.M. (1996). Population fluctuations, reproductive costs and life-history tactics in female Soay sheep. *Journal of Animal Ecology*, **65**, 675–89.

Cohen, D. (1966). Optimizing reproduction in a randomly varying environment. *Journal of Theoretical Biology*, **12**, 119–29.

Cohen, D. (1967). Optimizing reproduction in a randomly varying environment when a correlation may exist between the conditions at the time a choice has to be made and the subsequent outcome. *Journal of Theoretical Biology*, **16**, 1–14.

Cohen, D. (1970). A theoretical model for the optimal timing of diapause. *American Naturalist*, **104**, 389–400.

Cohen, D. (1971). Maximizing final yield when growth is limited by time or by limiting resources. *Journal of Theoretical Biology*, **33**, 299–307.

Cohen, D. (1976). The optimal timing of reproduction. *American Naturalist*, **110**, 801–7.

Collier, G.H. (1982). Determinants of choice. In *Nebraska Symposium on Motivation*, **29**, ed. D.J. Bernstein, pp. 69–127. Lincoln: University of Nebraska Press.

Collier, G.H. (1983). Life in a closed economy: the ecology of learning and motivation. In *Advances in Analysis of Behaviour*, **3**, eds. M.D. Zeiler and P. Harzem, pp. 223–74. Chichester: John Wiley.

Collins, E.J., and McNamara, J.M. (1993). The job search problem with competition: an evolutionarily stable dynamic strategy. *Advances in Applied Probability*, **25**, 314–33.

Cooch, E.G., and Ricklefs, R.E. (1994). Do variable environments significantly influence optimal reproductive effort in birds? *Oikos*, **69**, 447–59.

Cooper, W.S., and Kaplan, R.H. (1982). Adaptive 'coin-flipping': a decision-theoretic examination of natural selection for random individual variation. *Journal of Theoretical Biology*, **94**, 135–51.

Crowley, P.H., and Hopper, K.R. (1994). How to behave around cannibals: a density-dependent dynamic game. *American Naturalist*, **143**, 117–54.

Crowley, P.H., Travers, S.E., Linton, M.C., Cohn, S.L., Sih, A., and Sargent, R.C. (1991). Mate density, predation risk, and the seasonal sequence of mate choice: a dynamic game. *American Naturalist*, **137**, 567–96.

Daan, S., Dijkstra, C., Drent, R., and Meijer, T. (1989). Food supply and the annual timing of avian reproduction. In *Acta XIX International Congress of Ornithology*, ed. H. Oullet, pp. 392–407. Ottawa: University of Ottawa Press.

Daan, S., Dijkstra, C., and Tinbergen, J.M. (1990). Family planning in the kestrel (*Falco tinnunculus*): the ultimate control of covariation of laying date and clutch size. *Behaviour*, **114**, 83–116.

Daan, S., Deerenberg, C., and Dijkstra, C. (1996). Increased daily work precipitates natural death in the kestrel. *Journal of Animal Ecology*, **65**, 539–44.

Davison, M.C. (1969). Preference for mixed-interval versus fixed-interval schedules. *Journal of the Experimental Analysis of Behaviour*, **12**, 247–52.

Dawkins, M.S. (1990). From an animal's point of view: motivation, fitness and animal welfare. *Behavioral and Brain Sciences*, **13**, 1–61.

Dawkins, R. (1982). *The Extended Phenotype*. Oxford: W.H. Freeman.

Day, T., and Taylor, P.D. (1996). Evolutionarily stable versus fitness maximizing life histories under frequency-dependent selection. *Proceedings of the Royal Society, Series B*, **263**, 333–8.

Deerenberg, C., Apanius, V., Daan, S., and Bos, N. (1997). Reproductive effort decreases antibody responsiveness. *Proceedings of the Royal Society, Series B*, **264**, 1021–29.

Drent, R.H., and Daan, S. (1980). The prudent parent: energetic adjustments in avian breeding. *Ardea*, **68**, 225–52.

Elgar, M.A. (1989). Predator vigilance and group size in mammals and birds: a critical review of the empirical evidence. *Biological Reviews*, **64**, 13–33.

Ellner, S. (1985a). ESS germination strategies in randomly varying environments 1. Logistic-type models. *Theoretical Population Biology*, **28**, 50–79.

Ellner, S. (1985b). ESS germination strategies in randomly varying environments 2. Reciprocal yield-law models. *Theoretical Population Biology*, **28**, 80–116.

Ellner, S. (1986). Germination dimorphisms and parent–offspring conflict in seed-germination. *Journal of Theoretical Biology*, **123**, 173–85.

Emlen, J.M. (1984). Natural selection and population density feedback II. The evolution of functional response curves. *Theoretical Population Biology*, **25**, 62–77.

Emlen, J.M. (1973). *Ecology: An Evolutionary Approach*. New York: Addison Wesley.

Endler, J.A. (1995). Multiple-trait coevolution and environmental gradients in guppies. *Trends in Ecology and Evolution*, **10**, 22–9.

Engen, S., and Stenseth, N.C. (1984). A general version of optimal foraging theory: the effect of simultaneous encounters. *Theoretical Population Biology*, **26**, 192–204.

Enquist, M., and Leimar, O. (1983). Evolution of fighting behaviour: decision rules and assessment of relative strength. *Journal of Theoretical Biology*, **102**, 387–410.

Enquist, M., and Leimar, O. (1987). Evolution of fighting behaviour: the effect of variation in resource value. *Journal of Theoretical Biology*, **127**, 187–205.

Enquist, M., and Leimar, O. (1990). The evolution of fatal fighting. *Animal Behaviour*, **39**, 1–9.

Eshel, I. (1983). Evolutionary and continuous stability. *Journal of Theoretical Biology*, **103**, 99–111.

Eshel, I. (1996). On the changing concept of evolutionary population stability as a reflection of a changing point of view in the quantitative theory of evolution. *Journal of Mathematical Biology*, **34**, 485 510.

Eshel, I., and Sansone, E. (1995). Owner–intruder conflict, Grafen effect and self-assessment. The Bourgeois Principle re-examined. *Journal of Theoretical Biology*, **177**, 341–56.

Eshel, I., Motro, U., and Sansone, E. (1997). Continuous stability and evolutionary convergence. *Journal of Theoretical Biology*, **185**, 333–43.

Eshel, I., Feldman, F.W., and Bergman, A. (1998). Long-term evolution, short-term evolution and population genetic theory. *Journal of Theoretical Biology*, **191**, 391–6.

Ewens, W.J. (1969). *Population Genetics*. London: Methuen & Co.

Fisher, R.A. (1958). *The Genetical Theory of Natural Selection*. New York: Dover.

Fraser, D.F., and Gilliam, J.F. (1992). Nonlethal impacts of predator invasion: facultative suppression of growth and reproduction. *Ecology*, **73**, 959–70.

Fretwell, S.D., and Lucas, H.L. (1970). On territorial behaviour and other factors influencing habitat distribution in birds. I. Theoretical development. *Acta Biotheoretica*, **19**, 16–36.

Fuchs, S. (1982). Optimality of parental investment: the influence of nursing on reproductive success of mother and female young house mice. *Behavioral Ecology and Sociobiology*, **10**, 39–51.

Gadgil, M., and Bossert, W.H. (1970). Life historical consequences of natural selection. *American Naturalist*, **104**, 1–24.

Gantmacher, F.R. (1959). *Applications of the Theory of Matrices*. New York: Interscience.

Geritz, S.A.H., Kisdi, E., Meszena, G., and Metz, J.A.J. (1998). Evolutionarily singular strategies and the adaptive growth and branching of the evoutionary tree. *Evolutionary Ecology*, **12**, 35–57.

Getty, T., and Pulliam, H.R. (1991). Random prey detection with pause-travel search. *American Naturalist*, **138**, 1459–77.

Gibb, J. (1954). Feeding ecology of tits, with notes on treecreeper and goldcrest. *Ibis*, **96**, 513–43.

Gibbon, J. (1977). Scalar expectancy theory and Weber's law in animal timing. *Psychological Review*, **84**, 279–325.

Gibbon, J., Church, R.M., Fairhurst, S., and Kacelnik, A. (1988). Scalar expectancy theory and choice between delayed rewards. *Psychological Review*, **95**, 102–14.

Gillespie, J.H. (1977). Natural selection for variances in offspring numbers: a new evolutionary principle. *American Naturalist*, **111**, 1010–14.

Gilliam, J.F. (1982). Foraging under mortality risk in size-structured populations. Ph.D. thesis, Michigan State University.

Gilliam, J.F. (1990). Hunting by the hunted: optimal prey selection by foragers under predation hazard. In *Behavioural Mechanisms of Food Selection*, ed. R.N. Hughes, pp. 797–818. Berlin: Springer-Verlag.

Gilliam, J.F., and Fraser, D.F. (1987). Habitat selection under predation hazard: test of a model with foraging minnows. *Ecology*, **68**, 1856–62.

Gilliam, J.F., Green, R.F., and Pearson, N.E. (1982). The fallacy of the traffic policeman: a response to Templeton and Lawlor. *American Naturalist*, **119**, 875–8.

Gladstein, D.S., Carlin, N.F., and Austad, S.N. (1991). The need for sensitivity analyses of dynamic optimization models. *Oikos*, **60**, 121–6.

Godin, J.-G.J. (1990). Diet selection under the risk of predation. In *Behavioural Mechanisms of Food Selection*. ed. R.N. Hughes, pp. 737–69. Berlin: Springer-Verlag.

Gotthard, K., and Nylin, S. (1995). Adaptive plasticity and plasticity as an adaptation: a selective review of plasticity in animal morphology and life history. *Oikos*, **74**, 3–17.

Gould, S.J., and Lewontin, R.C. (1979). The spandrels of San Marco and the Panglossian paradigm: a critique of the adaptationist programme. *Proceedings of the Royal Society, Series B*, **205**, 581–98.

Grafen, A. (1984). Natural selection, kin selection and group selection. In *Behavioural Ecology: An Evolutionary Approach*, 2nd edition, eds. J.R. Krebs and N.B. Davies, pp. 62–84. Oxford: Blackwell Scientific Publications.

Grafen, A. (1987). The logic of divisively asymmetric contests: respect for ownership and the desperado effect. *Animal Behaviour*, **35**, 462–7.

Grey, D.R. (1980). Minimisation of extinction probabilities in reproducing populations. *Theoretical Population Biology*, **18**, 430–43.

Green, W.C.H., and Rothstein, A. (1991). Trade-offs between growth and reproduction in female bison. *Oecologia*, **86**, 521–7.

Gross, M.R. (1985). Disruptive selection for alternative life histories in salmon. *Nature*, **313**, 47–8.

Gross, M.R. (1996). Alternative reproductive strategies and tactics: diversity within sexes. *Trends in Ecology and Evolution*, **11**, 92–8.

Gustafsson, L., Nordling, D., Andersson, M.S., Sheldon, B.C., and Qvarnström, A. (1994). Infectious diseases, reproductive effort and the cost of reproduction in birds. *Philosophical Transactions of the Royal Society, Series B*, **346**, 323–31.

Gustafsson, L., and Sutherland, W.J. (1988). The costs of reproduction in the collared flycatcher *Ficedula albicollis*. *Nature*, **335**, 813–15.

Haccou, P., and Iwasa, Y. (1995). Optimal mixed strategies in stochastic environments. *Theoretical Population Biology*, **47**, 212–43.

Haccou, P., and Iwasa, Y. (1996). Establishment probability in fluctuating environments: a branching process model. *Theoretical Population Biology*, **50**, 254–80.

Haccou, P., and McNamara, J.M. (1998). Effects of parental survival on clutch size decisions in fluctuating environments. *Evolutionary Ecology*, **12**, 459–75.

Haftorn, S. (1992). The diurnal body weight cycle in titmice *Parus* spp. *Ornis Scandinavica*, **23**, 435–43.

Halliday, T.R., and Sweatman, H.P.A. (1976). To breathe or not to breathe; the newt's problem. *Animal Behaviour*, **24**, 551–61.

Hamilton, W.D. (1964). The genetical evolution of social behaviour I. *Journal of Theoretical Biology*, **7**, 1–16.

Hamilton, W.D. (1966). The moulding of senescence by natural selection. *Journal of Theoretical Biology*, **12**, 12–45.

Hamilton, W.D. (1967). Extraordinary sex ratios. *Science*, **156**, 477–88.

Hammerstein, P. (1981). The role of asymmetries in animal contests. *Animal Behaviour*, **29**, 193–205.

Hammerstein, P. (1996). Darwinian adaptation, population genetics and the streetcar theory of evolution. *Journal of Mathematical Biology*, **34**, 511–32.

Hammerstein, P., and Parker, G.A. (1987). Sexual selection: games between the sexes. In *Sexual Selection: Testing the Alternatives*, eds. J.W. Bradbury and M.B. Andersson, pp. 119–42. New York: Wiley.

Hammond, K.A., and Diamond, J. (1997). Maximal sustained energy budgets in humans and animals. *Nature*, **386**, 457–62.

Hanski, I. (1988). Four kinds of extra long diapause in insects: a review of theory and observations. *Annales Zoologici Fennici*, **25**, 37–53.

Harsanyi, J.C. (1973). Games with randomly distributed payoffs: a new rationale for mixed strategy equilibrium points. *International Journal of Game Theory*, **2**, 1–23.

Hedenström, A. (1992). Flight performance in relation to fuel load in birds. *Journal of Theoretical Biology*, **158**, 535–37.

Hedenström, A., and Alerstam, T. (1995). Optimal flight speed of birds. *Philosophical Transactions of the Royal Society of London, Series B*, **348**, 471–87.

Heino, M., Metz, J.A.J., and Kaitala, V. (1997). Evolution of mixed maturation strategies in semelparous life histories: the crucial role of dimensionality of feedback environment. *Philosophical Transactions of the Royal Society of London, Series B*, **352**, 1647–55.

Heller, R. (1980). On optimal diet in a patchy environment. *Theoretical Population Biology*, **17**, 201–14.

Heller, R., and Milinski, M. (1979). Optimal foraging of sticklebacks on swarming prey. *Animal Behaviour*, **27**, 1127–41.

Herrnstein, R.J. (1964). Aperiodicity as a factor in choice. *Journal of the Experimental Analysis of Behavior*, **7**, 179–82.

Herrnstein, R.J. (1970). On the law of effect. *Journal of the Experimental Analysis of Behavior*, **13**, 243–66.

Hines, W.G.S. (1987). Evolutionary stable strategies: a review of basic theory. *Theoretical Population Biology*, **31**, 195–272.

Hofbauer, J., and Sigmund, K. (1988). *The Theory of Evolution and Dynamical Systems*. Cambridge: Cambridge University Press.

Holekampe, K.E., and Smale, L. (1993). Ontogeny of dominance in free-living spotted hyaenas: juvenile rank relations with other immature individuals. *Animal Behaviour*, **46**, 451–66.

Holmgren, N., and Hedenström, A. (1995). The scheduling of molt in migrating birds. *Evolutionary Ecology*, **9**, 354–68.

Houston, A.I. (1987). The control of foraging decisions. In *Quantitative Analysis of Behaviour*, **6**, eds. M.L. Commons, A. Kacelnik and S.J. Shettleworth, pp. 41–61. Hillsdale: Lawrence Erlbaum.

Houston, A.I. (1995). Energetic constraints and foraging efficiency. *Behavioral Ecology*, **6**, 393–6.

Houston, A.I., and Davies, N.B. (1985). The evolution of cooperation and life history in the dunnock, *Prunella modularis*. In *Behavioural Ecology*, eds. R.M. Sibly and R.H. Smith, pp. 471–87. Oxford: Blackwell Scientific Publications.

Houston, A.I., and McNamara, J.M. (1982). A sequential approach to risk-taking. *Animal Behaviour*, **30**, 1260–1.

Houston, A.I., and McNamara, J.M. (1985a). The variability of behaviour and constrained optimization. *Journal of Theoretical Biology*, **112**, 265–73.

Houston, A.I., and McNamara, J.M. (1985b). A general theory of central-place foraging for single-prey loaders. *Theoretical Population Biology*, **28**, 233–62.

Houston, A.I., and McNamara, J.M. (1985c). The choice of two prey types that minimises the probability of starvation. *Behavioral Ecology Sociobiology*, **17**, 135–41.

Houston, A.I., and McNamara, J.M. (1986a). Evaluating the selection pressure on foraging decisions. In *Relevance of Models and Theories in Ethology*, eds. R. Campan and R. Zayan, pp. 61–75. Toulouse: Privat, I.E.C.

Houston, A.I., and McNamara, J.M. (1986b). The influence of mortality on the behaviour that maximizes reproductive success in a patchy environment. *Oikos*, **47**, 267–74.

Houston, A.I., and McNamara, J.M. (1987). Singing to attract a mate – a stochastic dynamic game. *Journal of Theoretical Biology*, **129**, 57–68.

Houston, A.I., and McNamara, J.M. (1988). Fighting for food: a dynamic version of the Hawk–Dove game. *Evolutionary Ecology*, **2**, 51–64.

Houston, A.I., and McNamara, J.M. (1989). The value of food: effects of open and closed economies. *Animal Behaviour*, **37**, 546–62.

Houston, A.I., and McNamara, J.M. (1991). Evolutionarily stable strategies in the repeated hawk–dove game. *Behavioral Ecology*, **2**, 219–27.

Houston, A.I., and McNamara, J.M. (1992). Phenotypic plasticity as a state-dependent life-history decision. *Evolutionary Ecology*, **6**, 243–53.

Houston, A.I., and McNamara, J.M. (1993). A theoretical investigation of the fat reserves and mortality levels of small birds in winter. *Ornis Scandinavica*, **24**, 205–19.

Houston, A.I., and McNamara, J.M. (1997). Patch choice and population size. *Evolutionary Ecology*, **11**, 703–22.

Houston, A.I., Kacelnik, A., and McNamara, J.M. (1982). Some learning rules for acquiring information. In *Functional Ontogeny*, ed. D.J. McFarland, pp. 140–91. London: Pitman.

Houston, A.I., McNamara, J.M., and Thompson, W.A. (1992). On the need for a sensitive analysis of optimisation models, or, 'This simulation is not as the former'. *Oikos*, **63**, 513–17.

Houston, A.I., McNamara, J.M., and Hutchinson, J.M.C. (1993). General results concerning the trade-off between gaining energy and avoiding predation. *Philosophical Transactions of the Royal Society of London, Series B*, **341**, 375–97.

Houston, A.I., Welton, N.J., and McNamara, J.M. (1997). Acquisition and maintenance costs in the long-term regulation of avian fat reserves. *Oikos*, **78**, 331–40.

Huey, R.B., and Kingsolver, J.G. (1989). Evolution of thermal sensitivity of ectotherm performance. *Trends in Ecology and Evolution*, **4**, 131–5.

Hursh, S.R. (1980). Economic concepts for the analysis of behavior. *Journal of the Experimental Analysis of Behavior*, **34**, 219–38.

Hursh, S.R. (1984). Behavioral economics. *Journal of the Experimental Analysis of Behavior*, **42**, 435–52.

Hutchinson, J.M.C., McNamara, J.M., and Cuthill, I.C. (1993). Song, sexual selection, starvation and strategic handicaps. *Animal Behaviour*, **45**, 1153–77.

Iwasa, Y. (1991). Sex change evolution and cost of reproduction. *Behavioral Ecology*, **2**, 56–68.

Iwasa, Y., and Cohen, D. (1989). Optimal growth schedule of a perennial plant. *American Naturalist*, **133**, 480–505.

Iwasa, Y., and Odendaal, F.J. (1984). A theory on the temporal pattern of operational sex ratio: the active-inactive model. *Ecology*, **65**, 886–93.

Iwasa, Y., and Haccou, P. (1994). ESS emergence patterns of male butterflies in stochastic environments. *Evolutionary Ecology*, **8**, 503–23.

Iwasa, Y., and Obara, Y. (1989). A game model for the daily activity schedule of the male butterfly. *Journal of Insect Biology*, **2**, 589–608.

Jansson, C., Ekman, J., and Brömssen, A. von (1981). Winter mortality and food supply in tits *Parus* spp. *Oikos*, **37**, 313–22.

Johns, M., and Miller, R.G. (1963). Average renewal loss rate. *Annals of Mathematical Statistics*, **34**, 396–401.

Johnstone, R.A. (1997). The tactics of mutual male choice and competitive search. *Behavioral Ecology and Sociobiology*, **40**, 51–9.

Johnstone, R.A., Reynolds, J.D., and Deutsch, J.C. (1996). Mutual mate choice and sex differences in choosiness. *Evolution*, **50**, 1382–91.

Juanes, F., and Hartwick, E.B. (1990). Prey size selection in Dungeness crabs: the effect of claw damage. *Ecology*, **71**, 744–58.

Kacelnik, A., and Bateson, M. (1996). Risky theories – the effects of variance on foraging decisions. *American Zoologist*, **36**, 402–34.

Kacelnik, A., and Bateson, M. (1997). Risk-sensitivity: crossroads for theories of decision-making. *Trends in Cognitive Sciences*, **1**, 304–9.

Kacelnik, A., and Krebs, J.R. (1983). The dawn chorus in the great tit (*Parus major*): proximate and ultimate causes. *Behaviour*, **83**, 287–309.

Kacelnik, A., Krebs, J.R., and Bernstein, C. (1992). The ideal free distribution and predator–prey populations. *Trends in Ecology and Evolution*, **7**, 50–5.

Kaitala, V., Lindström, K., and Ranta, E. (1989). Foraging, vigilance and risk of predation in birds – a dynamic game study of ESS. *Journal of Theoretical Biology*, **138**, 329–45.

Kamil, A.C., Lindstrom, F., and Peters, J. (1985). The detection of cryptic prey by blue-jays (*Cyanocitta cristata*). *Animal Behaviour*, **33**, 1068–79.

Katz, P.L. (1974). A long-term approach to foraging optimization. *American Naturalist*, **108**, 758–82.

Kawecki, T.J. (1993). Age and size at maturity in a patchy environment: fitness maximization versus evolutionary stability. *Oikos*, **66**, 309–17.

Kawecki, T.J., and Stearns, S.C. (1993). The evolution of life histories in spatially heterogeneous environments: optimal reaction norms revisited. *Evolutionary Ecology*, **7**, 155–74.

King, D.A. (1990). The adaptive significance of tree height. *American Naturalist*, **135**, 809–28.

King, D., and Roughgarden, J. (1982). Graded allocation between vegetative and reproductive growth for annual plants in growing seasons of random length. *Theoretical Population Biology*, **22**, 1–16.

Kirkpatrick, M. (1984). Demographic models based on size, not age, for organisms with indeterminate growth. *Ecology*, **65**, 1874–84.

Kirkwood, J.K. (1983). A limit to metabolisable energy intake in mammals and birds. *Comparative Biochemistry and Physiology*, **75A**, 1–3.

Kirkwood, T.B.L., and Rose, M.R. (1991). Evolution of senescence: late survival sacrificed for reproduction. *Philosophical Transactions of the Royal Society, Series B*, **332**, 15–24.

Kisdi, E., and Meszéna, G. (1993). Density dependent life history evolution in fluctuating environments. In *Adaptation in a Stochastic Environment*, eds. J. Yoshimura and C. Clark, pp. 26–62. Lecture Notes in Biomathematics. Berlin: Springer-Verlag.

Kisdi, E., Meszéna, G., and Pásztor, L. (1998). Individual optimization: mechanisms shaping the optimal reaction norm. *Evolutionary Ecology*, **12**, 211–21.

Konarzewski, M. (1993). The evolution of clutch size and hatching asynchrony in altricial birds: the effect of environmental variability, egg failure and predation. *Oikos*, **67**, 97–106.

Kozłowski, J. (1992). Optimal allocation of resources to growth and reproduction: implications for age and size at maturity. *Trends in Ecology and Evolution*, **7**, 15–19.

Kuhn, H.W. (1953). Extensive games and the problem of information. In *Contributions to the Theory of Games Vol. II*, eds. H.W. Kuhn and A.W. Tucker, pp. 193–216. Princeton: Princeton University Press.

Lack, D. (1950). The breeding seasons of European birds. *Ibis*, **92**, 288–316.

Lalonde, R.G. (1991). Optimal offspring provisioning when resources are not predictable. *American Naturalist*, **138**, 680–6.

Lande, R. (1982). Elements of a quantitative genetic model of life history evolution. In *Evolution and Genetics of Life Histories*, eds. H. Dingle and J.P. Hegmann, pp. 21–9. New York: Springer-Verlag.

Leimar, O. (1996). Life-history analysis of the Trivers and Willard sex-ratio problem. *Behavioral Ecology,* **7**, 316–25.

Leimar, O. (1997a). Repeated games: a state space approach. *Journal of Theoretical Biology,* **184**, 471–98.

Leimar, O. (1997b). Reciprocity and communication of partner quality. *Proceedings of the Royal Society, Series B,* **264**, 1209–15.

Leimar, O. Multidimensional convergence stability and the canonical adaptive dynamics (unpub.).

Leimar, O., and Enquist, M. (1984). Effects of asymmetries in owner–intruder conflicts. *Journal of Theoretical Biology,* **111**, 475–91.

León, J.A. (1976). Life histories as adaptive strategies. *Journal of Theoretical Biology,* **60**, 301–35.

Leonardsson, K. (1991). Predicting risk-taking behaviour from life-history theory using static optimization technique. *Oikos,* **60**, 149–54.

Leslie, P.H. (1945). On the use of matrices in certain population mathematics. *Biometrika,* **33**, 213–45.

Lessels, C.M. (1991). The evolution of life histories. In *Behavioural Ecology: An Evolutionary Approach,* 3rd edition, eds. J.R. Krebs and N.B. Davies, pp. 32–68. Oxford: Blackwell Scientific Publications.

Lewontin, R.C., and Cohen, D. (1969). On population growth in randomly varying environments. *Proceedings of the National Academy of Science,* **62**, 1056–60.

Lima, S.L. (1986). Predation risk and unpredictable feeding conditions: determinants of body mass in birds. *Ecology,* **67**, 377–85.

Lima, S.L. (1987a). Clutch size in birds – a predation perspective. *Ecology,* **68**, 1062–70.

Lima, S.L. (1987b). Vigilance while feeding and its relation to the risk of predation. *Journal of Theoretical Biology,* **124**, 303–16.

Lima, S.L. (1988). Vigilance and diet selection – the classical diet model reconsidered. *Journal of Theoretical Biology,* **132**, 127–43.

Lima, S.L. (1989). Iterated prisoner's dilemma: an approach to evolutionarily stable cooperation. *American Naturalist,* **134**, 828–34.

Lima, S.L. (1998). Stress and decision making under the risk of predation: recent development from behavioral, reproductive, and ecological perspectives. *Advances in the Study of Behavior*, **27**, 215–90.

Lima, S.L., and Dill, L.M. (1990). Behavioral decisions made under the risk of predation: a review and prospectus. *Canadian Journal of Zoology*, **68**, 619–40.

Liou, L.W., Price, T., Boyce, M.S., and Perrins, C.M. (1993). Fluctuating environments and clutch size evolution in great tits. *American Naturalist*, **141**, 507–16.

Lloyd, D.G. (1987). Selection on offspring size at independence and other size-versus-number strategies. *American Naturalist*, **129**, 800–17.

Logan, F.A. (1965). Decision making by rats: delay versus amount of reward. *Journal of Comparative and Physiological Psychology*, **59**, 1–12.

Logue, A.W. (1988). Research on self-control: an integrating framework. *Behavioural and Brain Sciences*, **11**, 665–709.

Lucas, J.R., and Howard, R.D. (1995). On alternative reproductive tactics in anurans: dynamic games with density and frequency dependence. *American Naturalist*, **146**, 365–97.

Lucas, J.R., Howard, R.D., and Palmer, J.G. (1996). Callers and satellites: chorus behaviour in anurans as a stochastic dynamic game. *Animal Behaviour*, **51**, 501–18.

Lucas, J.R., and Schmid-Hempel, P. (1988). Diet choice in patches: time-constraint and state-space solutions. *Journal of Theoretical Biology*, **131**, 307–32.

Lucas, J.R., and Walter, L.R. (1991). When should chickadees hoard food? Theory and experimental results. *Animal Behaviour*, **41**, 579–601.

Ludwig, D., and Rowe, L. (1990). Life-history strategies for energy gain and predator avoidance under time constraints. *American Naturalist*, **135**, 686–707.

Mäkelä, A. (1985). Differential games in evolutionary theory: height growth strategies of trees. *Theoretical Population Biology*, **27**, 239–67.

Mangel, M. (1987). Oviposition site selection and clutch size in insects. *Journal of Mathematical Biology*, **25**, 1–22.

Mangel, M. (1989). Evolution of host selection in parasitoids: does the state of the parasitoid matter? *American Naturalist*, **139**, 688–705.

Mangel, M. (1990). A dynamic habitat selection game. *Mathematical Biosciences*, **100**, 241–8.

Mangel, M. (1992). Rate maximizing and state variable theories of diet selection. *Bulletin of Mathematical Biology*, **54**, 413–22.

Mangel, M., and Clark, C.W. (1986). Towards a unified foraging theory. *Ecology*, **67**, 1127–38.

Mangel, M., and Clark, C.W. (1988). *Dynamic Modeling in Behavioral Ecology*. Princeton: Princeton University Press.

Mangel, M., and Ludwig, D. (1992). Definition and evaluation of the fitness of behavioural and developmental programs. *Annual Review of Ecology and Systematics*, **23**, 507–36.

March, J.G. (1996). Learning to be risk averse. *Psychological Review*, **103**, 309–19.

Masman, D., Daan, S., and Dijkstra, C. (1988). Time allocation in the kestrel (*Falco tinnunculus*), and the principle of energy minimization. *Journal of Animal Ecology*, **57**, 411–32.

Matessi, C., and Di Pasquale, C. (1996). Long-term evolution of multilocus traits. *Journal of Mathematical Biology*, **34**, 613–53.

Matsuda, H., and Abrams, P.A. (1994). Timid consumers: self-extinction due to adaptive change in foraging and anti-predator effort. *Theoretical Population Biology*, **45**, 76–91.

Maynard Smith, J. (1977). Parental investment: a prospective analysis. *Animal Behaviour*, **25**, 1–9.

Maynard Smith, J. (1978). Optimization theory in evolution. *Annual Review of Ecology and Systematics*, **9**, 31–56.

Maynard Smith, J. (1982). *Evolution and the Theory of Games*. Cambridge: Cambridge University Press.

Maynard Smith, J., and Brown, R.L.W. (1986). Competition and body size. *Theoretical Population Biology*, **30**, 166–79.

Maynard Smith, J., and Parker, G.A. (1976). The logic of asymmetric contests. *Animal Behaviour*, **24**, 159–75.

Maynard Smith, J., and Price, G.R. (1973). The logic of animal conflict. *Nature*, **246**, 15–18.

McFarland, D.J. (1971). *Feedback Mechanisms in Animal Behaviour*. London: Academic Press.

McGinley, M.A., Temme, D.H., and Geber, M.A. (1987). Parental investment in offspring in variable environments: theoretical and empirical considerations. *American Naturalist*, **130**, 370–98.

McNamara, J.M. (1982). Optimal patch use in a stochastic environment. *Theoretical Population Biology*, **21**, 269–88.

McNamara, J.M. (1984). Control of a diffusion by switching between two drift coefficient pairs. *SIAM Journal of Control*, **22**, 87–94.

McNamara, J.M. (1985). An optimal sequential policy for controlling a Markov renewal process. *Journal of Applied Probability*, **22**, 324–35.

McNamara, J.M. (1990a). The policy which maximises long-term survival of an animal faced with the risks of starvation and predation. *Advances in Applied Probability*, **22**, 295–308.

McNamara, J.M. (1990b). The starvation–predation trade-off and some behavioural and ecological consequences. In *Mechanisms of Food Selection*, ed. R.N. Hughes, pp. 39–58. Berlin: Springer-Verlag.

McNamara, J.M. (1991). Optimal life histories: a generalisation of the Perron–Frobenius theorem. *Theoretical Population Biology*, **40**, 230–45.

McNamara, J.M. (1993a). State-dependent life-history equations. *Acta Biotheoretica*, **41**, 165–74.

McNamara, J.M. (1993b). Evolutionary paths in strategy space: an improvement algorithm for life-history strategies. *Journal of Theoretical Biology*, **161**, 23–37.

McNamara, J.M. (1994a). Multiple stable age distributions in a population at evolutionary stability. *Journal of Theoretical Biology*, **169**, 349–54.

McNamara, J.M. (1994b). Timing of entry into diapause: optimal allocation to "growth" and "reproduction" in a stochastic environment. *Journal of Theoretical Biology*, **168**, 201–9.

McNamara, J.M. (1995). Implicit frequency dependence and kin selection in fluctuating environments. *Evolutionary Ecology*, **9**, 185–203.

McNamara, J.M. (1996). Risk-prone behaviour under rules which have evolved in a changing environment. *American Zoologist*, **36**, 484–95.

McNamara, J.M. (1997). Optimal life histories for structured populations in fluctuating environments. *Theoretical Population Biology*, **51**, 94–108.

McNamara, J.M. (1998). Phenotypic plasticity in flutuating environments: consequences of the lack of individual optimisation. *Behavioral Ecology*, **9**, 642–8.

McNamara, J.M., and Collins, E.J. (1990). The job search problem as an employer–candidate game. *Journal of Applied Probability*, **28**, 815–27.

McNamara, J.M., and Houston, A.I. (1980). The application of statistical decision theory to animal behaviour. *Journal of Theoretical Biology*, **85**, 673–90.

McNamara, J.M., and Houston, A.I. (1982). Short-term behaviour and life-time fitness. In *Functional Ontogeny*, ed. D.J. McFarland, pp. 60–87. London: Pitman.

McNamara, J.M., and Houston, A.I. (1985a). Optimal foraging and learning. *Journal of Theoretical Biology*, **117**, 231–49.

McNamara, J.M., and Houston, A.I. (1985b). A simple model of information use in the exploitation of patchily distributed food. *Animal Behaviour*, **33**, 553–60.

McNamara, J.M., and Houston, A.I. (1986). The common currency for behavioral decisions. *American Naturalist*, **127**, 358–78.

McNamara, J.M., and Houston, A.I. (1987a). Memory and efficient use of information. *Journal of Theoretical Biology*, **125**, 385–95.

McNamara, J.M., and Houston, A.I. (1987b). A general framework for understanding the effects of variablility and interruptions on foraging behaviour. *Acta Biotheoretica*, **36**, 3–22.

McNamara, J.M., and Houston, A.I. (1987c). Starvation and predation as factors limiting population size. *Ecology*, **68**, 1515–19.

McNamara, J.M., and Houston, A.I. (1987d). Partial preferences and foraging. *Animal Behaviour*, **35**, 1084–99.

McNamara, J.M., and Houston, A.I. (1990a). Starvation and predation in a patchy environment. In *Living in a Patchy Environment*, eds. B. Shorrocks and I.R. Swingland, pp. 23–43. Oxford: Oxford University Press.

McNamara, J.M., and Houston, A.I. (1990b). The value of fat reserves and the trade-off between starvation and predation. *Acta Biotheoretica*, **38**, 37–61.

McNamara, J.M., and Houston, A.I. (1990c). State-dependent ideal free distributions. *Evolutionary Ecology*, **4**, 298–311.

McNamara, J.M., and Houston, A.I. (1992a). Risk-sensitive foraging: a review of the theory. *Bulletin of Mathematical Biology*, **54**, 355–78.

McNamara, J.M., and Houston, A.I. (1992b). Evolutionarily stable levels of vigilance as a function of group size. *Animal Behaviour*, **43**, 641–58.

McNamara, J.M., and Houston, A.I. (1992c). State dependent life-history theory and its implications for optimal clutch size. *Evolutionary Ecology*, **6**, 170–85.

McNamara, J.M., and Houston, A.I. (1994). The effect of a change in foraging options on intake rate and predation rate. *American Naturalist*, **144**, 978–1000.

McNamara, J.M., and Houston, A.I. (1996). State-dependent life histories. *Nature*, **380**, 215–20.

McNamara, J.M., and Houston, A.I. (1997). Currencies for foraging based on energetic gain. *American Naturalist*, **150**, 603–17.

McNamara, J.M., Gasson, C., and Houston, A.I. Incorporating rules for responding into evolutionary games (unpub.).

McNamara, J.M., Mace, R.H., and Houston, A.I. (1987). Optimal daily routines of singing and foraging in a bird singing to attract a mate. *Behavioral Ecology and Sociobiology*, **20**, 399–405.

McNamara, J.M., Houston, A.I., and Krebs, J.R. (1990). Why hoard? The economics of food storing in tits, *Parus* spp. *Behavioral Ecology*, **1**, 12–23.

McNamara, J.M., Merad, S., and Houston, A.I. (1991). A model of risk-sensitive foraging for a reproducing animal. *Animal Behavior*, **41**, 787–92.

McNamara, J.M., Houston, A.I., and Weisser, W.W. (1993). Combining prey choice and patch use – what does rate-maximizing predict? *Journal of Theoretical Biology*, **164**, 219–38.

McNamara, J.M., Houston, A.I., and Lima, S.L. (1994a). Foraging routines of small birds in winter: a theoretical investigation. *Journal of Avian Biology*, **25**, 287–302.

McNamara, J.M., Houston, A.I., and Webb, J.N. (1994b). Dynamic kin selection. *Proceedings of the Royal Society, Series B*, **258**, 23–8.

McNamara, J.M., Webb, J.N., and Collins, E.J. (1995). Dynamic optimisation in fluctuating environments. *Proceedings of the Royal Society, Series B*, **261**, 279–84.

McNamara, J.M., Webb, J.N., Collins, E.J., Székely, T., and Houston, A.I. (1997). A general technique for computing evolutionarily stable strategies based on errors in decision-making. *Journal of Theoretical Biology*, **189**, 211–25.

McNamara, J.M., Forslund, P., and Lang, A. (1999). An ESS model for divorce strategies in birds. *Philosophical Transactions of the Royal Society of London Series B*, **354**, 223–36.

McNamara, J.M., Welham, R.K., and Houston, A.I. (1998). The timing of migration within the context of an annual routine. *Journal of Avian Biology*, **29**, 416-423.

Merad, S. (1991). Dynamic modelling of risk-taking and agressiveness of foraging animals: an evolutionary approach. Ph.D. thesis, University of Bristol.

Merad, S., and McNamara, J.M. (1994). Optimal foraging of a reproducing animal as a discounted reward problem. *Journal of Applied Probability*, **31**, 287–300.

Mesterton-Gibbons, M. (1992). Ecotypic variation in the asymmetric Hawk–Dove game: when is Bourgeois an evolutionarily stable strategy? *Evolutionary Ecology*, **6**, 198–222.

Metz, J.A.J., Nisbet, R.M., and Geritz, S.A.H. (1992). How should we define 'fitness' for general ecological scenarios? *Trends in Ecology and Evolution*, **7**, 198–202.

Milinski, M., and Parker, G.A. (1991). Competition for resources. In *Behavioural Ecology: An Evolutionary Approach*, 3rd edition, eds. J.R. Krebs and N.B. Davies, pp. 137–68. Oxford: Blackwell Scientific Publications.

Mitchell, W.A., and Brown, J.S. (1990). Density-dependent harvest rates by optimal foragers. *Oikos*, **57**, 180–90.

Mitchell, W.A., and Valone, T.J. (1990). The optimization research program: studying adaptations by their function. *Quarterly Review of Biology*, **65**, 43–52.

Moody, A.L., Houston, A.I., and McNamara, J.M. (1996). Ideal free distributions under predation risk. *Behavioral Ecology and Sociobiology*, **38**, 131–43.

Motro, U. (1994). Evolutionary and continuous stability in asymmetric games with continuous strategy sets: the parental investment conflict as an example. *American Naturalist*, **144**, 229–41.

Murphy, G.I. (1968). Patterns in life history and the environment. *American Naturalist*, **102**, 391–403.

Mylius, S.D., and Diekmann, O. (1995). On evolutionarily stable life histories, optimization and the need to be specific about density dependence. *Oikos*, **74**, 218–24.

Newman, J.A. (1991). Patch use under predation hazard: foraging behaviour in a simple stochastic environment. *Oikos*, **61**, 29–44.

Newton, I. (1980). The role of food in limiting bird numbers. *Ardea*, **68**, 11–30.

Nilsson, J.-Å., and Svensson, E. (1996). The cost of reproduction: a new link between current reproductive effort and future reproductive success. *Proceedings of the Royal Society, Series B*, **263**, 711–14.

Oaten, A. (1977). Optimal foraging in patches: a case for stochasticity. *Theoretical Population Biology*, **12**, 263–85.

Olsson, O. (1996). Seasonal effects of timing and reproduction in the King Penguin: a unique breeding cycle. *Journal of Avian Biology*, **27**, 7–14.

Orians, G.H., and Pearson, N.E. (1979). On the theory of central place foraging. In *Analysis of Ecological Systems*, eds. D.J. Horn, R.D. Mitchell and G.R. Stairs, pp. 155–77. Columbus: Ohio State University Press.

Orzack, S.H., and Sober, E. (1994). Optimality models and the test of adaptationism. *American Naturalist*, **143**, 361–80.

Orzack, S.H., and Tuljapurkar, S. (1989). Population dynamics in variable environments VII. The demography and evolution of iteroparity. *American Naturalist*, **133**, 901–23.

Osborne, M.J., and Rubinstein, A. (1994). *A Course in Game Theory*. Cambridge, *MA*: Massachusetts Institute of Technology.

Parker, G.A. (1978). Searching for mates. In *Behavioural Ecology: An Evolutionary Approach*, eds. J.R. Krebs and N.B. Davies, pp. 214–44. Oxford: Blackwell Scientific Publications.

Parker, G.A. (1982). Phenotype-limited evolutionarily stable strategies. In *Current Problems in Sociobiology*, ed. King's College Sociobiology Group, pp. 173–201. Cambridge: Cambridge University Press.

Parker, G.A. (1984). Evolutionarily stable strategies. In *Behavioural Ecology: An Evolutionary Approach*, 2nd edition, eds. J.R. Krebs and N.B. Davies, pp. 30–61. Oxford: Blackwell Scientific Publications.

Parker, G.A. (1993). Sperm competition games – sperm size and sperm number under adult control. *Proceedings of the Royal Society of London, Series B*, **253**, 245–54.

Parker, G.A., and Begon, M.E. (1993). Sperm competition games – sperm size and number under gametic control. *Proceedings of the Royal Society of London, Series B*, **253**, 255–62.

Partridge, L., and Barton, N.H. (1993). Optimality, mutation and the evolution of aging. *Nature*, **362**, 305–11.

Partridge, L., and Barton, N.H. (1996). On measuring the rate of ageing. *Proceedings of the Royal Society, Series B*, **263**, 1365–71.

Partridge, L., and Harvey, P.H. (1988). The ecological context of life history evolution. *Science*, **241**, 1449–55.

Perrin, N., and Sibly, R.M. (1993). Dynamic models of energy allocation and investment. *Annual Review of Ecology and Systematics*, **24**, 379–410.

Perrins, C.M. (1970). The timing of birds' breeding seasons. *Ibis*, **112**, 242–55.

Piersma, T., and Lindström, Å. (1997). Rapid reversible changes in organ size as a component of adaptive behaviour. *Trends in Ecology and Evolution*, **12**, 134–8.

Price, G.R. (1970). Selection and covariance. *Nature*, **227**, 520–1.

Price, K., and Boutin, S. (1993). Territorial bequethal by red squirrel mothers. *Behavioral Ecology*, **4**, 144–50.

Prince, P.A., Ricketts, C., and Thomas, G. (1981). Weight loss in incubating albatrosses and its implications for their energy and food requirements. *Condor*, **83**, 238–42.

Pyke, G.H. (1979). The economics of territory size and time budget in the golden-winged sunbird. *American Naturalist*, **114**, 131–45.

Real, L.A. (1980). Fitness, uncertainty, and the role of diversification in evolution and behavior. *American Naturalist*, **115**, 623–38.

Real, L. (1990). Search theory and mate choice. I. Models of single-sex discrimination. *American Naturalist*, **136**, 376–404.

Reboreda, J.C., and Kacelnik, A. (1991). Risk sensitivity in starlings: variability in food amount and food delay. *Behavioral Ecology*, **2**, 301–8.

Reeve, H.K., and Sherman, P.W. (1993). Adaptation and the goals of evolutionary research. *Quarterly Review of Biology*, **68**, 1–32.

Regelmann, K. (1986). Learning to forage in a variable environment. *Journal of Theoretical Biology*, **120**, 321–9.

Reiter, J., and Le Boeuf, B. (1991). Life history consequences of variation in age at primiparity in northern elephant seals. *Behavioral Ecology and Sociobiology*, **28**, 153–60.

Riechert, S.E., and Hammerstein, P. (1983). Game theory in the ecological context. *Annual Review of Ecology and Systematics*, **14**, 377–409.

Roff, D.A. (1992). *The Evolution of Life Histories*. New York: Chapman and Hall.

Rowe, L., Ludwig, D., and Schluter, D. (1994). Time, condition, and the seasonal decline of avian clutch size. *American Naturalist*, **143**, 698–722.

Ryan, M.J., Tuttle, M.D., and Rand, A.S. (1982). Bat predation and sexual advertisement in a neotropical anuran. *American Naturalist*, **119**, 136–9.

Saether, B.-E., and Heim, M. (1993). Ecological correlates of individual variation in age at maturity in female moose (*Alces alces*): the effects of environmental variability. *Journal of Animal Ecology*, **62**, 482–9.

Sasaki, A., and Ellner, S. (1995). The evolutionary stable phenotype distribution in a random environment. *Evolution*, **49**, 337–50.

Schaffer, W.M. (1974a). Selection for optimal life histories: the effects of age structure. *Ecology*, **55**, 291–303.

Schaffer, W.M. (1974b). Optimal reproductive effort in fluctuating environments. *American Naturalist*, **108**, 783–90.

Schaffer, W.M. (1983). The application of optimal control theory to the general life history problem. *American Naturalist*, **121**, 418–31.

Scheiner, S.M. (1993). Genetics and evolution of phenotypic plasticity. *Annual Review of Ecology and Systematics*, **24**, 35–68.

Schlichting, C.D. (1986). The evolution of phenotypic plasticity in plants. *Annual Review of Ecology and Systematics*, **17**, 667–93.

Schmitz, O.J., and Ritchie, M.E. (1991). Optimal diet selection with variable nutrient intake – balancing reproduction with risk of starvation. *Theoretical Population Biology*, **39**, 100–14.

Schoener, T.W. (1971). Theory of feeding strategies. *Annual Review of Ecology and Systematics*, **2**, 369–404.

Schultz, D.L. (1991). Parental investment in temporally varying environments. *Evolutionary Ecology*, **5**, 415–27.

Seger, J., and Brockmann, H.J. (1987). What is bet-hedging? In *Oxford Surveys in Evolutionary Biology*, eds. P.H. Harvey and L. Partridge, pp. 182–211. Oxford: Oxford University Press.

Selten, R. (1980). A note on evolutionarily stable strategies in asymmetric animal conflicts. *Journal of Theoretical Biology*, **84**, 93–101.

Selten, R. (1983). Evolutionary stability in extensive 2-person games. *Mathematical Social Sciences*, **5**, 269–363.

Sheldon, B.C., and Verhulst, S. (1996). Ecological immunology: costly parasite defences and trade-offs in evolutionary ecology. *Trends in Ecology and Evolution*, **11**, 317–21.

Sibly, R.M., Calow, P., and Nichols, N. (1985). Are patterns of growth adaptive? *Journal of Theoretical Biology*, **112**, 553–74.

Sibly, R.M., and McFarland, D.J. (1976). On the fitness of behavior sequences. *American Naturalist*, **110**, 601–17.

Sih, A. (1987) Predator and prey lifestyles : an evolutionary and ecological overview. In *Predation: Direct and Indirect Impacts on Acquatic Communities*, eds. W.C. Kerfoot, and A. Sih, pp. 203–24. Hanover. *NH*: University Press of New England.

Sinervo, B., and DeNardo, D.F. (1996). Costs of reproduction in the wild: path analysis of natural selection and experimental tests of causation. *Evolution*, **50**, 1299–313.

Sirot, E., and Bernstein, C. (1996). Time sharing between host searching and food searching in parasitoids: state-dependent optimal strategies. *Behavioral Ecology*, **7**, 189–94.

Skyrms, B. (1996). *Evolution of the Social Contract*. Cambridge: Cambridge University Press.

Smith, C.C., and Fretwell, S.D. (1974). The optimal balance between size and number of offspring. *American Naturalist*, **108**, 499–506.

Stearns, S.C. (1992). *The Evolution of Life Histories*. Oxford: Oxford University Press.

Stearns, S.C., and Koella, J.C. (1986). The evolution of phenotypic plasticity in life-history traits: predictions of reaction norms for age and size at maturity. *Evolution*, **40**, 893–913.

Stephens, D.W. (1981). The logic of risk-sensitive foraging preferences. *Animal Behaviour*, **29**, 628–29.

Stephens, D.W., and Charnov, E.L. (1982). Optimal foraging: some simple stochastic models. *Behavioral Ecology and Sociobiology*, **10**, 251–63.

Stephens, D.W., and Krebs, J.R. (1986). *Foraging Theory*. Princeton: Princeton University Press.

Sutherland, W.J., and Anderson, C.W. (1987). Six ways in which a foraging predator may encounter options with different variances. *Biological Journal of the Linnean Society*, **30**, 99–114.

Swennen, C., Leopold, M.F., and de Bruijn, L.L.M. (1989). Time-stressed oystercatchers, *Haematopus ostralegus*, can increase their intake rate. *Animal Behaviour*, **38**, 8–22.

Székely, T., Webb, J.N., McNamara, J.M., and Houston, A.I. A dynamic game-theoretic model of avian parental care and desertion (unpub.).

Taylor, H.M., Gourley, R.S., Lawrence, C.E., and Kaplan, R.S. (1974). Natural selection of life history attributes: an analytical approach. *Theoretical Population Biology*, **5**, 104–22.

Taylor, P.D. (1985). A general mathematical model for sex allocation. *Journal of Theoretical Biology*, **112**, 799–818.

Taylor, P.D. (1989). Evolutionary stability in one-parameter models under weak selection. *Theoretical Population Biology*, **36**, 125–43.

Taylor, P.D. (1990). Allele frequency change in a class-structured population. *American Naturalist*, **135**, 95–106.

Taylor, P.D. (1991). Optimal life histories with age dependent trade-off curves. *Journal of Theoretical Biology*, **148**, 33–48.

Taylor, P.D., and Frank, S.A. (1996). How to make a kin selection model. *Journal of Theoretical Biology*, **180**, 27–37.

Templeton, A.R., and Lawlor, L.R. (1981). The fallacy of the averages in ecological optimization theory. *American Naturalist*, **117**, 390–3.

Thomas, B. (1985a). On evolutionarily stable sets. *Journal of Mathematical Biology*, **22**, 105–15.

Thomas, B. (1985b). Evolutionarily stable sets in mixed-strategist models. *Theoretical Population Biology*, **28**, 332–41.

Tregenza, T. (1995). Building on the ideal free distribution. *Advances in Ecological Research*, **26**, 253–307.

Trivers, R.L., and Willard, D.E. (1973). Natural selection of parental ability to vary the sex ratio of offspring. *Science*, **179**, 90–2.

Tuljapurkar, S. (1990). *Population Dynamics in Variable Environments*. Berlin: Lecture Notes in Biomathematics, Springer-Verlag.

Turelli, M., Gillespie, J.H., and Schoener, T.W. (1982). The fallacy of the fallacy of the averages in ecological optimization theory. *American Naturalist*, **119**, 879–84.

Tuttle, M.D., and Ryan, M.J. (1981). Bat predation and the evolution of frog vocalization in the neotropics. *Science*, **214**, 677–8.

van Damme, E. (1991). *Stability and Perfection of Nash Equilibria*. Berlin: Springer-Verlag.

van der Werf, E. (1992). Lack's clutch size hypothesis: an examination of the evidence using meta-analysis. *Ecology*, **73**, 1699–705.

Vincent, T.L., and Brown, J.S. (1988). The evolution of ESS theory. *Annual Review of Ecology and Systematics*, **19**, 423–43.

Walters, C.J., and Juanes, F. (1993). Recruitment limitation as a consequence of natural selection for use of restricted feeding habitats and predation risk-taking by juvenile fishes. *Canadian Journal of Fisheries and Aquatic Sciences*, **50**, 2058–70.

Webb, J.N., Houston, A.I., McNamara, J.M., and Székely, T. Multiple patterns of parental care (unpub.).

Weber, T.P., Ens, B.J., and Houston, A.I. (1998). Optimal avian migration: a dynamic model of fuel stores and site use. *Evolutionary Ecology*, **12**, 377–401.

Weibull, J.W. (1995). *Evolutionary Game Theory*. Cambridge *MA*: Massachusetts Institute of Technology.

Weissing, F. (1991). Evolutionary stability and dynamic stability in a class of evolutionary normal form games. In *Game Equilibrium Models. I. Evolution and Game Dynamics*, ed. R. Selten, pp. 29–97. Berlin: Springer-Verlag.

Weissing, F.J. (1996). Genetic versus phenotypic models of selection: can genetics be neglected in a long-term perspective? *Journal of Mathematical Biology*, **34**, 533–55.

Werner, E.E. (1991). Nonlethal effects of a predator on competitive interactions between two anuran larvae. *Ecology*, **72**, 1709–20.

Werner, E.E., and Anholt, B.R. (1993). Ecological consequences of the trade-off between growth and mortality rates mediated by foraging activity. *American Naturalist*, **142**, 242–72.

Werner, E.E., and Gilliam, J.F. (1984). The ontogenetic niche and species interactions in size-structured populations. *Annual Review of Ecology and Systematics*, **15**, 393–425.

Werner, E.E., and Hall, D.J. (1988). Ontogenetic habitat shifts in bluegill: the foraging rate-predation risk trade-off. *Ecology*, **69**, 1352–66.

Werner, E.E., Gilliam, J.F., Hall, D.J., and Mittlebach, G.G. (1983). An experimental test of the effects of predation on habitat use in fish. *Ecology*, **64**, 1540–47.

Westoby, M. (1981). How diversified seed-germination behavior is selected. *American Naturalist*, **118**, 882–5.

Williams, G.C. (1966). Natural selection, the costs of reproduction, and a refinement of Lack's principle. *American Naturalist*, **100**, 687–90.

Willmer, P.G. (1986). Foraging patterns and water balance: problems of optimization for a xerophilic bee, *Chalicodoma sicula*. *Journal of Animal Ecology*, **55**, 941–62.

Witter, M.S., and Cuthill, I.C. (1993). The ecological costs of avian fat storage. *Philosophical Transactions of the Royal Society of London, Series B*, **340**, 73–92.

Ydenberg, R.C. (1989). Growth-mortality trade-offs and the evolution of juvenile life histories in the Alcidae. *Ecology*, **70**, 1494–506.

Ydenberg, R.C., Welham, C.V.J., Schmid-Hempel, R., Schmid-Hempel, P., and Beauchamp, G. (1994). Time and energy constraints and the relationships between currencies in foraging theory. *Behavioral Ecology*, **5**, 28–34.

Index

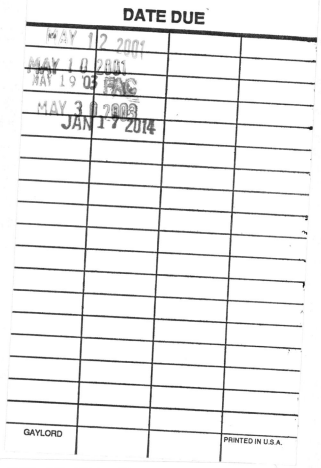